- CLEAR AND CONCISE
- GRADED EXERCISES
- COMPREHENSIVE
- VERY USER FRIENDLY
- SOLUTIONS AT BACK

UNDERSTANDING YEAR 10 MATHS

Authors:

Warwick Marlin B.Sc., Dip.Ed.

Ian Bull B.Sc., Dip.Ed., BFA, Grad Dip

Author/Editor:
Warwick Marlin B.Sc., Dip.Ed.
Ian Bull B.Sc., Dip.Ed., BFA, Grad Dip

Publisher:
Five Senses Education Pty Ltd
ABN: 16 001 414437
2/195 Prospect Highway
Seven Hills NSW Australia 2147
sevenhills@fivesenseseducation.com.au
www.fivesenseseducation.com.au

Trade Enquiries:
Phone (02) 9838 9265
Fax (02) 9838 8982
Email: fsonline@fivesenseseducation.com.au

ISBN: 978-1-76032-361-5
1st Edition: February 2021

AUTHOR'S ACKNOWLEDGEMENTS

To **Jones** my typesetter

I have been now working with Jones for more than 8 years on the 'Understanding Maths Series' of books. He has done so much excellent and brilliant work in translating my thoughts and ideas, and sometimes my untidy hand written pages, into the final high standard of overall presentation shown in this new series of books. A patient and very professional man, he aims for, and achieves excellence in his work at all times.

To **Edgar O'Neill** my proof reader

Many thanks Edgar for patiently working through all the exercises and problems and triple checking all the solutions. I am extremely appreciative of your excellent professional services.

To **Ian Bull** my co-author

Many thanks Ian for the excellent set of extension questions that you have produced throughout the book. Your unique and interesting questions test the problem solving abilities of students and also their skills to apply their knowledge to practical everyday problems and for completing the final proof reading before publication.

To **Roger Furniss** my publisher

I have had an association with Roger for well over 15 years now, and I thank him most sincerely for publishing, distributing and recommending my books to teachers, parents and other educators. He has been in the publishing and retail industry for many years, and he knows much more about educational books than any other person that I know. Also a very special thanks to all the staff at **Five Senses Education** who have always proved over the years to be so cooperative, friendly and supportive.

I started writing the original 'Understanding Maths Series' well over 20 years ago, and during that time many hundreds of parents have either phoned us, or written in to tell us how these summary guides have helped their children in terms of better understanding and improved results. This is a special thanks for all those positive comments, and also to the many Mathematics teachers and other educators who are recommending these books to their students. I hope that you will embrace this new advanced and superior series with even more enthusiasm and positivity.

Teachers

This book summarises the steps and stages by which Australian children acquire necessary mathematical skills and concepts. It eliminates the need to wade through lengthy curriculum documents, and it provides a clear, easy to follow summary for teachers to use, which they can also confidently recommend to parents, as it supports classroom activities and exercises.

B.Ed. and Dip.Ed. Student Primary Teachers

This book provides a concise record of the steps to be taken and the activities suggested to support the acquisition of mathematical concepts. It is the perfect reference book for teaching practice, and during the early years of teaching, when there is so much to learn.

'Education is the most powerful weapon which you can use to change the world.'

Nelson Mandela

AVAILABILITY OF BOOKS IN AUSTRALIA.....by the same Author/or Editor

All of the books below incorporate the same high presentation, format and philosophy. They can be purchased directly from Five Senses Education, but they are also available in many educational bookshops throughout NSW and Australia (and also some selected bookshops in New Zealand).

Understanding Maths Series

Understanding Year 1 Maths
Understanding Year 2 Maths
Understanding Year 3 Maths
Understanding Year 3 Maths (Advanced Edition)
Understanding Year 4 Maths
Understanding Year 4 Maths (Advanced Edition)
Understanding Year 5 Maths
Understanding Year 5 Maths (Advanced Edition)
Understanding Year 6 Maths
Understanding Year 6 Maths (Advanced Edition)
Understanding Year 7 Maths
Understanding Year 7 Maths (Advanced Edition)
Understanding Year 8 Maths
Understanding Year 9 & 10 Intermediate Maths
Understanding Year 9 & 10 Advanced Maths
Understanding Year 9 Maths
Understanding Year 9 Maths (Advanced Edition)
Understanding Year 10 Maths
Understanding Year 10 Maths (Advanced Edition)
Understanding Year 11 & 12 General Maths
Understanding Year 11 2 & 3 Unit Maths
Understanding Year 12 2 & 3 Unit Maths

Essential Exercises Series

Essential Exercises Year 1 Maths
Essential Exercises Year 2 Maths
Essential Exercises Year 3 Maths
Essential Exercises Year 4 Maths
Essential Exercises Year 5 Maths
Essential Exercises Year 6 Maths
Year 5 & 6 Scholarship Tests

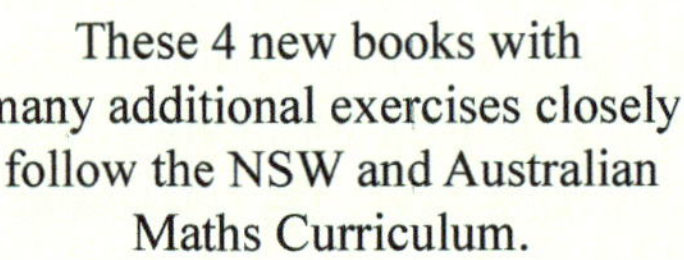

Important Facts and Formulas Series

Important Facts and Formulas Year 7
Important Facts and Formulas Year 8
Important Facts and Formulas Year 9&10 Intermediate
Important Facts and Formulas Year 9&10 Advanced
Important Facts and Formulas Year 11 2/3 unit
Important Facts and Formulas Year 12 2/3 unit

Understanding Comprehension Series

Understanding Year 3 Comprehension
Understanding Year 4 Comprehension
Understanding Year 5 Comprehension
Understanding Year 6 Comprehension
Understanding Year 7 Comprehension
Understanding Year 8 Comprehension

For availability and all other information regarding the above titles:

Five Senses Education
2/195 Prospect Highway
Seven Hills Sydney NSW 2147

www.fivesenseseducation.com.au

www.understandingmaths.com

For all educational needs, Five Senses Education is probably the largest supplier of educational books and learning aids in Australia.

CONTENTS

INTRODUCTION

Note (i): The Australian National Curriculum has been split into 3 major strands:

1. Number & Algebra (NA)
2. Measurement & Geometry (MG)
3. Statistics & Probability (SP)

Note (ii): The student may think that the last 5 chapters have already been explained in the 'Understanding Year 9 Maths' book by the same authors. However it is very important to note that although these final chapters do certainly revise earlier work, they also explain many new concepts and ideas that have not been included before. In addition, the exercises are more difficult and challenging. Some schools may choose not to teach ALL of the last 5 topics, because they are primarily intended for students who wish to study higher levels of Maths in later years.

WHY ARE YEARS 9 AND 10 SO IMPORTANT?

This book has been written to broadly follow both the Australian and the NSW Year 10 Mathematics Curriculum (5.2). It is intended to be a very thorough and concise summary of most major core topics which are taught and covered in Year 10 classes throughout Australia. Please keep in mind that this curriculum is based on the work covered in 2 years of high school (Year 9 and Year 10) which implies that different schools may teach the topics in a different order throughout the two year period. Therefore if you cannot find a particular topic summary in this Year 10 book (for example: measurement), then you will almost certainly find it explained in the Understanding Year 9 book. This book will prove to be very beneficial to students, teachers and coaches in all states. It has primarily been developed to be used as an additional learning aid to the conventional school text book, but it could also be used as a class text on its own account.

Years 9 and 10 are arguably the two most important years in the subject area of mathematics, because many completely new topics and ideas are introduced at a fairly rapid rate in order to complete the demanding syllabus in preparation for senior study. It is also essential for students to thoroughly know all the work covered in this two year course, because it will have a great influence on their later years choices and results. Teachers usually have little or no time to revise, and each new topic builds on theory covered in previous topics. With new ideas being introduced every day, it is easy for even good students to suddenly start falling behind — especially if they miss out on one or two key ideas.

This book is intended to overcome these problems by summarising all key formulae, rules and ideas in each major topic, as well as providing hundreds of fully worked examples. In addition, the graded topic review exercises at the end of each chapter will give students ample practice at typical exam and test questions.

Warwick Marlin

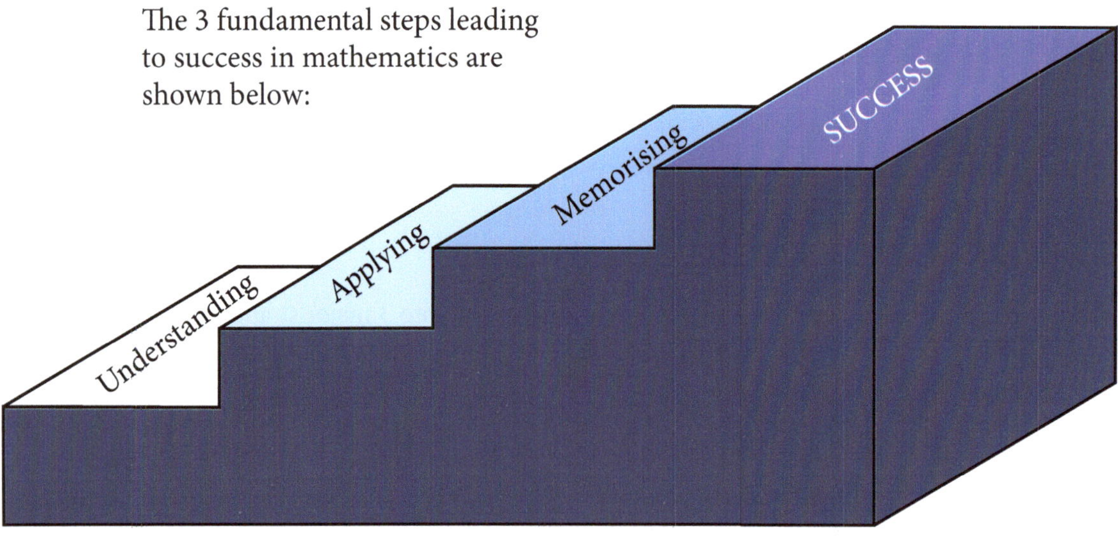

'The challenge of a good summary book is not to make simple ideas more complicated, but to make complicated ideas more simple.'

FEATURES AND BENEFITS OF THIS BOOK

1) Firstly, the book has been split up into **11 MAJOR IMPORTANT TOPICS.**
2) Each of these major topics has been broken down into a number of simpler ideas and rules, thus saving educators and students valuable **TIME** in research.
3) Most pages explain only one idea or rule, thus giving students **CLARITY**.
4) Each concept is thoroughly, but simply, explained for **UNDERSTANDING**.
5) Each formula, rule or set of steps is highlighted in larger print for ease of **MEMORISING**.
6) Each formula and page contains at least one fully worked example. This not only reinforces understanding, but also shows the student how to **APPLY** each formula to typical questions.
7) At the end of each chapter, there are usually at least 4 comprehensive graded exercises for **PRACTICE**, which cover all the ideas in the topic. If a student is not sure how to do a particular problem, all they need to do is turn back to the page number shown in order to find a very similar example. (Please read more about these exercises in the next section.)
8) In addition, the text presentation is well spaced out, in order to make the subject matter more **APPEALING** and **USER FRIENDLY** to this age group.
9) The opening contents page of each chapter contains **CURRICULUM REFERENCES** which will prove helpful and beneficial to teachers.
10) At the end of each chapter there is a **CONDENSED SUMMARY** of the complete chapter. These chapter summaries give all the formulas and rules that students must memorise before sitting for a test or exam.

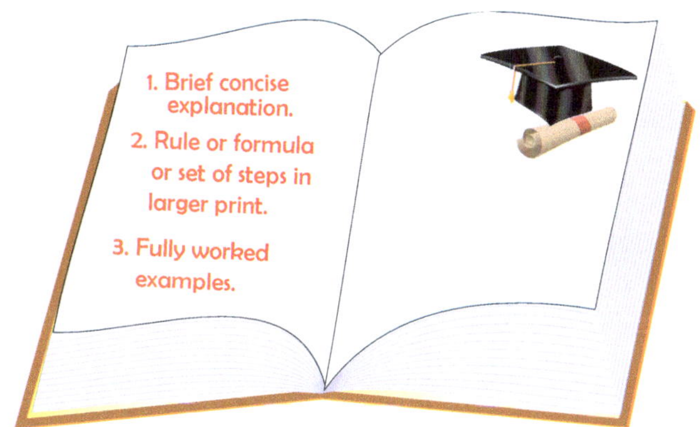

As you can see this book should prove to be an invaluable teaching aid for teachers, coaches and tutorial centres, because it thoroughly summarizes the 11 major topics which are the basis of the Year 10 Australian and NSW (5.2) Mathematics Curriculum.

Most important of all, it will prove to be an excellent reference for students of different ability groups. The user friendly format and layout makes it very much faster for a pupil to thoroughly master one major topic in a relatively short period of time, because it is so easy to see how each idea is linked to the previous one. It has all the rules and corresponding examples clearly set out topic by topic, page by page. In addition it teaches the pupils to read explanations, as well as to look back and research similar problems. And of course the graded exercises at the end of each topic chapter will help students to practise and apply their knowledge. It is so easy for students to work through the book by themselves with the minimum of supervision and help.

HOW TO USE THIS BOOK EFFECTIVELY?

As stated in the previous section, each chapter has been carefully split up into all the important ideas relating to that particular major topic. After carefully reading through and understanding a chapter, the next stage is for the student to work through some of the revision exercises given at the end. These have been graded into different levels of difficulty, and most Year 10 students should try to complete the first 3 levels. Remember that completing a level involves careful marking of your answers from the solutions given at the back of the book. If you got any questions wrong, then it is very important to find out why you got them wrong, before you move onto the next level.

The more difficult questions have partly or fully worked solutions.

EASIER QUESTIONS These questions given in level 1 (sometimes also in level 2) are intended to build up confidence and ensure that all the basic concepts have been understood. Most of the questions are almost identical to the examples given throughout the chapter – except with the numbers changed. If the student does have difficulty, then he/she can easily refer back to a similar example on the page number shown in the margin.

AVERAGE QUESTIONS The questions in Level 2 are of average difficulty level, and will give all students (average & gifted) a good opportunity to practise and consolidate the ideas and rules given throughout most of the chapter. Once again, if a student does have difficulty with a particular problem, then he/she can quickly refer back to a similar example on the page number shown. All students should try to complete and understand the questions in the first 2 levels.

HARDER QUESTIONS The questions in this level are usually more difficult than the ones in level 1 and 2 and some of the more difficult ideas in the topic are tested. Reference page numbers have NOT been included in the margin, in order to make students who reach this level more skilful in researching through the chapter for their own information. Level 3 should be done by more capable students, or those students who have found earlier levels straightforward.

PROBLEM SOLVING EXTENSION QUESTIONS This more difficult level has been included to challenge those students who are more gifted at Maths. Usually the questions are more sentence and problem oriented, and therefore they involve more reading and comprehension skills. It is unlikely that any of the questions in this level can be done mentally, because several different ideas, rules or steps are usually required. Because 'problem solving' is becoming very important in examinations, we have tried to include at least 1 exercise at the end of each chapter.

Note: Most chapters have 4 graded exercises at the end.
The reference pages work in the following way.
For example: 177, 178 in the margin means turn back to page 177 and page 178.
However 173-176 means turn back to pages 173, 174, 175 and 176.

All students should try to complete at least the first 2 or 3 levels. Some of the questions in the extension exercise are very challenging.

REVISION OF EARLIER YEARS

As already stated earlier, it is important to always keep in mind that the syllabus follows a spiral pattern. This means that each year there is some repetition and revision of the previous years work in order to reinforce and consolidate ideas.

For Example:

A topic such as geometry (which every parent and student has heard of) is taught from Year 7 right through to Year 12. But in each higher Grade, new ideas are added and built onto the ideas taught in previous years. And of course the geometry and theory becomes more and more complicated and difficult as the student gets older ..... just like the building bricks and foundations of a building.

Consequently, at the beginning of each chapter, we always start with some revision of work from previous years before introducing brand new work. Therefore students should be prepared to revisit and read some old theory and ideas.

You will see me on many of the pages... I will be trying to give you some reminders and advice.

Coincidence or Not?

If

A	B	C	D	E	F	G	H	I	J	K	L	M	N	O	P	Q	R	S	T	U	V	W	X	Y	Z
1	2	3	4	5	6	7	8	9	10	11	12	13	14	15	16	17	18	19	20	21	22	23	24	25	26

K + N + O + W + L + E + D + G + E
11+ 14 + 15 + 23 + 12 + 5 + 4 + 7 + 5 = 96%

H + A + R + D + W + O + R + K
8 + 1 + 18 + 4 + 23 + 15 + 18 + 11 = 98%

Both are very important but the total falls just short of 100%. However,

A + T + T + I + T + U + D + E
1 + 20 + 20 + 9 + 20 + 21 + 4 + 5 = 100%

'For success in any subject or discipline, attitude is just as important as ability.'

...Walter Scott

CALCULATORS AND COMPUTERS

The use and understanding of scientific **CALCULATORS** are essential in Years 9 and 10. They will be used extensively in most of the topics (Financial Maths, Indices, Measurement, Surds, Statistics and Trigonometry). There are several different manufacturers and models available on the market, and the type of calculator that you purchase will in most cases be the one recommended by your school or Mathematics teacher. Most calculators will include a user's guide, and it will be most beneficial for students to learn as much about their particular scientific calculator as possible. Calculators allow students to take risks and experiment with numbers. Playing with calculators gives opportunities for self discovery, often stimulating learning and interest in mathematical processes. Keep in mind that the most important part of 'problem solving' is understanding which operations or processes are used to get the answer.

Due to the large range and diversity of calculators, each with their own different function keys, it is almost impossible in this type of summary book to incorporate and explain their use in the calculation required for the solutions to the exercises. However, in some topics requiring the use of a calculator (namely: statistics and trigonometry), we have used the Casio fx-82 AU PLUS 11.

There are however four facts which must be made.

1. With simple problems, the brain is often quicker than the calculator.
2. It is easy to press a wrong key, so it is essential to estimate the answer to ensure the calculator's answer is a reasonable one to a particular problem.
3. If a student does not know which operation, or series of operations is needed, a calculator is useless. It does as it's told!
4. Calculators do not eliminate the need to thoroughly know addition/subtraction facts or times tables.

Also if one cannot understand how to apply the 4 operations (+, –, ×, ÷) to ordinary numerical fractions, then it will be impossible to apply these operations to algebraic fractions.

COMPUTERS provide access to many websites, thus giving unlimited practice with a particular mathematical concept, and allowing students to explore Mathematics at a greater depth and to develop problem solving skills and special awareness.

Many programs allow students to explore alternative strategies. They involve decision making and interpretation of information, often involving analysis and evaluation. Word processing and design programs can be used effectively to improve presentation. Graphing, data bases and spread sheets provide links with other curriculum areas.

FURTHER FINANCIAL MATHEMATICS

The 'Australian Curriculum Mathematics' (ACM) references for this sub-strand of 'Number and Algebra' (NA) are given below. This chapter may contain additional extension work which the author feels will be beneficial to the student.

- *Earning an income (NSW).*
- *Income tax (NSW).*
- *Simple interest (ACMNA 211).*
- *Compound interest and the formula (ACMNA 229).*
- *Depreciation (NSW).*
- *Time payments (NSW).*

Gottfried Wilhelm Leibnitz (1646 – 1716)

Leibnitz (sometimes referred to as 'The last Universal Genius') was another famous and highly acclaimed German mathematician, logician and natural philosopher who was prolifically active during the 'Age of Reason' or 'Enlightenment' (1685 – 1815). In mathematics his most famous contributions were in the fields of differential and integral calculus which he conceived independently of Isaac Newton, and arguably even before Newton. In fact, most maths works favour the Leibnitz notation as the conventional expression of calulus. He was one of the most ingenious inventors of mechanical calculators and the 'Leibnitz Wheel' was the first mass produced calculator that was ever developed. In addition, he refined the binary counting system which is the foundation of all digital modern day computers and silicon microchips. He also made significant contributions to physics, technology, probability theory, biology, medicine, geology, psychology, linguistics and computer science. If that wasn't enough, he also wrote many works on philosophy, politics, law, ethics, theology, history and philology. His numerous journals were written primarily in Latin, French and German but he also wrote some of his works in English, Dutch and Italian.

WAYS OF EARNING MONEY

Salary: A person on a 'salary' earns a fixed amount of money per year, usually paid on a fortnightly basis — and these people do not usually receive extra pay for working overtime. A good example is a school teacher.

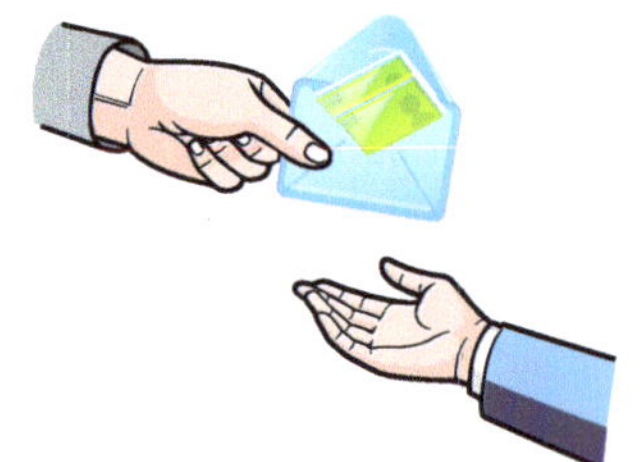

Wages: Usually paid weekly, and this is calculated on an hourly basis. For many jobs, the Industrial Award allows 'overtime rates' for any additional hours of work beyond the normal working week. These are calculated as follows:

Normal wages = rate of pay per hour × regular hours
Time-and-a-half = rate of pay per hour × overtime hours × 1.5
Double-time = rate of pay per hour × overtime hours × 2

Commission: Many people involved in sales are paid on a commission basis.
Examples include car sales, real estate sales, computer sales, swimming pool sales, kitchen sales, etc. etc.
The commission is expressed as a percentage of the value of the goods sold.

Casual: The person is not employed permanently, but works when required on a fixed rate per hour.
Examples include casual secretaries, casual teachers, gym instructors, etc.

Piece work: This is any type of employment in which a worker is paid a fixed piece rate for each unit produced or action performed regardless of time.
Examples include brick layers, fruit pickers, factory workers.

ALLOWANCES AND TERMS

HOLIDAY LOADING

People are often paid an extra amount during their annual holiday. This amount is usually calculated as a percentage of the person's usual average wage or salary.

SPECIAL RATES

These are usually paid to wage earners, who obtain an extra rate per hour, for working in dangerous or unpleasant conditions. Examples are: working with poisonous substances, working in confined spaces, working in sewerage, etc.

WEEKEND RATES

People on wages are often paid 'time-and-a-half or 'double-time' for working on Saturdays, Sundays or public holidays.

SUPERANNUATION

This is a lump sum amount of money paid out to a person when they retire. Usually the person and the employer contribute money for many years before retirement to a superannuation fund, and this earns interest on their contributions.

BONUS PAYMENTS

If a company makes a high end of year profit, they sometimes split a percentage of this profit up amongst the employees as a reward for efficient and loyal service.

GROSS PAY

This refers to a person's total weekly wage or salary before any deductions have been made. Net pay refers to the actual 'take home' pay after deductions have been made. These deductions could include income tax, superannuation, union membership, insurance, etc.

Example 1: Below is a typical time sheet for Mr. F. Jones.
Calculate his wage at the end of the week.

Name	Rate per hour	Regular hours	Overtime		Wage
			Time-and-a-half	Double time	
F. JONES	$17.62	40	7	3	

Normal wages = $\$17.62 \times 40 =$ $704.80
Time-and-a-half = $17.62 \times 7 \times 1.5 =$ $185.01
Double-time = $\$17.62 \times 3 \times 2 =$ $105.72
Total weekly wage = $995.53

Note: Time-and-a-half simply means 'multiply the respective overtime amount by 1.5'.
Double-time simply means 'multiply the respective overtime amount by 2.'

Example 2: A tiler is paid piecework rates at $12.75 per square metre. How much will he earn for tiling a rectangular kitchen floor measuring 3.2 m wide by 4.7 m long?

For 1 m^2 he gets paid $12.75

Area of floor = length × breadth
$= 4.7 \times 3.2 = 15.04 \text{ m}^2$
$\therefore$ For 15.04 m^2 he gets paid $15.04 \times \$12.75 = \$191.76$

FURTHER EXAMPLES

Example 1: John Ravenswood works as an architect and earns a salary of \$84 728 per year. He is entitled to 4 weeks annual leave per year with an $18\frac{1}{2}\%$ holiday loading. How much would he receive for the 4 week holiday period?

Solution:

One year = 52 weeks
∴ He earns = \$84 728 ÷ 52
= \$1 629.38 per week
= \$6 517.54 per 4 weeks.
Holiday loading = 18.5% × \$6 517.54
= \$1 205.74
Total holiday pay = \$6 517.54 + \$1 205.74
= \$7 723.28

Example 2: A labourer who normally gets paid \$14.31 per hour, also gets a special rate of \$2.57 cents/hour when he works in wet conditions.
How much will he earn for 8 hours of work digging trenches on a rainy day?

Solution:

Total amount per hour = \$14.31 + \$2.57
= \$16.88
In 8 hours he earns = 8 × \$16.88
= \$135.04

Example 3: A real estate salesman charges 5% commission on the first \$80 000 of the value of a house, and 2% commission on the remaining balance. How much commission will he earn if he sells a house for \$560 000?

Solution:

5% of \$80 000 = 0.05 × 80 000 = \$4 000
The remaining balance is \$560 000 – \$80 000 = \$480 000
He obtains 2% commission on this remaining balance
2% of \$480 000 = 0.02 × 480 000 = \$9 600.00
Therefore his total commission is \$13 600.

Example 4: Vicki Chan receives a weekly gross salary of \$780. If she pays 26% of this in income tax, 5% in superannuation and \$41.40 in medical insurance, what is her net weekly pay?

Solution:

Income tax = 26% × \$780 = \$202.80
Superannuation = 5% × \$780 = \$39.00
Medical insurance = \$41.40
Total deduction = \$283.20
Therefore her net weekly pay is \$780 – \$283.20 = \$496.80

INCOME TAX

The Government has to raise money to pay for its considerable expenses: hospitals, roads, schools, parks, salaries to ministers, etc. One of the ways of raising this money is to tax the yearly income of every individual wage earner — this is called **INCOME TAX**.
It is designed so that people with larger incomes pay more tax than people with smaller incomes. Every person with a certain minimum income must by law complete an income tax return at the end of the financial year (June 30th).

When all the legal deductions have been subtracted from a person's gross earnings, then the 'tax payable' is worked out using a tax table. This table changes from time to time and shouldn't be memorised. However, students must understand how to calculate the tax using a table, because this is a very frequently asked exam question.

Typical Tax Table

Taxable income		Tax on taxable income
From	**To**	
$1	$5 250	Nil
$5 251	$17 650	Nil plus 32 cents for each $1 in excess of $5 250
$17 651	$34 000	$3 968 plus 39 cents for each $1 in excess of $17 650
$34 001 and over		$10 344.50 plus 47 cents for each $1 in excess of $34 000

As you can see, the table has been divided into 4 major categories of incomes, with each of these on a separate line. Once you have been given a person's taxable income, simply read the information given on the corresponding line, and then do the calculations as stated on that line.

Example: Find the tax payable on the following taxable incomes:
a) $4876 b) $24 700 c) $69 460 d) $15 476

Solutions: a) Nil tax because it lies between $1 and $5 250.

b) This must be worked out using only line 3 on the tax table because it lies in the range $17 651 and $34 000.

i.e. $3968 plus 39 cents for each $1 in excess of $17 650
= $3968 + 0.39 × 7050
= $3968 + $2749.50
= $6717.50

Note: $24 700 is $7050 in excess of $17 650.

FURTHER EXAMPLES

c) This must be worked out using only the final line on tax table because it lies in the range of \$34 001 and over.

i.e. \$10 344.50 plus 47 cents for each \$1 in excess of \$34 000
= \$10 344.50 + 0.47 × 35 460
= \$10 344.50 + \$16 666.20
= \$27 010.70

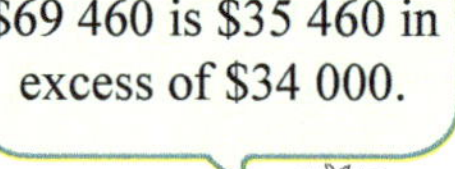

d) This must be worked out using only line 2, because it lies between \$5 251 and \$17 650.

i.e. Nil plus 32 cents for each \$1 in excess of \$5 250
= \$0.00 + 0.32 × 10 226
= \$0.00 + \$3 272.32
= \$3 272.32

Note: \$15 476 is \$10 226 in excess of \$5 250.

Example: Mr. Chen is a teacher who earns an annual salary of \$62 400. From private tutoring after school hours, he earns a further \$8 750 per year. He is allowed various tax deductions (for resource books etc.) amounting to \$3 843, and he has paid in advance tax installments of \$18 760 during the year.

Find: (i) his total gross income
(ii) his taxable income
(iii) tax payable from the table
(iv) balance owed or to be refunded.

Solutions: (i) Total gross income = \$62 400 + \$8 750
= \$71 150

(ii) Taxable income = gross income – deductions
= \$71 150 – \$3 843
= \$67 307

(iii) Tax payable = \$10 344.50 + \$0.47 × 33 307
= \$25 998.79

(iv) Balance owed
to taxation office = tax payable – tax instalments
= \$25 998.79 – \$18 760
= \$7 238.79

Note: In many instances, people get refunds from the Taxation Office at the end of June, because their tax instalments are larger than the tax payable.

SIMPLE INTEREST

When money is deposited into a bank, or other savings institution, then one can earn a certain amount of interest on the capital or principal invested. In reverse, if a person borrows money (for a car, house, etc.) then they have to pay the bank interest on the amount borrowed. The interest paid is expressed as a percentage rate per year. The words 'per year' are usually shortened to p.a. (per annum from latin). There are two methods of calculating the simple interest.

> i) Find the interest generated after 1 year.
>
> ii) Multiply this answer by the number of years the money has been invested.

Example: Find the simple interest if \$12 000 is invested in a savings account at 3% p.a. interest for 6 years.

Solution:

Interest per year $= 3\% \times \$12\,000$
$= 0.03 \times \$12\,000$
$= \$360.00$

$\therefore$ Interest for 6 years $= 6 \times \$360.00$
$= \$2\,160.00$

Multiply the interest rate by the amount of money invested.

ALTERNATIVELY: We could use the simple interest formula:

$$I = P \times R \times N$$

Where I = Interest
P = Principal invested
R = Rate of interest in decimal form
N = Time in years

By using the simple interest formula the above example could be done in one step.

$$I = P \times R \times N$$
$$\therefore I = 12\,000 \times \frac{3}{100} \times 6$$
$$\therefore I = 12\,000 \times 0.03 \times 6$$
$$\therefore I = \$2\,160.00$$

Where P = \$12 000
R = 3% = 0.03
N = 6 years

Note: You can use R as a fraction $\frac{3}{100}$, but usually it is easier to change the percentage into the decimal equivalent of 0.03.

FURTHER EXAMPLES

Example 1: Find the simple interest if \$12 000 is invested in a savings account at $2\frac{1}{2}\%$ per annum interest for 7 months.

Solution: The interest rate R is $2\frac{1}{2}\%$ per year.
Therefore we first have to change 7 months to $\frac{7}{12}$ of a year.

$I = P \times R \times N$ Where $P = \$12\ 000$

$R = 2.5\% = 0.025$

$N = \frac{7}{12}$

$\therefore\ I = 12\ 000 \times 0.025 \times \frac{7}{12}$

$= \$175.00$

In the simple interest formula $I = P \times R \times N$ there are 4 variables I, P, R and N. Most questions will ask you to evaluate the interest (I), but you could also be given the interest and asked to calculate one of the other variables in the formula.

Example 2: Joy deposited \$25 000 into a savings account. After 2 years her new balance was \$26 200. What is the rate of interest per year?

Solution: $I = P \times R \times N$ Where $I = \$1\ 200$

$P = \$25\ 000$

$R = ?$

$N = 2$ years

$\therefore\ 1\ 200 = 25\ 000 \times R \times 2$

$\therefore\ R = \frac{1\ 200}{50\ 000} = 0.024 = 2.4\%$

Note: We could also just transpose the formula to $R = \frac{I}{P \times N}$

Example 3: I deposit some money into a savings account for 3 years at an interest rate of 1.5% p.a. If my interest after 3 years is \$1 350, how much did I deposit?

Solution: $I = P \times R \times N$ Where $I = \$1\ 350$

$P = ?$

$R = 1.5\% = 0.015$

$N = 3$ years

$\therefore\ 1\ 350 = P \times 0.015 \times 3$

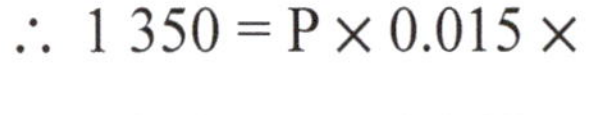

$\therefore\ 1\ 350 = P \times 0.045$

$\therefore\ P = 1\ 350 \div 0.045$

$\therefore\ = \$30\ 000$

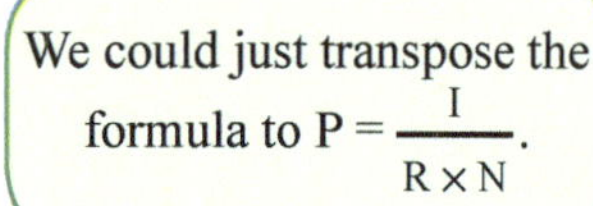

COMPOUND INTEREST

This is different from 'simple interest' because at the end of each year the interest earned is added to the principal. The interest for the next year is therefore calculated on the 'principal plus added interest'. This means that compound interest is always increasing year by year, unlike simple interest which remains the same.

Example: Find to the nearest cent the compound interest on a principal of $3400 invested for 3 years at 9% p.a.

Principal for 1st year	$3 400.00
Interest for 1st year	$306.00
Principal for 2nd year	$3 706.00
Interest for 2nd year	$333.54
Principal for 3rd year	$4 039.54
Interest for 3rd year	$363.56
Total amount	$4 403.10
∴ Compound interest	$1 003.10

$3\,400 \times .09 = 306.00$

$3\,706 \times .09 = 333.54$

$4\,039.54 \times .09 = 363.56$

To calculate the actual compound interest, we have to subtract the original principal from the total amount accrued.
i.e. $4 403.10 – $3 400 = $1 003.10

Note: Compare the simple interest earned on $3400 at 9% p.a. for 3 years.

$I = P \times R \times N$

$= 3\,400 \times 0.09 \times 3$

$= \$918$

The simple interest remains constant each year.

The difference between the compound interest and the simple interest in the example above is $1 003.10 – $918 = $85.10. It can be clearly seen that one can earn more if the interest is compounded.

COMPOUND INTEREST FORMULA

On the previous page it was explained how to work out the compound interest on a year to year calculation using the steps shown. However, there is a faster method using the formula below:

$$A = P\left(1 + \frac{r}{100}\right)^n$$

where A = Total amount after n time intervals
P = Principal invested
r = Interest rate for 1 time interval
n = Number of time intervals

Note: (i) When using the above formula, it is important to note that 'A' is the total amount (principal + interest) after n time periods. To find the actual compound interest it is necessary to subtract the original principal (P) from the amount accrued (A). Therefore read the question carefully to find out what is actually required.

(ii) The time interval 'n' is usually in years. However it could be in half years or months as shown in the second example.

Example 1: Find to the nearest cent the compound interest on a principal of \$3400 invested for 3 years at 9% p.a.

Using $A = P\left(1 + \frac{r}{100}\right)^n$

$\therefore \quad A = 3\,400\left(1 + \frac{9}{100}\right)^3$

$\therefore \quad A = 3\,400 \times 1.09^3$

$\therefore \quad A = 4\,403.10$

$\therefore \quad$ Compound interest = \$4 403.10 – \$3 400.00

= \$1 003.10

This is the same example as on the previous page, except we are now using the faster formula method.

Example 2: Find the amount of compound interest earned if \$6 000 is invested at 6% p.a. for 5 years and the interest is compounded monthly.

Using $A = P\left(1 + \frac{r}{100}\right)^n$

$\therefore \quad A = 6\,000\left(1 + \frac{0.5}{100}\right)^{60}$

$\therefore \quad A = 6\,000 \times 1.005^{60}$

$\therefore \quad A = \$8\,093.10$

$\therefore \quad$ Compound interest = \$8 093.10 – \$6 000.00

= \$2 093.10

Note: 'n' and 'r' must first be converted to monthly time intervals.

$n = 60$ months

$r = \frac{6}{12}\%$ p.m. $= \frac{1}{2}\%$ p.m.

DEPRECIATION

An article is said to be depreciating when it loses value year by year. Many businesses have to calculate the depreciation on their machinery, cars, office equipment, etc. for tax purposes. They can claim this as a tax deduction. The amount of depreciation is calculated using the same methods shown in compound interest. The only difference is that the amount of depreciation is subtracted each year, while in compound interest the amount of interest was added each year.

Example: A photocopying machine originally costing \$7 000.00 depreciates at 20% p.a. Find the depreciation after 3 years.

Value at beginning of 1st year	\$7 000
Depreciation for 1st year	\$1 400
Value at beginning of 2nd year	\$5 600
Depreciation for 2nd year	\$1 120
Value at beginning of 3rd year	\$4 480
Depreciation for 3rd year	\$896
Value after 3 years	\$3 584
Total depreciation over 3 years	\$3 416

$$\begin{array}{r} 7\,000 \times \\ .20 \\ \hline 1\,400 \end{array} \qquad \begin{array}{r} 5\,600 \times \\ .20 \\ \hline 1\,120 \end{array} \qquad \begin{array}{r} 4\,480 \times \\ .20 \\ \hline 896 \end{array}$$

The total depreciation over 3 years can be calculated by subtracting the value after 3 years (\$3 584) from the original value (\$7 000).

$\therefore$ Depreciation = \$3 416.00

Note: It is very important to read the question carefully, because in some instances you could be asked to find the 'value' of the article after so many years, rather than the 'depreciation'.

DEPRECIATION FORMULA:

It was explained above how to work out the depreciation on a year to year calculation using the steps shown. As with compound interest, there is a faster method of calculating depreciation using the formula below:

$$A = P\left(1 - \frac{r}{100}\right)^n$$

where A = Value after n time intervals
P = Initial value of the item
r = Interest rate for 1 time interval
n = Number of time intervals

Note: Once again, the time interval 'n' is usually measured in years, but it could also be in 'half years' or 'months'.

EXAMPLES

Example 1: A photocopying machine originally costing \$7 000.00 depreciates at 20% p.a. Find the depreciation after 3 years.

Solution: a) Using $A = P\left(1 - \frac{r}{100}\right)^n$

$\therefore \quad A = 7\,000\left(1 - \frac{20}{100}\right)^3$

$\therefore \quad A = 7\,000 \times 0.8^3$

$\therefore \quad A = \$3\,584$

Its value after 3 years is \$3 584.00.

b) The depreciation is \$7 000 – \$3 584 = \$3 416.

Example 2: A car now worth \$8000 has been depreciating at a rate of 15% per year for the last 4 years. What was its original value 4 years ago to the nearest dollar?

Solution: Using $A = P\left(1 - \frac{r}{100}\right)^n$

$\therefore \quad 8\,000 = P\left(1 - \frac{15}{100}\right)^4$

$\therefore \quad 8\,000 = P \times 0.85^4$

$\therefore \quad P = \frac{8\,000}{0.85^4}$

$= \$15\,325.00$

The original value was \$15 325.00.

Sometimes you will have to find other variables in the formula besides '*A*'.

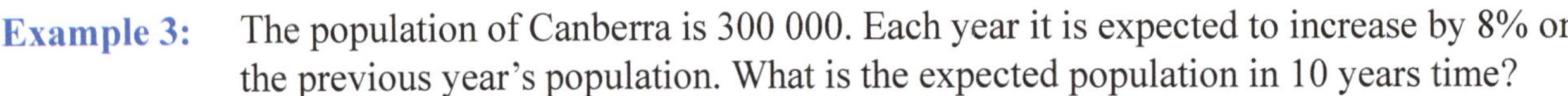

Example 3: The population of Canberra is 300 000. Each year it is expected to increase by 8% on the previous year's population. What is the expected population in 10 years time?

Solution: This is an example of 'Growth' using the compound interest formula which doesn't involve money.

$A = P\left(1 + \frac{r}{100}\right)^n$

$\therefore \quad A = 300\,000\left(1 + \frac{8}{100}\right)^{10}$

$\therefore \quad A = 300\,000 \times 1.08^{10}$

$\therefore \quad A = 647\,677$

Expected population is 647 677 people.

Not all problems are necessarily related to money. Some problems involve 'Growth' or 'Decay'.

TIME PAYMENT

The process of paying for an article over an extended period of time is often called TIME PAYMENT or HIRE PURCHASE. There is always an extra cost to cover the interest on the outstanding money still to be paid. This 'extra cost' can be quite high.

IMPORTANT TERMS:

CASH PRICE	—	this is the amount an article is worth if it is paid for immediately,
DEPOSIT	—	this is a 'part payment' of the full amount owing.
BALANCE	—	this is the amount still owing after the deposit has been paid.
MONTHLY INSTALMENT	—	this is the monthly amount the purchaser will have to pay over a certain period of time.

FINDING THE MONTHLY INSTALMENT:

Step (i): Calculate the deposit if it is expressed as a percentage.
Step (ii): Calculate the balance owing, i.e. Cash price – Deposit.
Step (iii): Calculate the interest on the balance using:

$$I = P \times R \times N$$

Step (iv): Calculate the total amount still owing.
i.e. Balance + Interest.
Step (v): Divide total amount owing by the number of months over which it is repaid.

Note: If a bank offers a loan at 'Reducible Interest' it means that one only pays interest on the balance of money owing at the end of each month.
If a bank offers a loan at a 'Flat rate' it means that interest is calculated on the original loan — even towards the end of the borrowing period when most of the loan has been repaid.
A flat interest rate is approximately equal to twice the reducible interest rate.

EXAMPLE ON TIME PAYMENT

Students should be aware of the difference between 'reducible rate' and 'flat rate'. However, to make calculations simpler in Year 10, most problems and questions involve only a 'flat interest rate'.
The following type of example occurs frequently in exams. You should check all the working with your calculator and memorise the different steps.

Example: The cash price of a new car is \$34 700. A person pays a deposit of 20% and agrees to pay the balance over 3 years at a flat interest rate of 14% p.a. What is his monthly instalment?

Solution:

Step (i): Calculate the deposit.

Deposit = 20% of \$34 700
= \$6 940

Step (ii): Calculate the balance.

Balance = \$34 700 – \$6 940
= \$27 760

Step (iii): Calculate the interest on the balance.

$$\text{Interest} = \frac{P \times R \times N}{100}$$

$$= \frac{27\,760 \times 14 \times 3}{100}$$

= \$11 659.20

Step (iv): Calculate the total amount still owing.
(i.e. the balance + interest)
Total amount owing = \$27 760 + \$11 659.20
= \$39 419.20

Step (v): Divide total amount owing by number of months.

$$\text{Monthly instalments} = \frac{34\,419.20}{36}$$

= \$1 094.98

Note: The extra amount paid, using time payment, is the interest, i.e. \$11 659.20 in the above example.

CHAPTER SUMMARY

PERCENTAGE PROFIT

It is useful in business to know how much profit is being made as a percentage of cost price.

Step 1: Find the profit = selling price – cost price

Step 2: Express the profit as a fraction of the cost price.

Step 3: Convert to a percentage by multiplying by $\frac{100\%}{1}$

ie. Percentage profit $= \frac{\text{Profit}}{\text{Cost price}} \times \frac{100\%}{1}$

Percentage loss $= \frac{\text{Loss}}{\text{Cost price}} \times \frac{100\%}{1}$

SALARIES

A person on a 'salary' earns a fixed amount of money per year, usually paid on a fortnightly basis — and these people do not usually receive extra pay for working overtime. A good example of a person on a salary is a school teacher.

WAGES

People who earn 'wages' get paid weekly, and this is calculated on an hourly basis. For many jobs, the Industrial Award allows 'overtime rates' for any additional hours of work beyond the normal working week.

Normal wages	=	**rate of pay per hour × regular hours**
Time-and-a-half	=	**rate of pay per hour × overtime hours × 1.5**
Double-time	=	**rate of pay per hour × overtime hours × 2**

Note: Time-and-a-half simply means multiply the respective overtime amount by 1.5. Double-time simply means multiply the respective overtime amount by 2.

COMMISSION

There are many occupations in the workforce which involve some type of selling (e.g. car sales, computer sales, real estate sales, clothes sales, swimming pool sales, etc.)

A sales person is usually paid a commission. This commission is expressed as a percentage of the value of goods sold.

CALCULATING INCOME TAX FROM A TABLE

When all the legal deductions have been subtracted from a person's gross earnings, then the 'tax payable' is worked out using a tax table. This table changes from time to time and shouldn't be memorised.

SIMPLE INTEREST

When money is deposited into a Bank or Building Society, then one can earn a certain amount of interest on the capital or principal invested. In earlier years you learned one method of working out the interest in 2 stages. There is however a formula which calculates the interest directly.

The Simple Interest Formula:

$$I = P \times R \times N$$

where I – Interest
P – Principal
R – Rate of interest
N – Number of years

COMPOUND INTEREST

This is different from 'simple interest' because at the end of each year any interest earned is added to the principal. The interest for the next year is therefore calculated on the 'principal plus added interest'. This means that compound interest is always increasing year by year, unlike simple interest which remains the same. Compound interest problems can be worked out on a year by year basis.

However, the following formula calculates the compound interest more directly:

$$A = P\left[1 + \frac{r}{100}\right]^n$$

where A = total amount after n time intervals
P = principal invested
r = interest rate for 1 time interval
n = number of time intervals

Note: When using the above formula, it is important to note that 'A' is the total amount (principal + interest) after n time periods. To find the actual compound interest it is necessary to subtract the original principal (P) from the amount accrued (A). Therefore read the question carefully to find out what is actually required.
The time interval 'n' is usually in years. However it could be in half years or months.

DEPRECIATION

An article is said to be depreciating when it loses value year by year. Many businesses have to calculate the depreciation on their machinery, cars, office equipment, etc. for tax purposes. The amount of depreciation is calculated using the same methods shown in compound interest. The only difference is that the amount of depreciation is subtracted each year, while in compound interest the amount of interest was added each year.

$$A = P\left[1 - \frac{r}{100}\right]^n$$

where A = value after n time intervals
P = initial value of the item
r = interest rate for 1 time interval
n = number of time intervals

TIME PAYMENT OR HIRE PURCHASE

The process of paying for an article over an extended period of time is often called **TIME PAYMENT** or **HIRE PURCHASE**. There is always an extra cost to cover the interest on the outstanding money still to be paid. This 'extra cost' can be quite high.

Step 1: Calculate the deposit if it is expressed as a percentage.

Step 2: Calculate the balance i.e. Cash price – Deposit.

Step 3: Calculate the interest on the balance using: $I = P \times R \times N$

Step 4: Calculate the total amount still owing, i.e. Balance + Interest.

Step 5: Divide total amount owing by the number of months over which it is repaid.

LEVEL 1 — FURTHER FINANCIAL MATHS

EASIER QUESTIONS

Note: Only turn back to page number shown if you have difficulty. — Page

Q1. Randy deposits \$12 000 into a savings account and he earns simple interest of $2\frac{1}{2}\%$ p.a. — 7, 8

a) What interest does he earn after 1 year?
b) What interest does he earn after 6 years?
c) What interest does he earn after 4 years and 9 months?

Q2. Find the simple interest earned on the following amounts: — 7, 8

a) \$8 000 at $3\frac{1}{2}\%$ p.a. for 4 years b) \$12 000 at $3\frac{1}{4}\%$ p.a. for 3 years
c) \$13 400 at 2% p.a. for $3\frac{1}{2}$ years d) \$18 000 at $2\frac{3}{4}\%$ p.a. for 18 months

Q3. Jessica deposits \$8 000 into a savings account for 4 years and earns \$960 interest. What is the interest rate per year? — 8

Q4. Zac invests \$12 000 in a savings account at 2% p.a. and earns \$1 800. For how long did Zac invest his money? — 8

Q5. Judy is working at a dress factory. If she works a normal 40 hour week, then she gets paid at \$22.00 per hour. For the next 10 hours she gets paid 'time-and-a-half'. And if she works more than 50 hours, she gets paid 'double-time'. She also gets taxed 26% of her gross earnings each week to get her net wages. — 2

a) Calculate her gross earnings if she worked 43 hours during the week.
b) Calculate her net earnings during the same week.
c) Calculate her net earnings if she works for 54 hours in one week.
d) Calculate how many hours she worked in one particular week, when her gross income was \$1 342.00.

Q6. Kim-Sun bought a new car, which costs \$38 000, if paid for in cash. He paid a $17\frac{1}{2}\%$ deposit, and repaid the balance owing over a 4 year period at a flat interest rate of $9\frac{1}{4}\%$. — 13, 14

a) What was his deposit?
b) How much interest did he pay over the 4 years?
c) What is his monthly instalment?

Q7. A savings account on March 1st is started with \$60 000. The account pays 3.2% interest per annum (on the minimum monthly balance) at the end of the month and \$4 000 is deposited on the last day of each month into the account. Complete the table to find the balance in the account on June 30th. — 9

Date	Balance	Contribution	Interest for the Month
March 1st	\$60 000		
March 31st			
April 1st			
April 30th			
May 1st			
May 30th			
June 1st			
June 30th			

 AVERAGE QUESTIONS

Note: Only turn back to page number shown if you have difficulty.

	Page
Q1. Find the simple interest earned for the following: a) \$15 000 at 3% p.a. for 6 years. b) \$25 000 at $2\frac{1}{4}$ p.a. for 5 years. c) \$8 500 at 1.7% p.a. for $4\frac{1}{2}$ years. d) \$42 800 at $1\frac{3}{4}$% p.a. for 2 years 3 months.	7, 8
Q2. Find the compound interest earned for the following: a) \$8 000 at 5% p.a. for 3 years. b) \$20 000 at $3\frac{1}{2}$% p.a. for 5 years. c) \$18 350 at 1.35% p.a. for 4 years. d) \$9 400 at $2\frac{1}{4}$% p.a. for 8 years.	9, 10
Q3. The initial cost price of a new car is \$42 500. If it depreciates at 20% p.a. what is its value after 5 years?	11
Q4. Find the compound interest on a principal of \$20 000 borrowed at 6% p.a. for 4 years compounded monthly.	10
Q5. Using the tax table on page 5, calculate the tax payable on the following taxable incomes: a) \$7 825 b) \$24 053 c) \$59 968 d) \$78 124	5, 6
Q6. a) How much more will Ingrid pay using time payment for a mountain bike worth \$940 if she pays $13\frac{1}{4}$% deposit and \$64.75 per month for 18 months. b) Greg wants to buy a car worth \$7 760. He can either pay a deposit of \$1 000 and then \$215 per month for 36 months, or he can borrow the entire amount at $6\frac{1}{2}$% p.a. simple interest payable over 24 months. Which is the better choice and by how much?	13, 14
Q7. a) Joydeep receives a normal hourly wage of \$21.40 and time-and-a-half for any overtime. On public holidays he receives double-time with a minimum of 4 hours paid. What is his wage for a week in which he works 40 normal hours, 7.5 hours overtime and 2 hours on a public holiday? b) Karen earns a weekly gross salary of \$820. Tax of \$148.90 is deducted, 6.5% is paid into a superannuation fund and 8% is paid directly towards her mortgage. Karen aims to save 12% of her remaining income. What amount will she save each week?	2, 3
Q8. a) Thomas received \$602 as a holiday loading payment ($17\frac{1}{2}$% on 4 weeks' pay). What is his weekly wage? b) Kim receives time-and-a-half for any hours worked exceeding the normal 38 hours per week. The normal hourly rate is \$18.40. In a particular week Kim receives \$795.80. How many hours overtime did he work in that week?	2, 3
Q9. Becky earns \$860 for a 40-hour week. Any overtime is paid at double-time. a) Using the tax schedule on page 20, what is the tax payable on Becky's normal wage assuming she works for 48 weeks during the year? b) How many hours overtime can she work in a year and still remain in the same tax category?	2, 3

Q1. Find the simple interest on a principal of:

a) \$5 893 at 23% p.a. for 3 months.

b) \$16 330 at $9\frac{1}{2}\%$ p.a. for 32 months.

c) \$27 500 at 4.25% p.a. for 44 days.

Q2. What simple interest rate would allow \$10 110 to grow to \$14 027 in 5 years? [Answer to 2 d.p.]

Q3. Find the compound interest on a principal of:

a) \$6 540 at 6% p.a. for 4.5 years.

b) \$15 430 at $5\frac{1}{4}\%$ p.a. for 42 months.

For this question, the interest is compounded bi-annually.

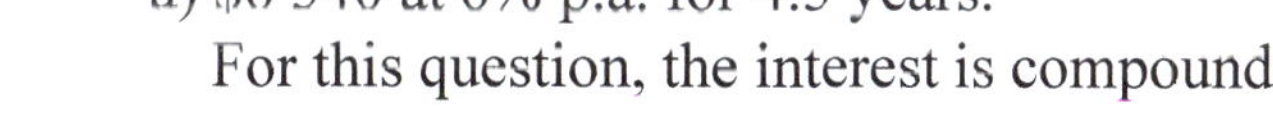

Q4. Find the compound interest on a principal of:

a) \$1 040 invested for 20 years at 15% p.a. compounded half yearly.

b) \$30 600 invested for 2 years at 6% compounded quarterly.

c) \$11 920 borrowed at 9% p.a. for 5 years compounded monthly.

Q5. I want to invest \$3 550 for 10 years. I have a choice of investing at a simple interest rate of 15.5% p.a. or a compound rate of 10.25% p.a. Which is the better option and by how much?

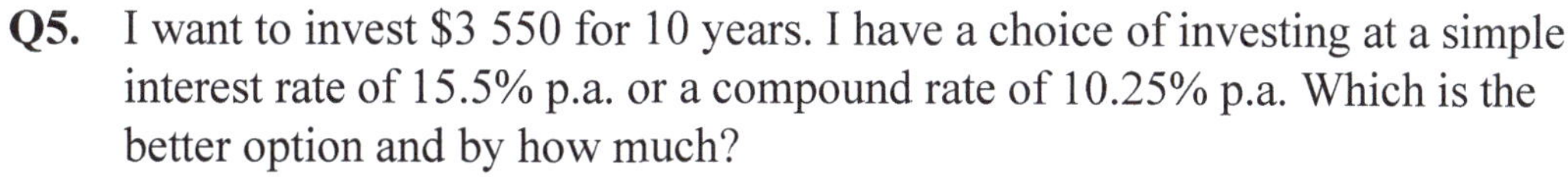

Q6. \$7 500 is invested for 3 years with interest compounded biannually. If at the end of the 3 years the investment is worth \$9 767, what is the applied interest rate?

Q7. Find the value after 8 years of:

a) a car costing \$29 000 depreciated at 15% p.a.

b) a printer costing \$569 depreciated at 6.5% p.a.

c) a mobile phone costing \$799 depreciated at $33\frac{1}{3}\%$ p.a.

Q8. After being depreciated at 7% p.a. for 12 years, a laptop computer is valued at \$413. What was its value 12 years ago?

Q9. Joe and Wendy bought a house costing \$210 000. They paid a 35% deposit and borrowed the remainder at a flat interest rate of 7.75% p.a. payable over 25 years. What is the amount of their total monthly instalment?

Q10. A \$25 000 loan is repaid over 8 years with monthly instalments of \$391. What was the interest rate (p.a.) charged on the loan?

Q11. The tax of Wendy-Li for 1 year was \$21 549.30. Use the tax table shown on page 3 to calculate her gross wage for the year.

LEVEL 4 — FURTHER FINANCIAL MATHS

EXTENSION QUESTIONS

Q1. The Tax Scale for the 2018 – 2019 financial year is given.

Tax Scale: 2018 – 2019

0 – $18 200	Nil
$18 201 – 37 000	19 cents for each dollar over $18 200
$37 001 – $90 000	$3 572 plus 32.5 cents for each $1 over $37 000
$90 001 – $180 000	$20 797 plus 37 cents for every dollar over $90 000
$180 001 and over	$54 097 plus 45 cents for each dollar over $180 000

Use the Tax Scale to find the annual amount of tax that each person needs to pay.

a) Thomas earns $86 560 per annum with 4 weeks Holiday Leave Loading at 17.5% and a $5 000 bonus.

b) Lucinda works at the rate of $24.50 per hour. Each week she works 28 hours at that rate and for 4 hours at time-and-a-half. She takes four weeks' unpaid holiday leave and is paid the 17.5% Holiday Leave Loading at the end of the year.

Q2. Employers are required to pay 9.5% of an employee's Gross Salary into their Superannuation Fund which doesn't affect the employee's taxable income. An employee can choose to make 'voluntary contributions' into their Superannuation Fund, known as Salary Sacrifice. The amount contributed in this way lowers an employee's taxable income. At the moment the taxation law sets the amount that can be paid into an employee's superannuation fund both as a result of contribution by the employer and through Salary Sacrifice at $25 000.

Use the 2018 – 2019 Tax Scale to find the annual amount of tax that each person needs to pay.

a) Katy earns $135 600 per annum with 4 weeks Holiday Leave Loading at $17\frac{1}{2}$%. Her employer makes the compulsory 9.5% payment into her Superannuation Fund and she makes the maximum salary sacrifice possible.

b) Fred works in a fast-food outlet for $24 per hour for 38 hours per week. He works for 40 weeks for the year and his employer makes the compulsory 9.5% amount and he makes the maximum salary sacrifice possible.

Q3. According to current depreciation rates, the value of a new vehicle drops by 20 percent after the first 12 months of ownership. Then, for the next four years, a car loses 10 percent of its value annually.

Find the value each year for the first five years for the following cars and state the percentage overall depreciation for the first five years.

(i) Lamborghini Huracan Peformante Spyder: $308 859

(ii) Porsche Turbo S E-hybrid Sport Turismo: $469 500

Q4. Calculate the difference between the interest earned on the following investments:

a) $20 000 is invested for five years with the interest compounding annually compared to $20 000 invested for five years with simple interest, both at the rate of 6% p.a.

b) $50 000 is invested for four years with the interest compounding annually compared to $50 000 invested for four years with the interest compounding monthly, both at the rate of 3% p.a.

c) $120 000 is invested at simple interest compared to $120 000 invested with interest compounding monthly, both at the rate of 8% p.a. for ten years.

SURFACE AREA AND VOLUME

The 'Australian Curriculum Mathematics' (ACM) references for this sub-strand of 'Measurement and Geometry' (MG) are given below. This chapter may contain additional extension work which the author feels will be beneficial to the student.

- *Surface area of a cylinder (ACMMG 217).*
- *Surface area of prisms (ACMMG 218).*
- *Volumes of prisms and cylinders (ACMMG 242).*
- *Volumes of pyramids, cones and spheres (ACMMG 271).*
- *Surface areas of pyramids, cones and spheres (ACMMG 271).*
- *Surface areas and volumes of composite solids (ACMMG 271).*

Srinivasa Ramanujan (1887 – 1920)

Ramanujan was an Indian who lived during the time of the British Rule in India. Although he had almost no formal training in mathematics, he made significant contributions in a number of fields, including mathematical analysis, number theory, infinite series and continued fractions. In addition, he found the solutions to many problems that were at that time considered unsolvable. He initially worked and developed his theories in relative isolation, even though he had tried unsuccessfully to interest the leading professors and mathematicians in India regarding his work. Eventually an English professor at Cambridge University (G.H.Hardy) recognized him as a mathematical genius and invited him over to further his research and studies. Hardy commented that Ramanujan had produced some ground breaking new theorems, unlike anything he had ever seen before. During his short life (only 33 years) he independently compiled more than 3 900 results and identities some of which are named after him, such as the 'Ramanujan prime' and the 'Ramanujan theta function'. He became one of the youngest Fellows of the Royal Society and the first Indian to be elected a Fellow of Trinity College, Cambridge. Unfortunately he died at a young age because of complications resulting from dysentery, but he is often compared to other great mathematical geniuses such as Euler and Gauss.

SURFACE AREA OF COMMON SOLIDS

To find the surface area of a solid means to add up the areas of all the surfaces which make up or surround that solid. From earlier years, you should already by able to calculate the surface area of prisms (i.e. rectangular, triangular and other prism shapes).

NAME	SHAPE	SURFACE AREA
Rectangular Prism	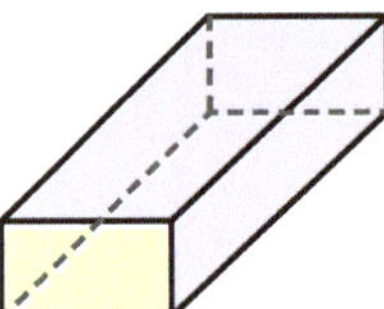	Add up the areas of the 6 rectangular faces. Opposite faces are congruent (i.e. identical in shape and size).
Triangular Prism	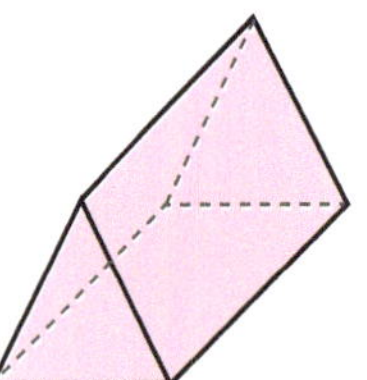	Add up the areas of the 5 faces. 2 congruent triangles plus 3 rectangles.
Square Pyramid	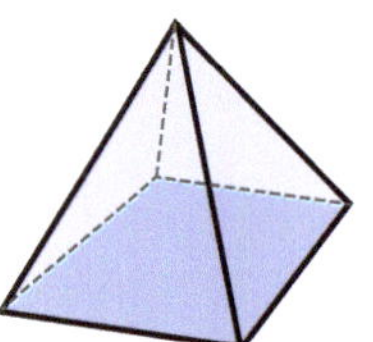	Add up the areas of the 5 faces. 1 square plus 4 congruent triangles.
Rectangular Pyramid	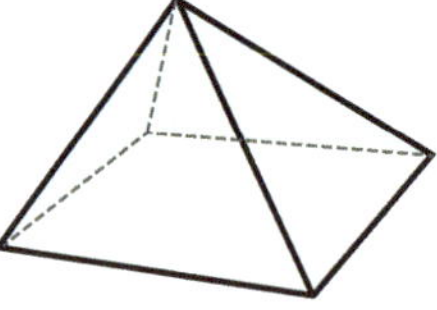	Add up the areas of the 5 faces. 1 rectangle plus 4 triangles. Opposite facing triangles are congruent.
Cylinder	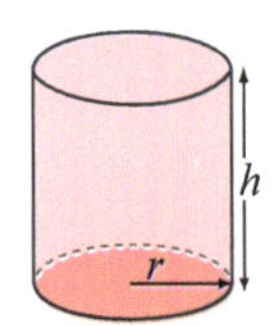 **Note:** This will be further explained later in the chapter.	Curved surfaced area = $2\pi rh$ Total surface area (including the 2 circles) $= 2\pi rh + 2\pi r^2$
Cone	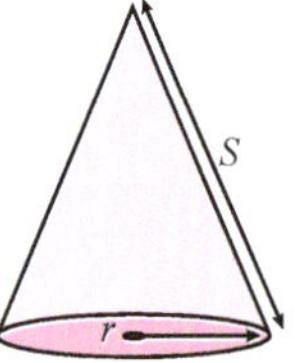	Surface area of a cone $= \pi rS + \pi r^2$ where r = radius and S = slant height
Sphere	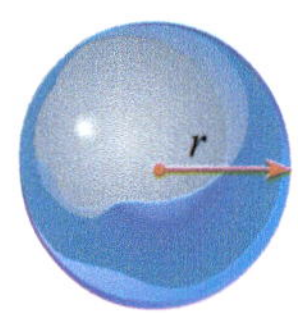	Surface area of a sphere $= 4\pi r^2$ where r = radius

SURFACE AREA OF PRISMS

Example 1: Find the surface area of the triangular prism with dimensions as shown.

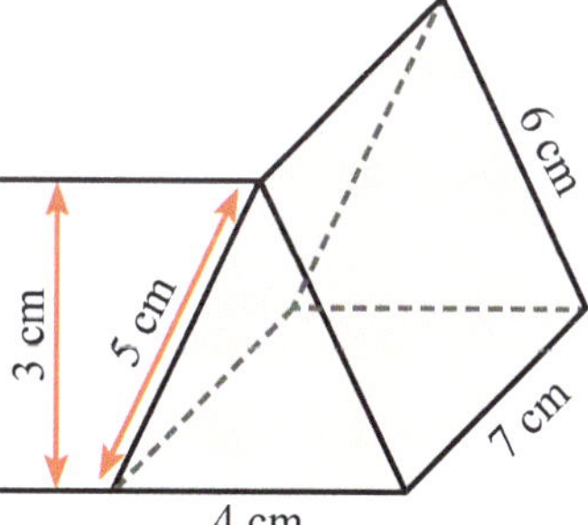

Solution:

$$\begin{aligned}\text{Front triangle} &= \tfrac{1}{2}bh \\ &= \tfrac{1}{2} \times 4 \times 3 \\ &= 6 \text{ cm}^2 \\ \text{Back triangle} &= 6 \text{ cm}^2 \\ \text{Rectangle (i)} &= 6 \times 7 \\ \text{(facing upwards)} &= 42 \text{ cm}^2 \\ \text{Rectangle (ii)} &= 5 \times 7 \\ \text{(unseen behind)} &= 35 \text{ cm}^2 \\ \text{Rectangle (iii)} &= 4 \times 7 \\ \text{(on base)} &= 28 \text{ cm}^2 \\ \text{Total surface area} &= 6 + 6 + 42 + 35 + 28 \\ &= 117 \text{ cm}^2\end{aligned}$$

Example 2: Find the surface area of the trapezoidal prism with dimensions as shown.

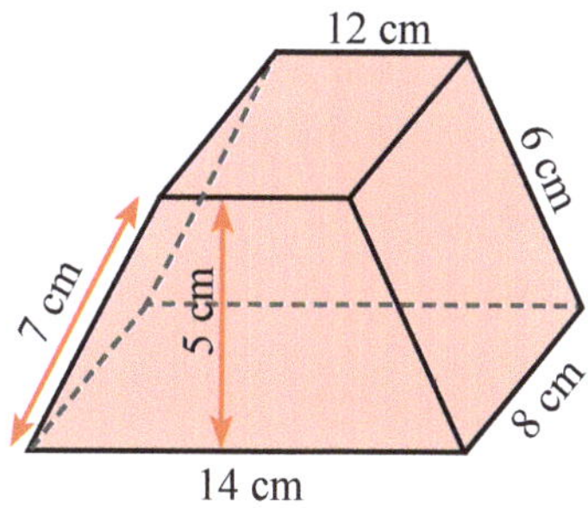

Solution:

$$\begin{aligned}\text{Front trapezium} &= \tfrac{1}{2}h(a+b) \\ &= \tfrac{1}{2} \times 5 \times (14 + 12) \\ &= 65 \text{ cm}^2 \\ \text{Back trapezium} &= 65 \text{ cm}^2 \\ \text{Rectangle (i)} &= 8 \times 6 \\ \text{(facing upwards)} &= 48 \text{ cm}^2 \\ \text{Rectangle (ii)} &= 8 \times 7 \\ \text{(unseen behind)} &= 56 \text{ cm}^2 \\ \text{Rectangle (iii)} &= 12 \times 8 \\ \text{(on the top)} &= 96 \text{ cm}^2 \\ \text{Rectangle (iv)} &= 14 \times 8 \\ \text{(on the bottom)} &= 112 \text{ cm}^2 \\ \text{Total surface area} &= 65 + 65 + 48 + 56 + 96 + 112 \\ &= 442 \text{ cm}^2\end{aligned}$$

APPLYING PYTHAGORAS' THEOREM

In some problems, Pythagoras' theorem needs to be applied first in order to calculate the length of one of the unknown sides. The examples below illustrate this idea.

Example 1: Find the surface area of the triangular prism shown.

Solution: As you can see, one of the dimensions (the vertical height of the triangle) has not been given.

Let the height = h

$$\therefore \quad 10^2 = h^2 + 6^2$$
$$\therefore \quad 100 = h^2 + 36$$
$$\therefore \quad h^2 = 64$$
$$\therefore \quad h = \sqrt{64}$$
$$= 8 \text{ cm}$$

Applying Pythagoras' theorem.

It is now easy to calculate the surface area using the same methods shown on the previous pages.

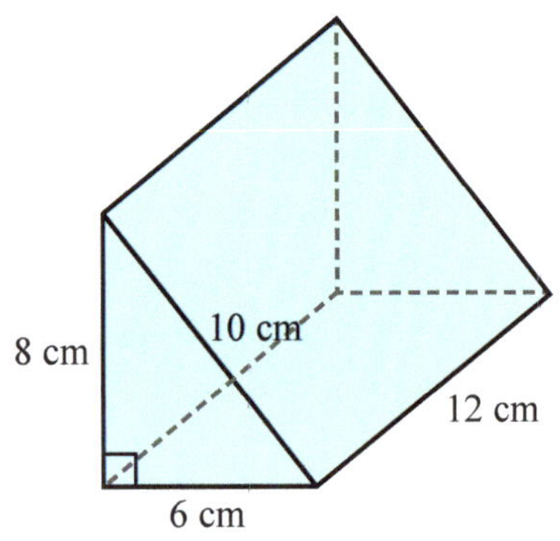

$$\text{Front triangle} = \frac{1}{2}bh$$
$$= \frac{1}{2} \times 6 \times 8$$
$$= 24 \text{ cm}^2$$
$$\text{Back triangle} = 24 \text{ cm}^2$$
$$\text{Rectangle (i)} = 12 \times 10$$
$$\text{(facing upwards)} = 120 \text{ cm}^2$$
$$\text{Rectangle (ii)} = 12 \times 6$$
$$\text{(on base)} = 72 \text{ cm}^2$$
$$\text{Rectangle (iii)} = 12 \times 8$$
$$\text{(unseen behind)} = 96 \text{ cm}^2$$
$$\text{Total surface area} = 24 + 24 + 120 + 72 + 96$$
$$= 336 \text{ cm}^2$$

The above prism is made up of 2 identical triangles and 3 differently seized rectangles.

SURFACE AREA OF A CYLINDER

If a closed cylinder is opened out flat, it will look like the figure below. This is called the net of the cylinder.

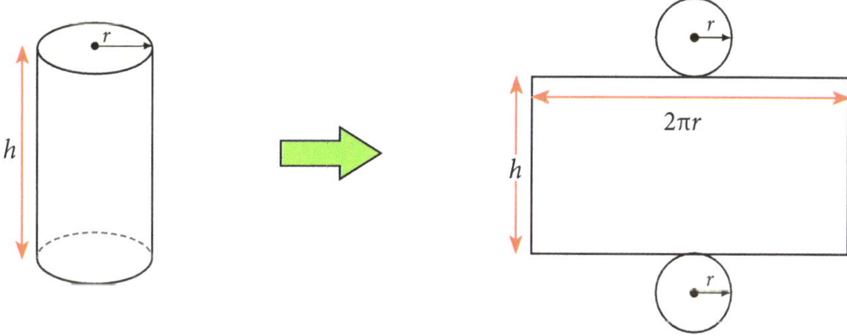

You will already see that the curved surface of the cylinder opens up to form a rectangle whose area is given by:

Length height = circumference of base × height

$$\text{Curved surface area} = 2\pi rh$$
$$\text{Total surface area} = \text{area of curved surface} + \text{area of 2 circular ends}$$
$$= 2\pi rh + 2\pi r^2$$

Note: It is important to read questions carefully as some problems do not require you to find the total surface area, including both circular ends.

Example: A cylinder of height 25 cm and radius 5 cm is closed at one end only. Find the surface area correct to 2 d.p.

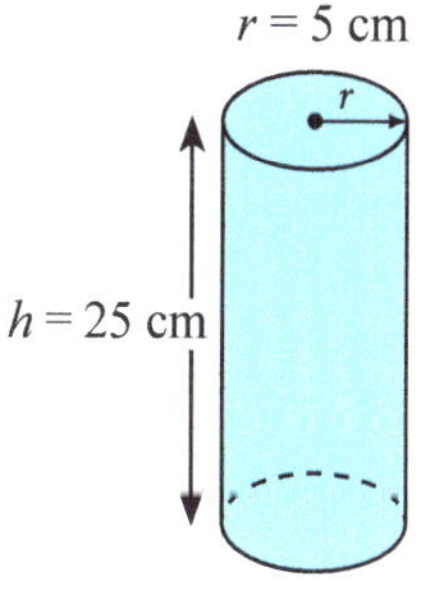

Solution: Total surface area $= \text{curved surface area} + 1 \text{ circular end}$

$$= 2\pi rh + \pi r^2$$
$$= 2 \times 3.142 \times 5 \times 25 + 3.142 \times 5 \times 5$$
$$= 864.05 \text{ cm}^2$$

EXAMPLES OF PYRAMIDS, CONES AND SPHERES

In these types of problems, you often have to use Pythagoras' Theorem to calculate unknown heights and other dimensions.

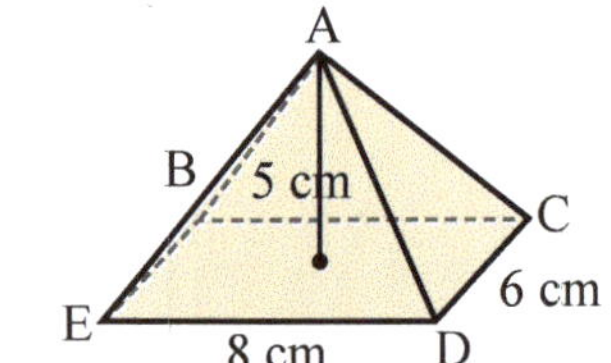

Example 1: Find the surface area of the rectangular pyramid correct to 2 decimal places.

Solution: It is first necessary to calculate the vertical heights of the slanting triangles using Pythagoras' Theorem:

In ΔAPR:

$$AR^2 = 5^2 + 3^2$$

$$\therefore \quad AR = \sqrt{34}$$

In ΔAQP:

$$AQ^2 = 5^2 + 4^2$$

$$\therefore \quad AQ = \sqrt{41}$$

Surface area = area of rectangle + 2 × area ΔAED + 2 × area ΔADC

$$= 8 \times 6 + 2\left(\tfrac{1}{2} \times 8 \times \sqrt{34}\right) + 2\left(\tfrac{1}{2} \times 6 \times \sqrt{41}\right)$$

$$= 133.07 \text{ cm}^2$$

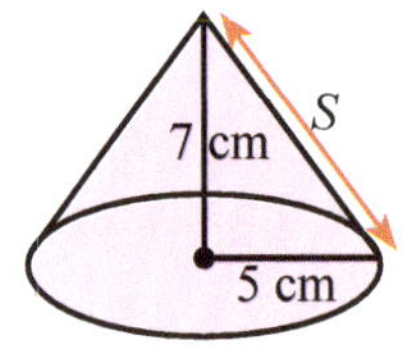

Example 2: Find the surface area of the cone which has a radius of 5 cm and a height of 7 cm. Give the answer correct to 4 significant figures.

Solution: Must first calculate the slant height using Pythagoras' Theorem.

$$S^2 = 7^2 + 5^2$$

$$\therefore \quad S = \sqrt{74}$$

Surface area $= \pi r S + \pi r^2$

$$= \pi \times 5 \times \sqrt{74} + \pi \times 5^2$$

$$= 213.7 \text{ correct to 4 s.f.}$$

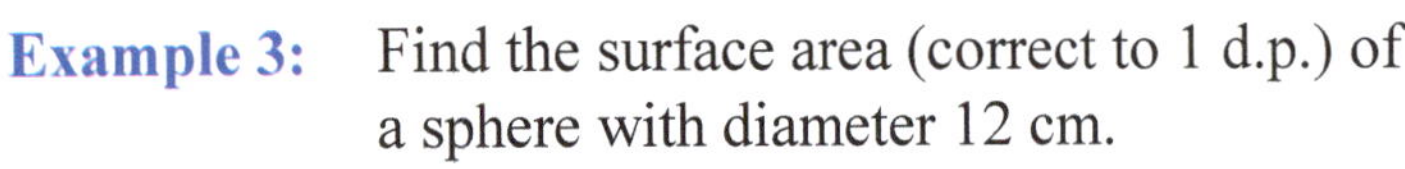

Example 3: Find the surface area (correct to 1 d.p.) of a sphere with diameter 12 cm.

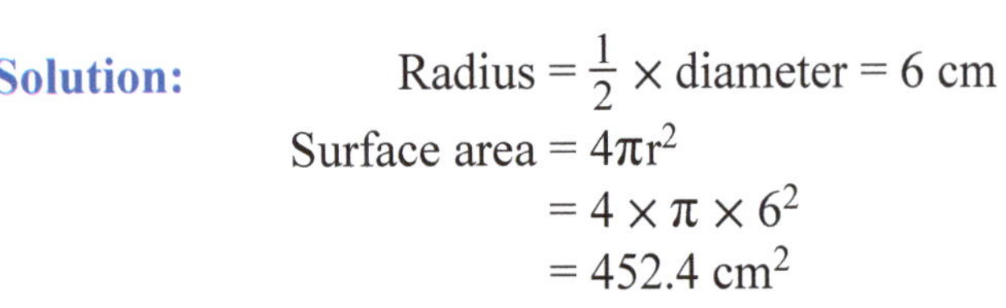

Solution: Radius $= \frac{1}{2} \times$ diameter = 6 cm

Surface area $= 4\pi r^2$

$$= 4 \times \pi \times 6^2$$

$$= 452.4 \text{ cm}^2$$

VOLUMES OF PRISMS

In earlier years you will have already been taught how to find the volume of a basic rectangular prism (i.e. box shape).

$$\begin{aligned}\text{Volume} &= \text{length} \times \text{breadth} \times \text{height}\\ &= \ell \times b \times h\\ &= \ell b h\end{aligned}$$

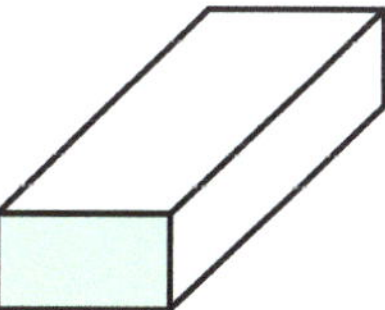

FIRSTLY, WHAT IS THE DEFINITION OF A PRISM?

> A prism is any solid which has a constant or uniform area of cross section.

For example, consider the solid shown. You will see that the shaded triangular shape at the front runs throughout the length of the solid. In other words, if you were to slice the solid in any place, the same triangular shape would be facing you. Hence, it is called a TRIANGULAR PRISM.

VOLUME OF ANY PRISM

Most of this work should be revision.

> (i) Find the constant area of cross section (A)
> (ii) Multiply this area by the length of the prism (h)
> i.e. Volume = A × h.

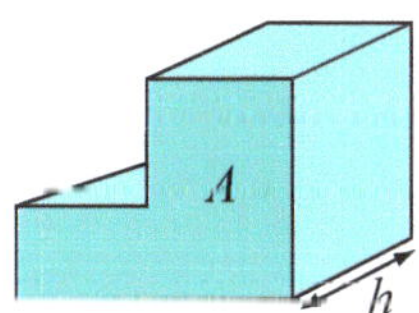

Volume = A × h

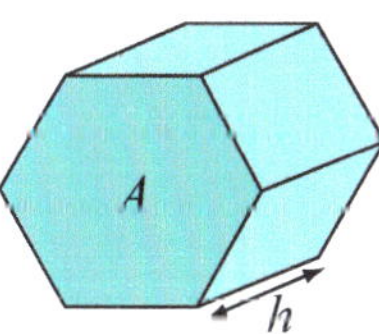

Volume = A × h

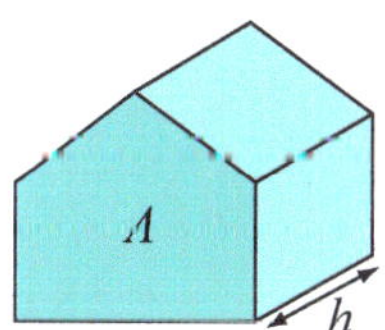

Volume = A × h

All the above figures are prisms, because they each have a constant cross section which has been shaded.

Example 1: The area of cross section of a hexagonal prism is 32 cm^2. If its length is 9 cm, find the volume.

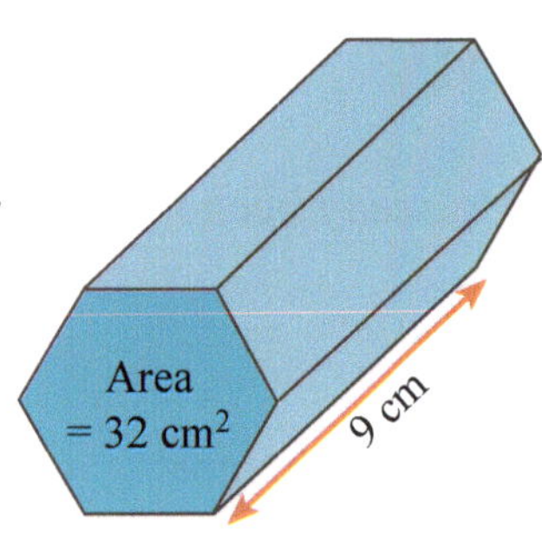

This is an easy question because the area of cross section has already been worked out for you.

Solution:

$$\begin{aligned} \text{Volume} &= \text{A} \times h \\ &= 32 \times 9 \\ &= 288 \text{ cm}^3 \end{aligned}$$

Example 2: Find the volume of the triangular prism with dimensions shown.

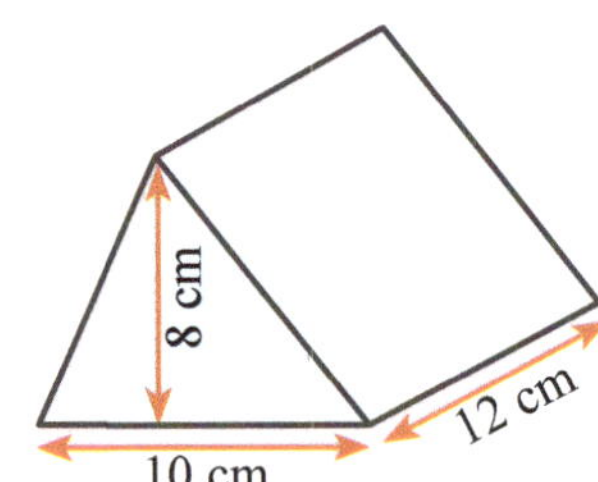

Solution: Step (i): Find cross sectional area of the triangle

$$\begin{aligned} \therefore \quad \text{Area of a triangle} &= \frac{1}{2} \times \text{base} \times \text{height} \\ &= \frac{1}{2} \times 10 \times 8 \\ &= 40 \text{ cm}^2 \end{aligned}$$

Step (ii): Multiply this area by the length of the prism.

$$\begin{aligned} \therefore \quad V &= A \times h \\ &= 40 \times 12 \\ &= 480 \text{ cm}^3 \end{aligned}$$

Example 3: Find the volume of the trapezoidal prism with dimensions shown.

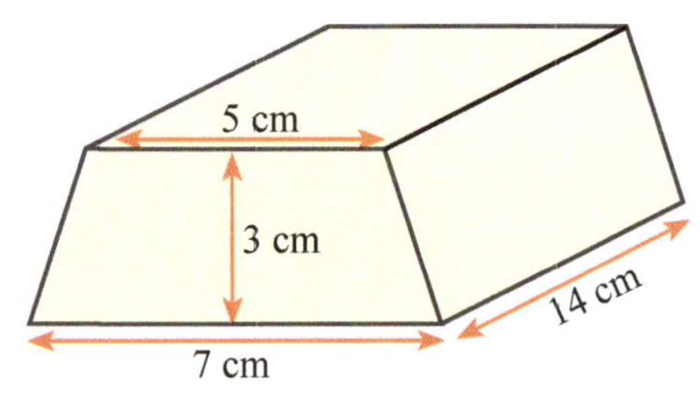

Solution: Step (i): Find cross sectional area of the trapezium.

$$\begin{aligned} \therefore \quad A &= \frac{\text{height} \times \text{sum of parallel sides}}{2} \\ &= \frac{3 \times (5 + 7)}{2} \\ &= 18 \text{ cm}^2 \end{aligned}$$

Step (ii): Multiply this area by the length of the prism.

$$\begin{aligned} \therefore \quad V &= A \times h \\ &= 18 \times 14 \\ &= 252 \text{ cm}^3 \end{aligned}$$

VOLUME OF A CYLINDER

A cylinder can be thought of as a prism which has the uniform cross section of a circle. The volume would therefore be obtained in the same way as for the prism.

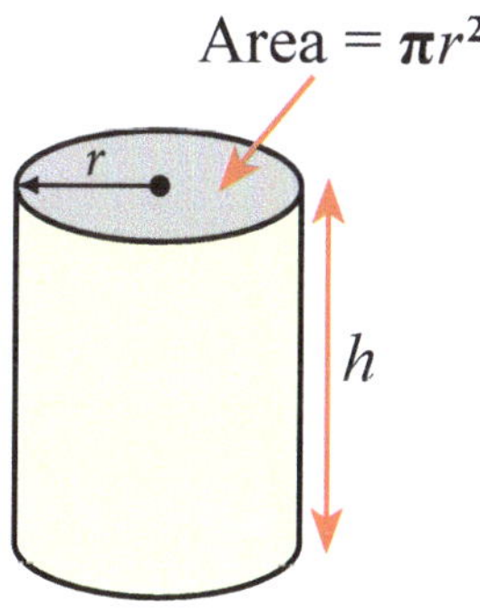

$$\begin{aligned} \text{Volume} &= \text{area of cross-section} \times \text{height} \\ &= Ah \qquad \text{where } A = \pi r^2 \\ &= \pi r^2 h \end{aligned}$$

Volume of cylinder $= \pi r^2 h$
where r = radius of the cylinder
and h = height of the cylinder

Example: Find the volume of a cylinder of radius 10 cm and height 20 cm
Take $\pi = 3.142$

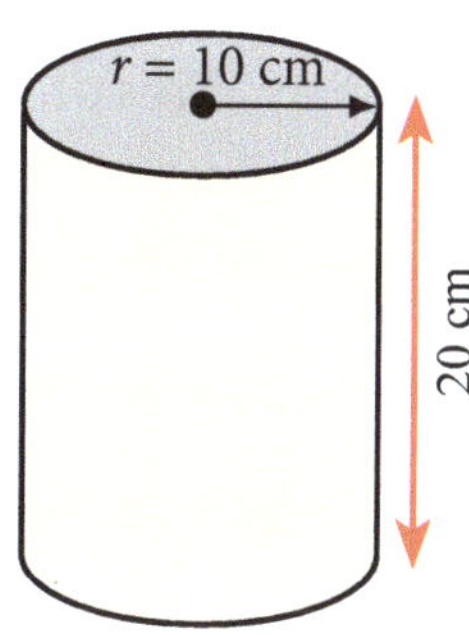

Solution:

$$\begin{aligned} \text{Volume} &= \pi r^2 h \\ &= 3.142 \times 10 \times 10 \times 20 \\ &= 6\,284 \text{ cm}^3 \end{aligned}$$

Note: It is often worth remembering the following conversions when doing volume problems involving capacity and weights of fluids.

$1 \text{ cm}^3 = 1 \text{ mL}$
$\therefore \quad 1\,000 \text{ cm}^3 = 1 \text{ L}$

1 cm^3 or 1 mL water weighs 1 g
$\therefore \quad 1\,000 \text{ cm}^3$ or 1 L water weighs 1 kg

Example 1: A tin shed for storage is made in the shape of half a cylinder with dimension as shown in the diagram. Find the volume correct to 1 d.p.

Take: $\pi = 3.142$

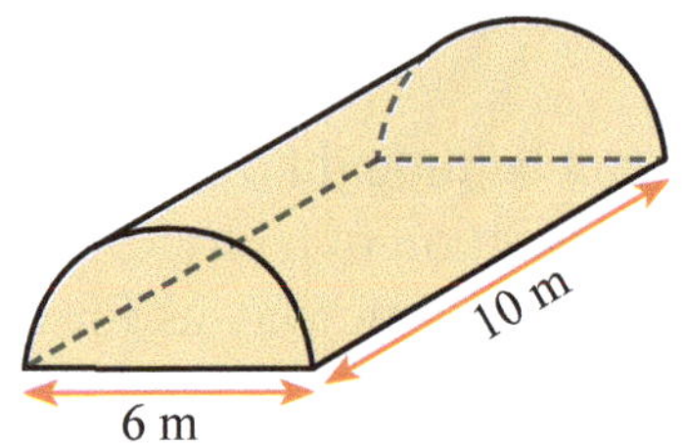

Solution:

$$\text{Volume of cylinder} = \pi r^2 h$$

$$\therefore \text{ Volume of } \frac{1}{2} \text{ cylinder} = \frac{1}{2}\pi r^2 h$$

$$= \frac{1}{2} \times 3.142 \times 3^2 \times 10$$

$$= 141.4 \text{ m}^3$$

Remember to halve the diameter in order to find the radius.

Example 2: An empty glass jar weighing 1.54 kg is in the shape of a cylinder which has a radius of 14 cm and a height of 20 cm. Calculate how many litres of water it will hold when filled to the top. Also calculate its weight when it is filled with water.

Take: $\pi = \frac{22}{7}$

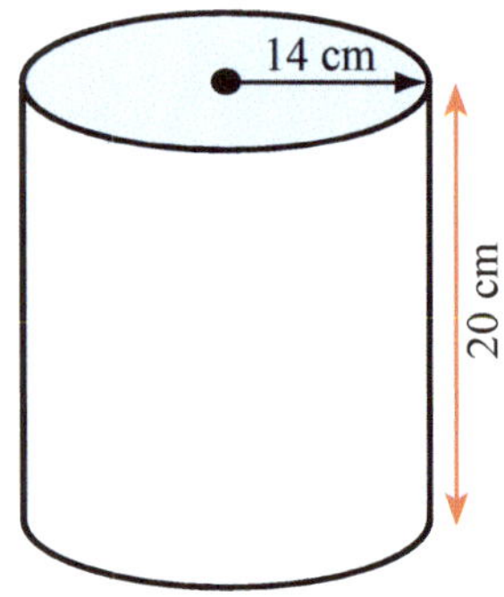

Solution: Volume $= \pi r^2 h$

$$= \frac{22}{7} \times \frac{14}{1} \times \frac{14}{1} \times \frac{20}{1}$$

$= 12\,320 \text{ cm}^3$

$= 12\,320$ mL Since $1 \text{ cm}^3 = 1$ mL

$= 12.32$ L Since 1 000 mL = 3 L

The glass jar will hold 12.32 litres.

1 litre of water weighs 1 kg

$\therefore$ 12.32 litres of water weighs 12.32 kg

$\therefore$ Weight of jar glass with water $= 1.54 + 12.32$

$= 13.86$ kg

OTHER VOLUMES

Most questions in tests and exams relate to volume and surface area of prisms and cylinders. Therefore, although you must still know this section of work, keep in mind that it is not examined as frequently as the ideas mentioned above. In some questions, you may have to use Pythagoras' Theorem to calculate the vertical height.

PYRAMID

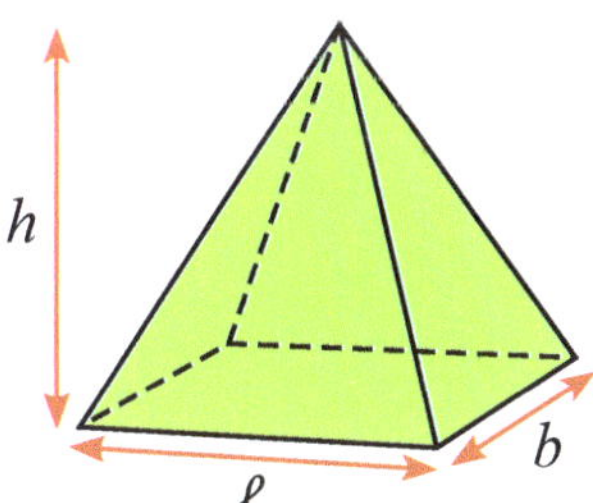

$$V = \frac{1}{3}\ell \times b \times h$$

where 'ℓ' and 'b' are the dimensions of the rectangular base and 'h' is the perpendicular or vertical height.

CONE

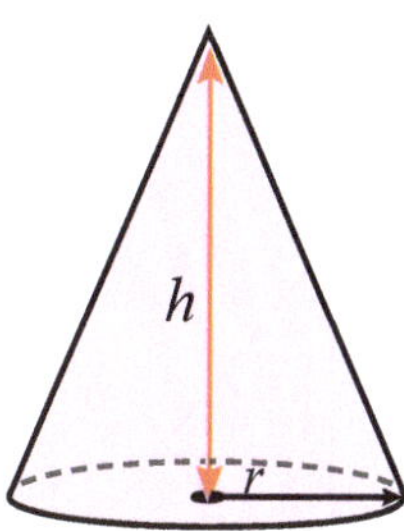

$$V = \frac{1}{3}\pi r^2 h$$

where r = radius of the base
h = perpendicular height

SPHERE

$$V = \frac{4}{3}\pi r^3$$

where r = radius of the sphere

OTHER EXAMPLES

Find the volumes of the 3 solids correct to 2 decimal places.

(i)

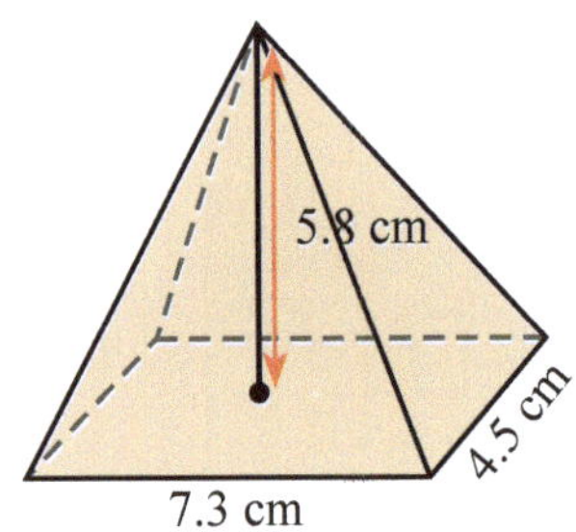

(ii)

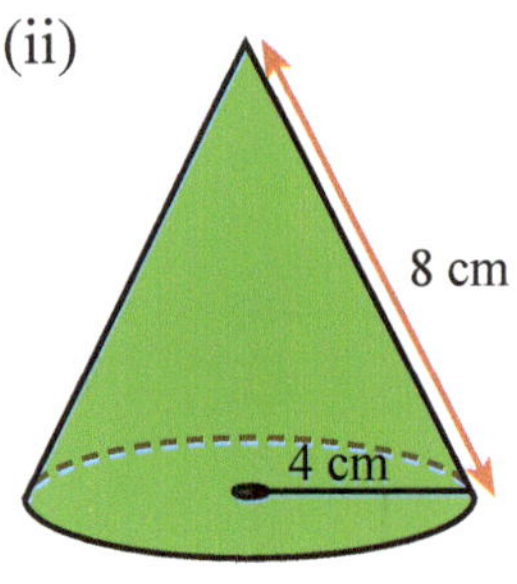

(iii)

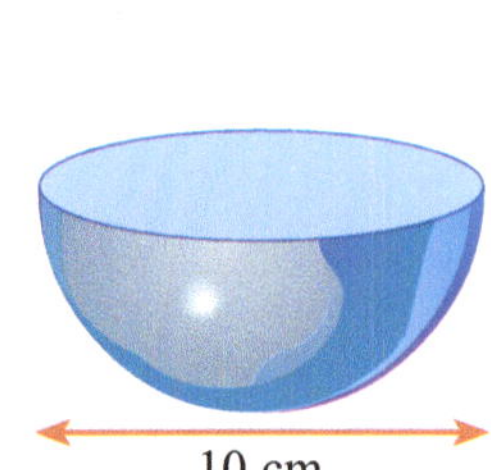

Solutions: (i) Volume of pyramid $= \frac{1}{3} \times \ell \times b \times h$

$$= \frac{1}{3} \times 7.3 \times 4.5 \times 5.8$$

$$= 63.51 \text{ cm}^3$$

(ii) Must first calculate the vertical height using Phytagoras' Theorem.

$$8^2 = h^2 + 4^2$$

$$\therefore \quad h^2 = 64 - 16$$

$$\therefore \quad h = \sqrt{48}$$

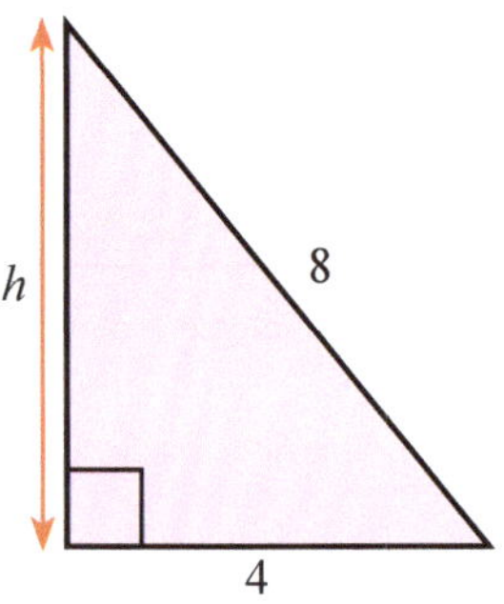

Volume of cone $= \frac{1}{3}\pi r^2 h$

$$= \frac{1}{3} \times \pi \times 4^2 \times \sqrt{48}$$

$$= 116.08 \text{ cm}^3$$

(iii) Volume of sphere $= \frac{4}{3}\pi r^3$

$\therefore$ Volume of hemisphere $= \frac{1}{2} \times \left(\frac{4}{3}\pi r^3\right)$

$$= \frac{1}{2} \times \frac{4}{3} \times \pi \times 5^3$$

$$= 261.80 \text{ cm}^3$$

COMPOSITE VOLUMES

The volumes of more complicated solids can be calculated by splitting the figure up into 2 or more simpler solids. The volume of each of these solids is calculated separately — and then finally added together to give the total volume.

Example 1: Find the volume of the solid shown correct to 1 decimal place.

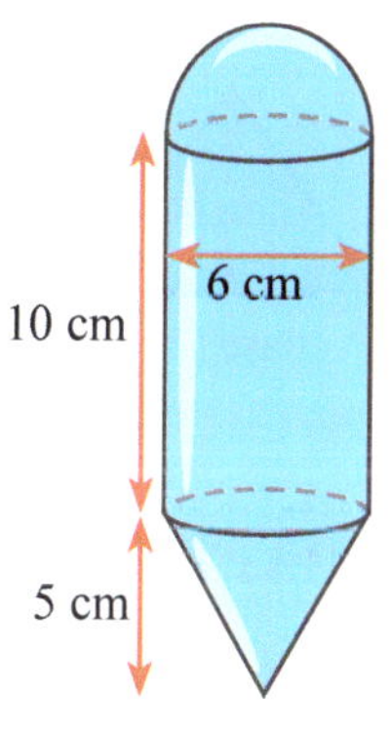

Solution: The solid can be split up into 3 simpler shapes: hemisphere, cylinder, cone

$$\text{Vol. of hemisphere} = \frac{1}{2} \times \left(\frac{4}{3}\pi r^3\right)$$
$$= \frac{1}{2} \times \left(\frac{4}{3}\pi \times 3^3\right)$$
$$= 56.55 \text{ cm}^3$$

$$\text{Vol. of cylinder} = \pi r^2 h$$
$$= \pi \times 3^2 \times 10$$
$$= 282.74 \text{ cm}^3$$

$$\text{Vol. of cone} = \frac{1}{3}\pi r^2 h$$
$$= \frac{1}{3} \times \pi \times 3^2 \times 5$$
$$= 47.12$$

$$\therefore \text{ Total volume} = 56.55 + 282.74 + 47.12$$
$$= 386.4 \text{ cm}^3$$

Example 2: The solid shown has a hole of diameter 4 cm drilled through it. Find the volume of the remaining solid correct to 2 decimal places.

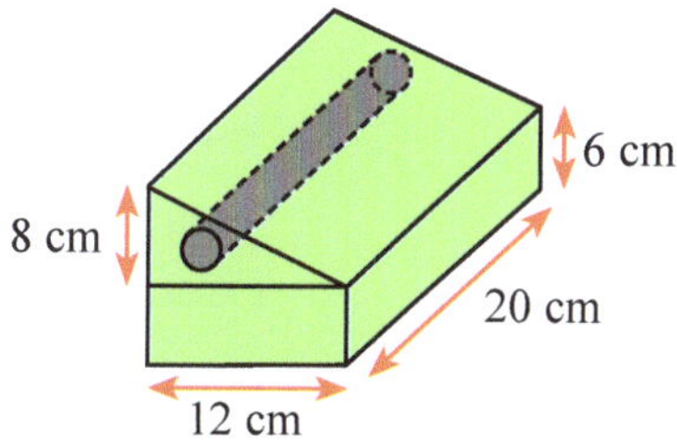

Solution: We have to find the volume of a rectangular prism plus the volume of a triangular prism, then subtract the volume of a cylinder.

$$\text{Volume of rectangular prism} = \ell \times b \times h$$
$$= 12 \times 20 \times 6$$
$$= 1\,440 \text{ cm}^3$$

$$\text{Volume of triangular prism} = \left(\frac{1}{2} \times b \times h\right) \times \text{length}$$
$$= \left(\frac{1}{2} \times 12 \times 8\right) \times 20$$
$$= 960 \text{ cm}^3$$

$$\text{Volume of cylinder} = \pi r^2 h$$
$$= \pi \times 2^2 \times 20$$
$$= 251.33 \text{ cm}^3$$

$$\therefore \text{ Volume of composite solid} = 1440 + 960 - 251.33$$
$$= 2\,148.67 \text{ cm}^3$$

CHAPTER SUMMARY

AREAS OF COMMON FIGURES

TRIANGLE

$A = \frac{1}{2}bh$

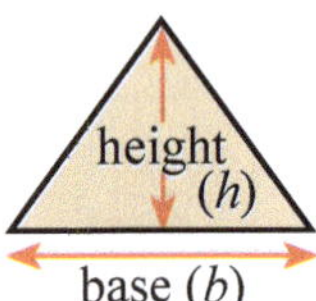

RECTANGLE

$A = l \times b$

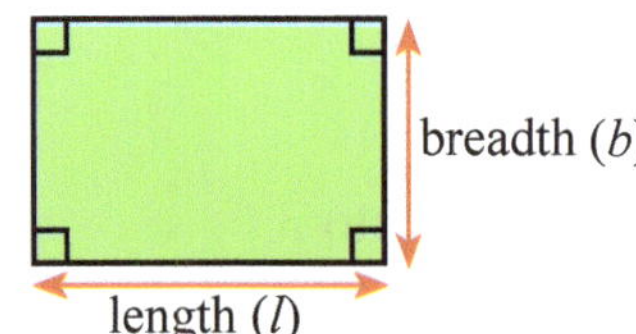

SQUARE

$A = l^2$

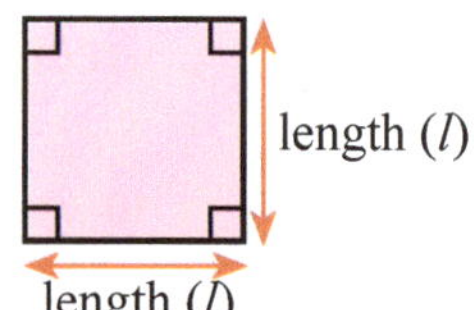

TRAPAZEUM

$A = \frac{1}{2}(a + b) \times h$

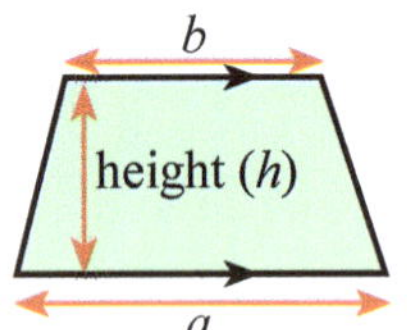

PARALLELOGRAM

$A = b \times h$

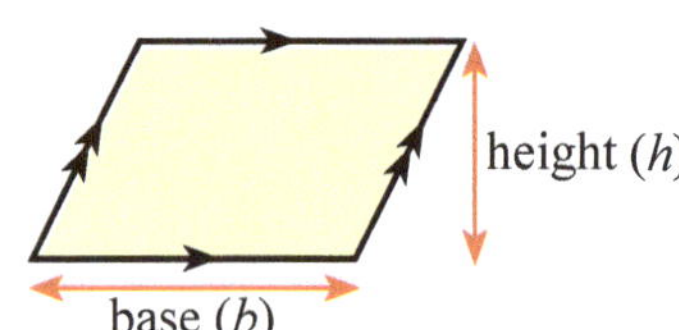

RHOMBUS

$A = \frac{1}{2} \times x \times y$

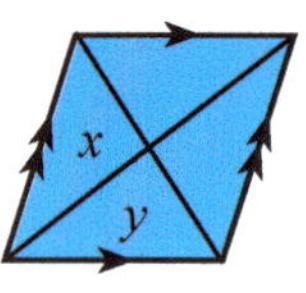

CIRCLES, SECTORS AND ARC LENGTHS

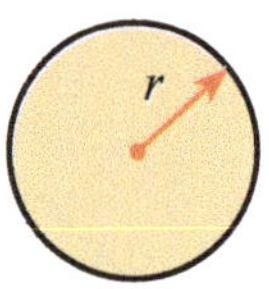

Area $= A = \pi r^2$
Circumference $= 2\pi r$

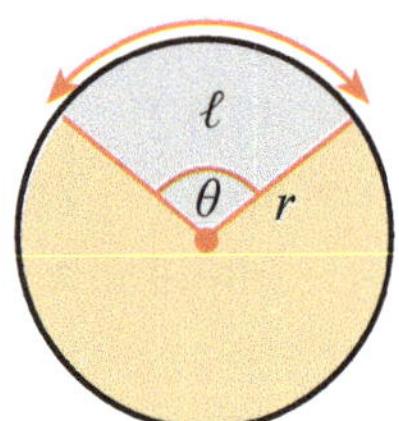

Area of sector $= \frac{\theta}{360} \times \pi r^2$

Arc length $(\ell) = \frac{\theta}{360} \times 2\pi r$

SURFACE AREA

To find the surface area of a solid means to add up the areas of all the surfaces which make up or surround that particular solid.

Example:

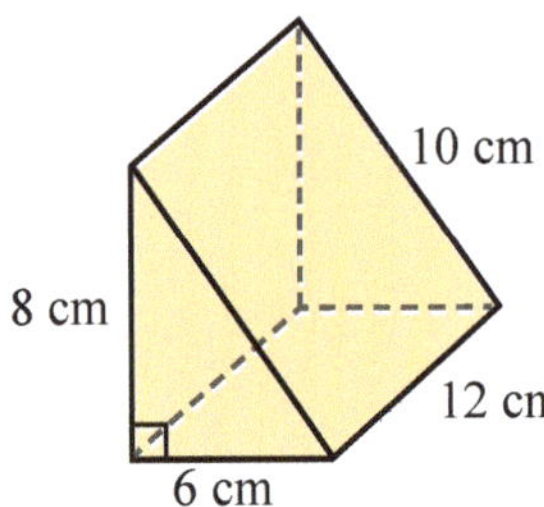

This triangular prism is made up of 2 identical triangles and 3 different sized rectangles.

Solution:

$$\begin{aligned}
\text{Front triangle} &= \frac{1}{2}bh \\
&= \frac{1}{2} \times 6 \times 8 \\
&= 24 \text{ cm}^2 \\
\text{Back triangle} &= 24 \text{ cm}^2 \\
\text{Rectangle (i)} &= 12 \times 10 \\
\text{(facing upwards)} &= 120 \text{ cm}^2 \\
\text{Rectangle (ii)} &= 12 \times 6 \\
\text{(on base)} &= 72 \text{ cm}^2 \\
\text{Rectangle (iii)} &= 12 \times 8 \\
\text{(unseen behind)} &= 96 \text{ cm}^2 \\
\text{Total surface area} &= 24 + 24 + 120 + 72 + 96 \\
&= 336 \text{ cm}^2
\end{aligned}$$

VOLUME OF ANY PRISM

Step 1: Find the constant area of cross section (A).

Step 2: Multiply this by the length of the prism (h).
i.e. **Volume =** $\boldsymbol{A \times h}$

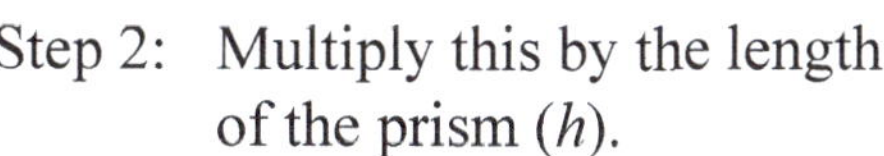

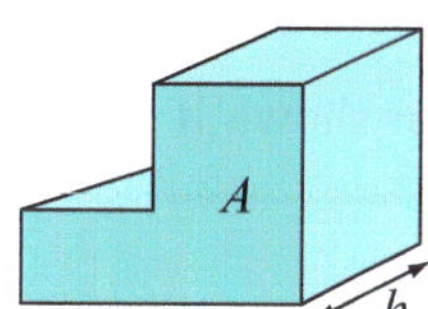

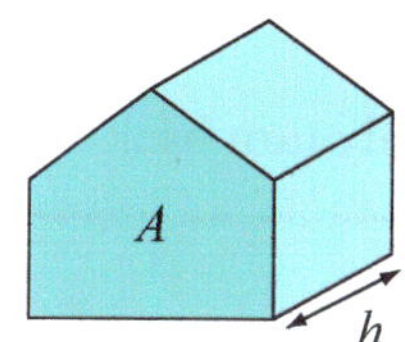

Volume = $\boldsymbol{A \times h}$

CYLINDERS

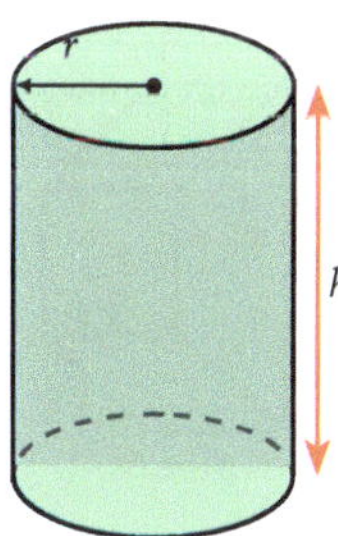

Curved surface area = $2\pi rh$
Total surface area closed cylinder = $2\pi r^2 + 2\pi rh$
Total surface area open cylinder = $\pi r^2 + \pi rh$
Volume = $\pi r^2 h$

CONES

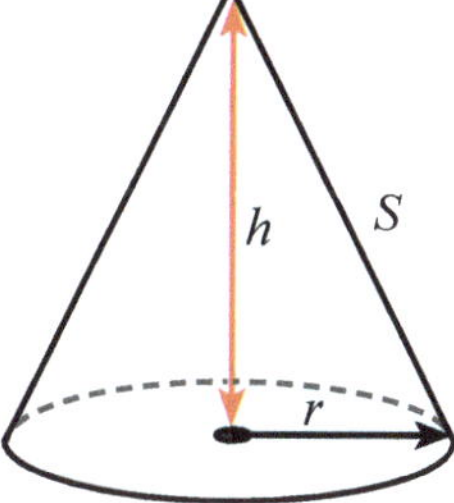

Surface Area = $\pi rS + \pi r^2$

Volume = $\frac{1}{3}\pi r^2 h$

PYRAMIDS

Triangular Pyramid  or **Rectangular Pyramid**

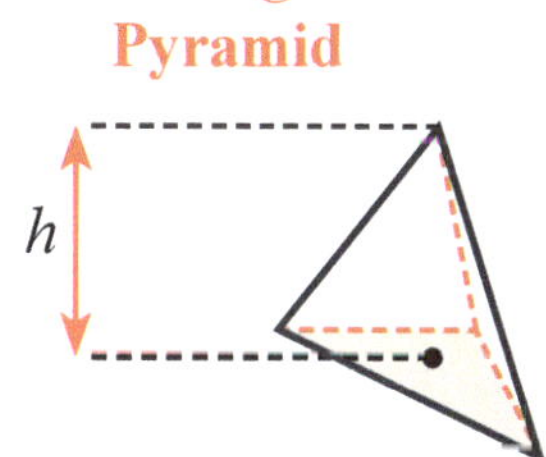

Volume = $\frac{1}{3}A \times h$

i.e. Volume = $\frac{1}{3}$(area of base) × vertical height

SPHERES

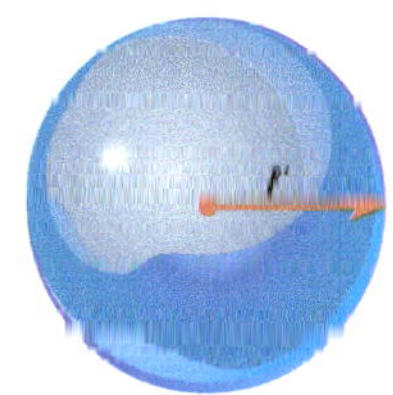

Surface area = $4\pi r^2$

Volume = $\frac{4}{3}\pi r^3$

LEVEL 1 — SURFACE AREA AND VOLUME

EASIER QUESTIONS

Note: Only turn back to page number shown if you have difficulty.	Page
Q1. Find the surface area of each prism or cylinder below:	22 - 25

a)

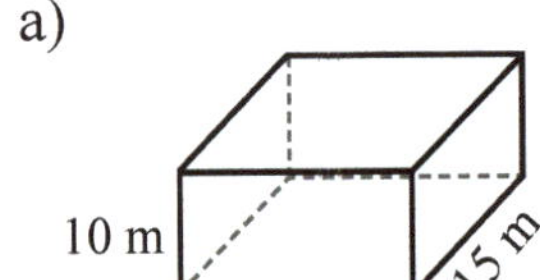

b)

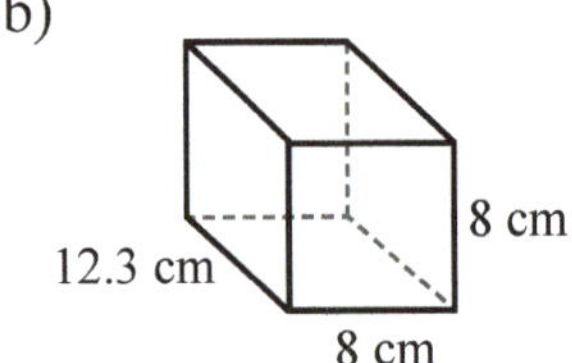

c)

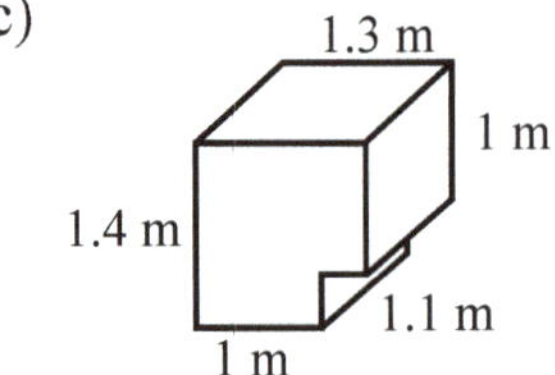

d)

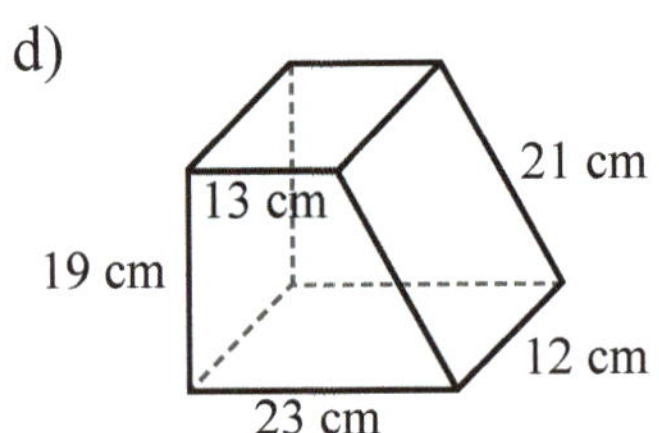

e)

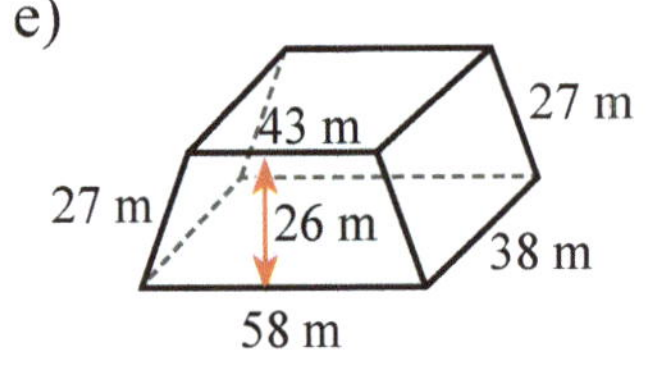

f)

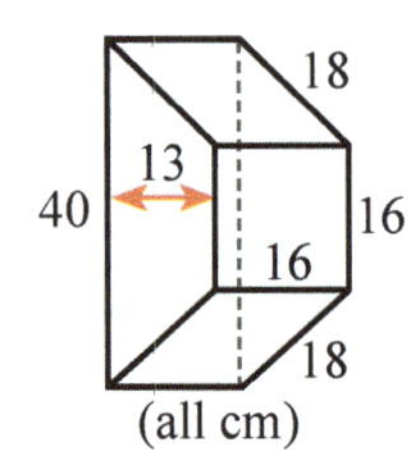

g)

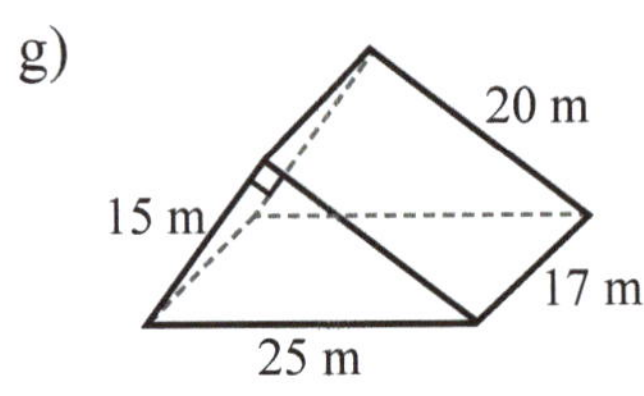

h)

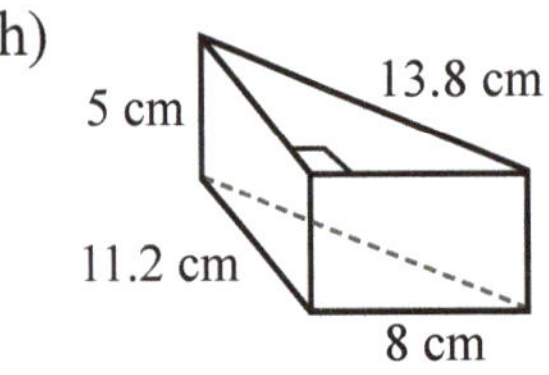

i)

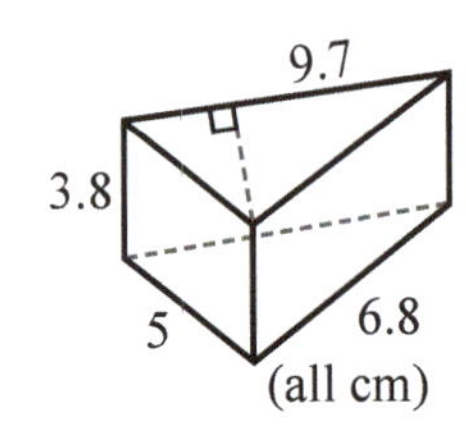

j)

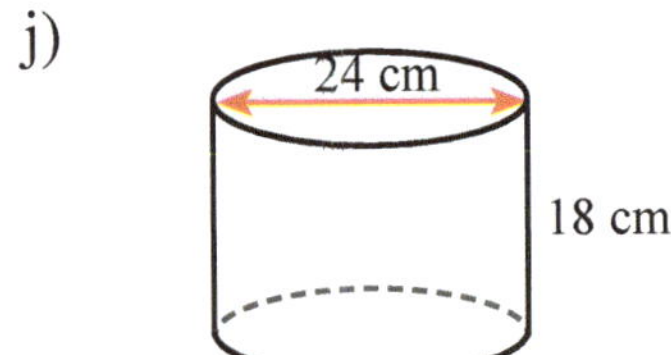

k)

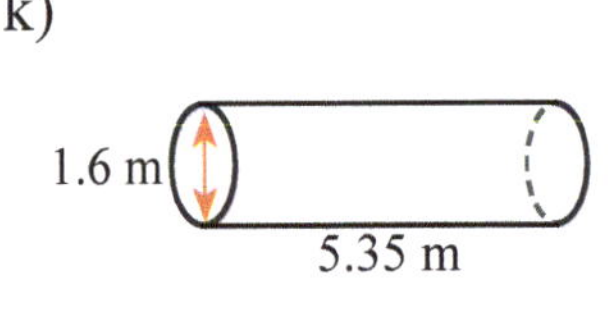

l)

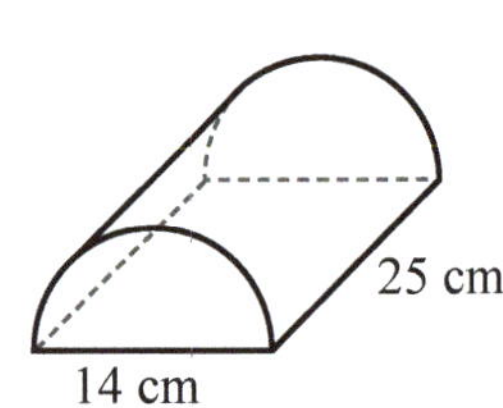

Q2. Calculate the surface area (correct to the nearest cm^2) of the solids: 26

a)

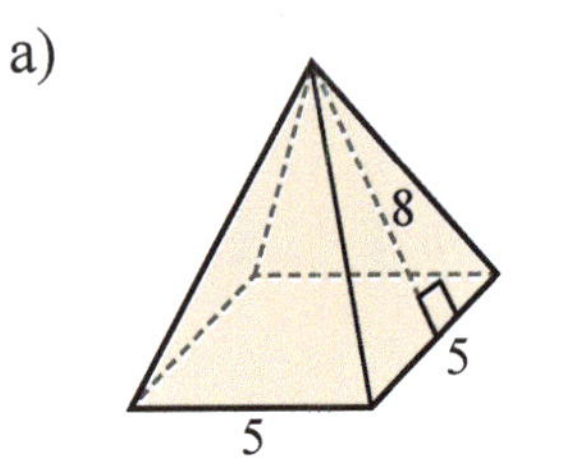

b)

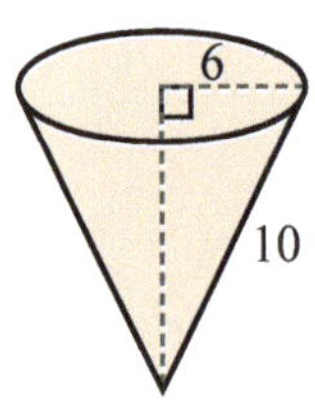

c)

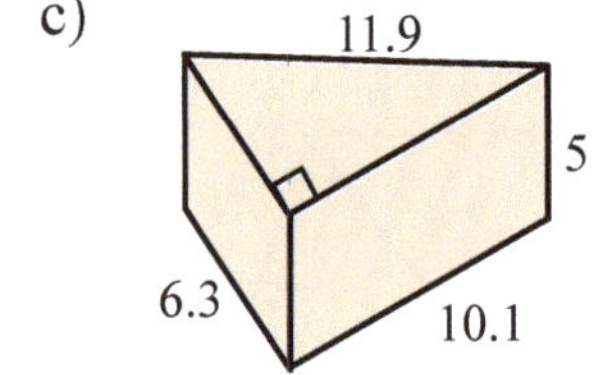

d)

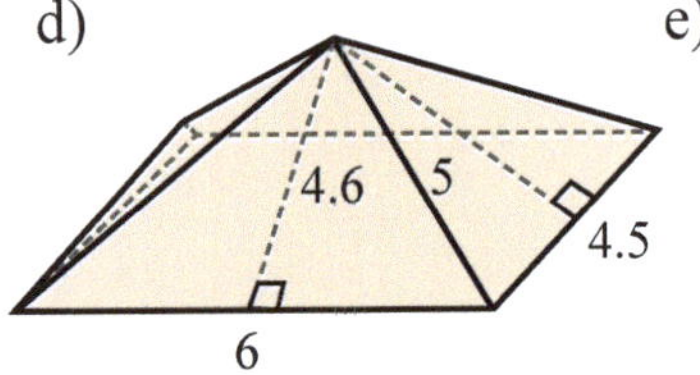

e)

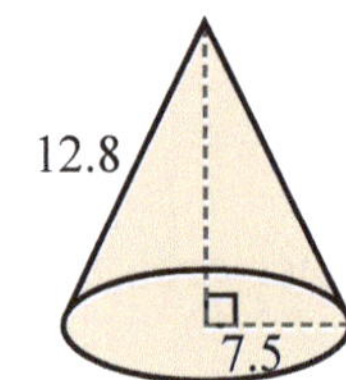

f)

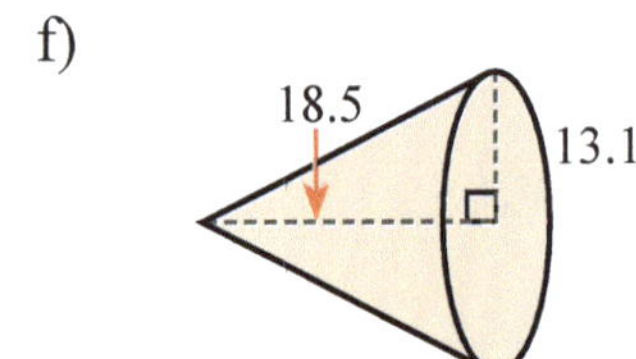

Q3. Calculate the surface area (correct to 1 d.p.) of: 26

a) a sphere of radius 10 cm
b) a sphere of diameter 15 cm
c) a sphere of diameter 9 cm
d) a hemisphere of diameter 21 cm

Note: Only turn back to page number shown if you have difficulty. | Page

Q1. Find the volume of these prisms: [units in cm] 27, 28

a)
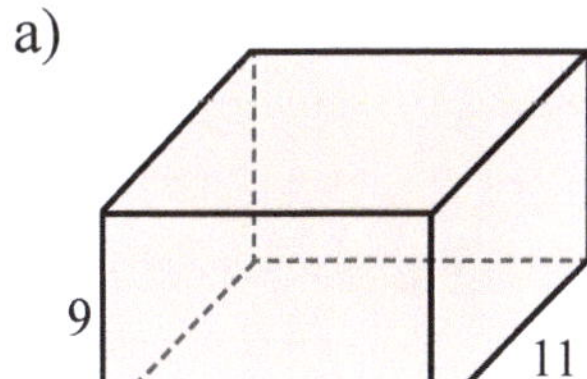

b)
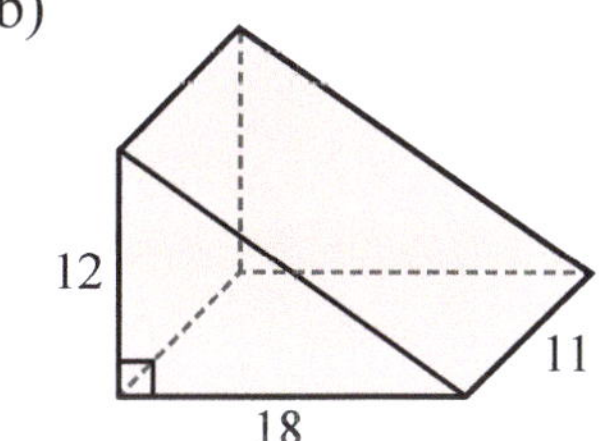

c)
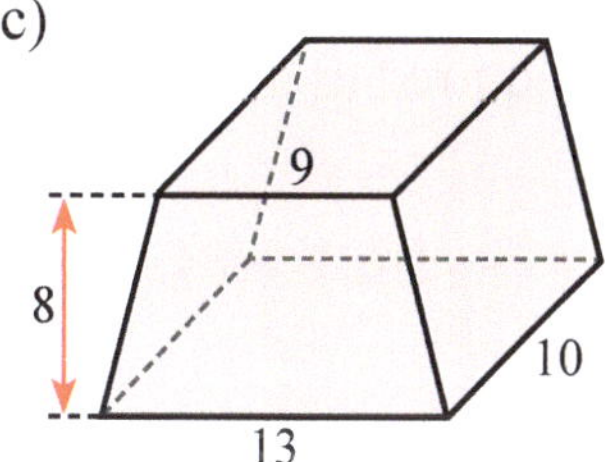

d)
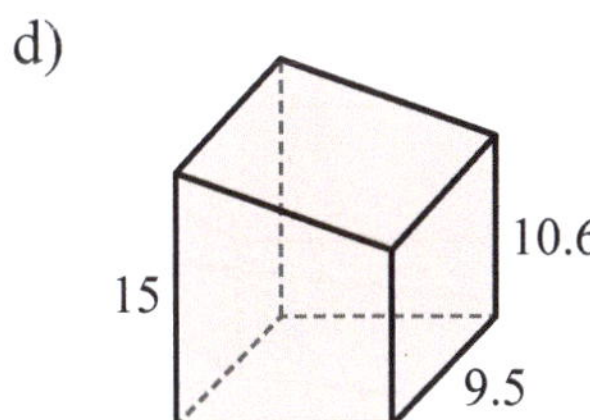

e)
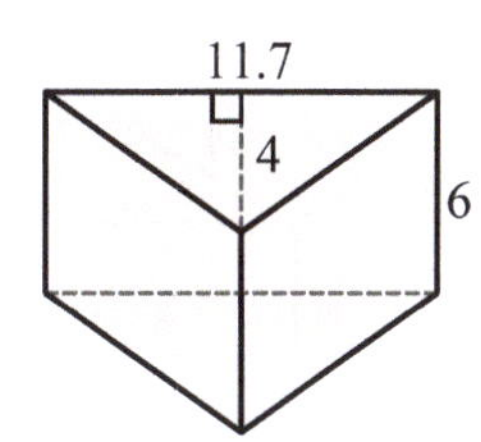

f)
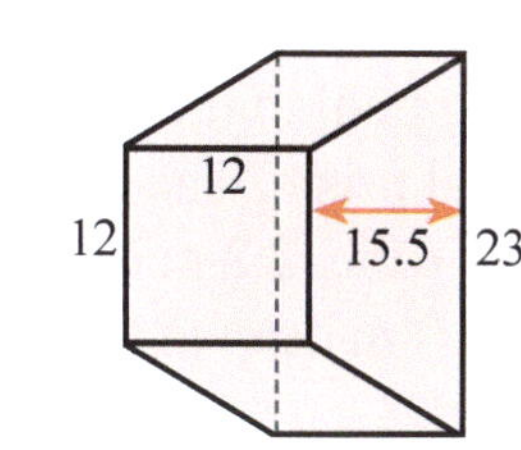

Q2. Find the volume of prism using $\pi \approx 3.142$: [units in m] 29, 30

a)
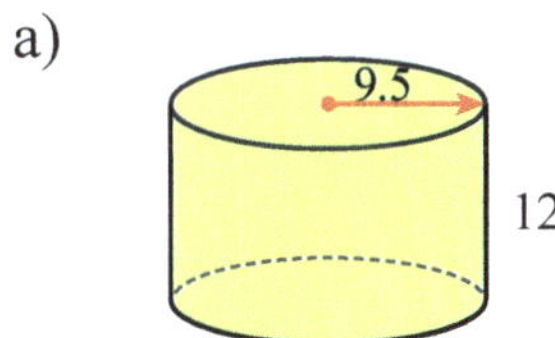

b)
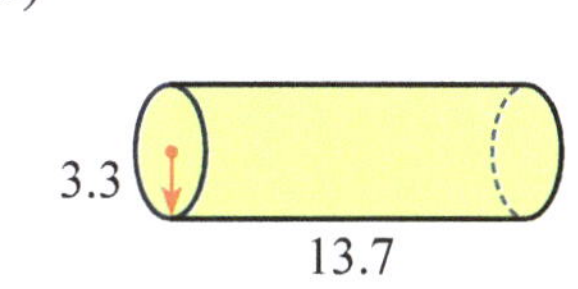

c)
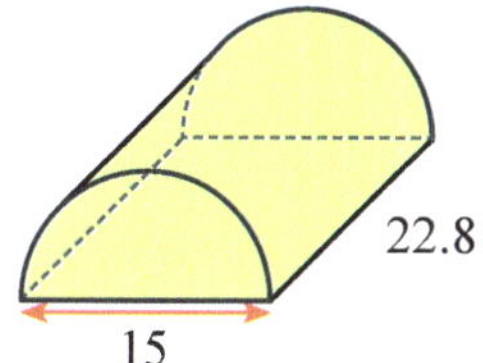

Q3. Find the volume, correct to 2 d.p. of each figure: [Units in cm] 31, 32

a)
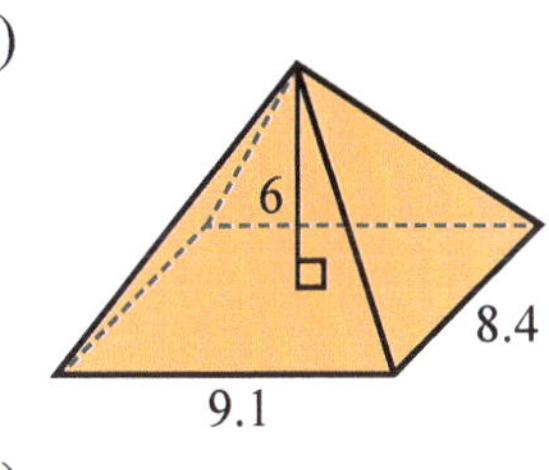

b)
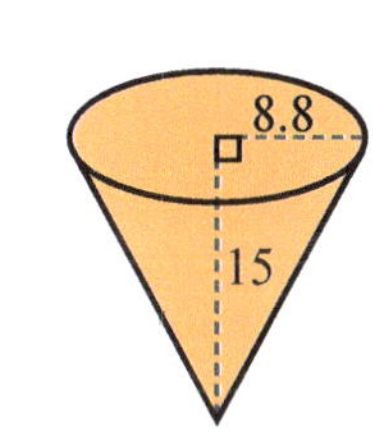

c)
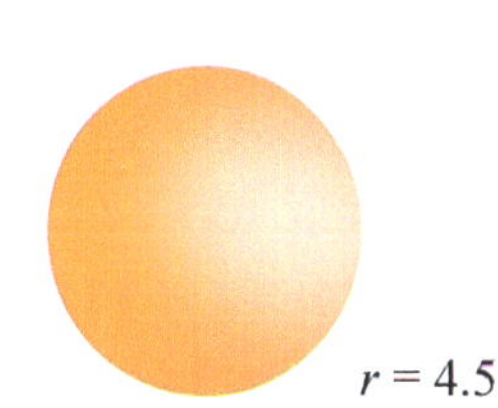

d)
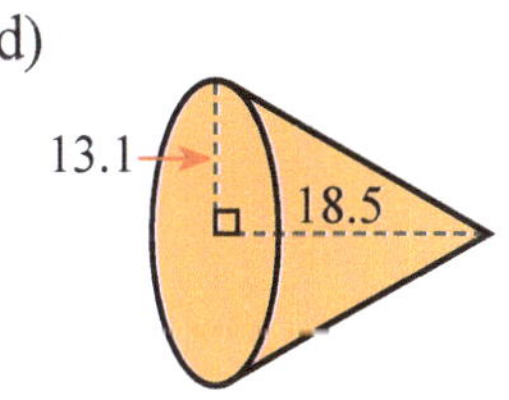

e)
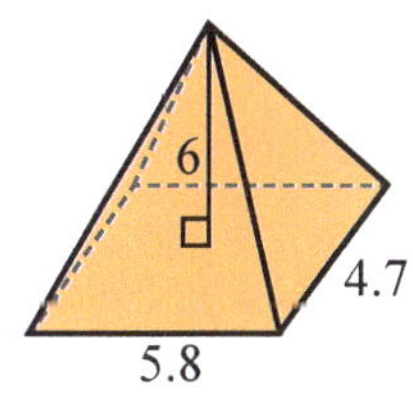

f)

Q4. Calculate the volume of each figure correct to 2 d.p.: [Units in m] 33

a)
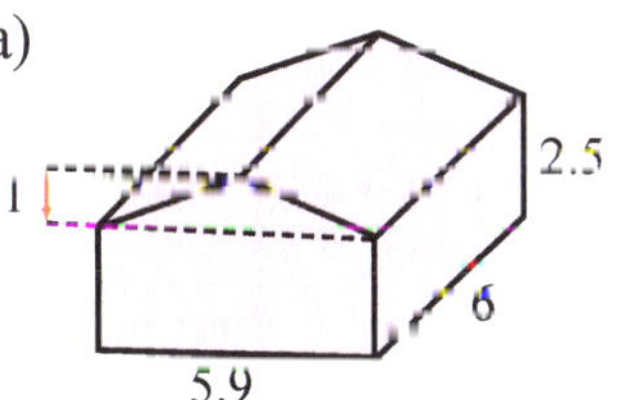

b)
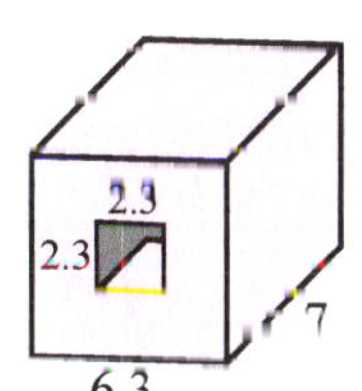

c)
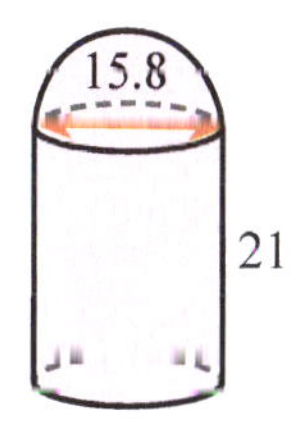

Q1. A hollow concrete pipe has an outside diameter of 1.1 m and an inside diameter of 0.9 m (i.e. the thickness of the pipe's walls is 10 cm). These pipes are made in 3 m lengths. Find the amount of concrete, in m^3 correct to 2 decimal places, required to make one pipe.

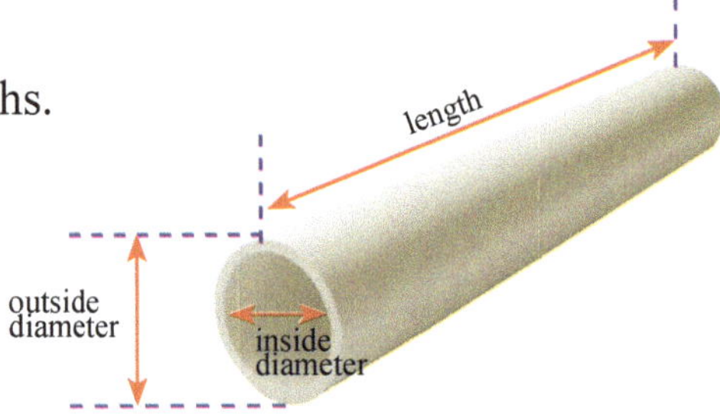

Q2. Paul the Pool Man builds a swimming pool in the shape of a trapezoidal prism. The pool is 10 m long, 5 m wide, 2.2 m deep at one end and 1.2 m deep at the other (as shown).

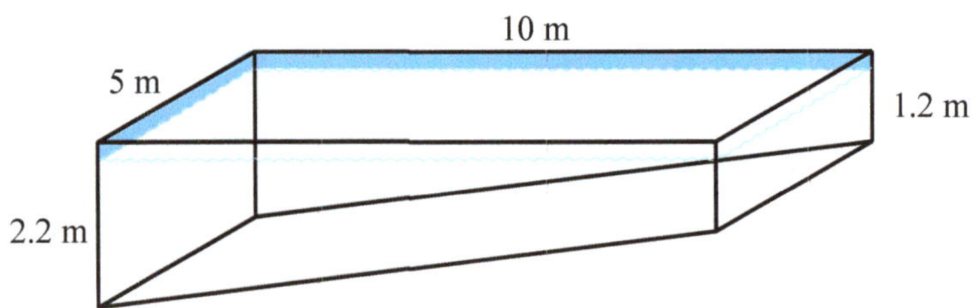

a) Find the volume of the pool in (i) m^3 and (ii) litres.

b) Show that the depth of water at the deep end is 1.3 metres when it contains 40 000 litres of water.

Q3. A wheat silo is made up from a cylinder with a cone on top. The silo has a diameter of 4 metres and is 6 metres high with the cone of height 3 metres on top of the cylinder. The diagram to the right shows the silo.

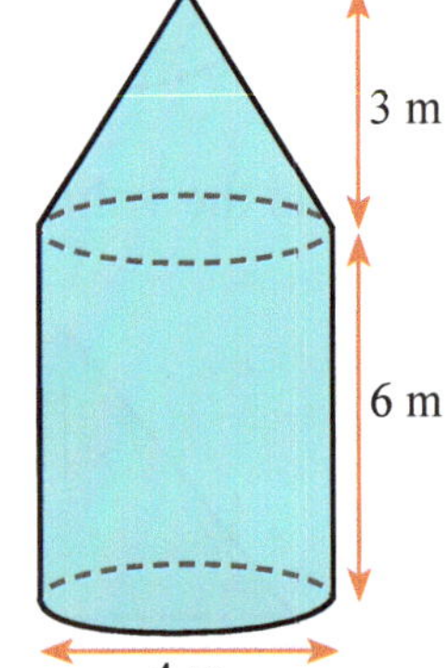

a) Find the volume of the silo to the nearest (i) cubic metre (ii) litre

b) Rather than have a conical top, the owners decide to have a semi-spherical top.

Would it hold more wheat? Show calculations to support your answer.

Q4. Four tennis balls are packed into the cylindrical container as shown.

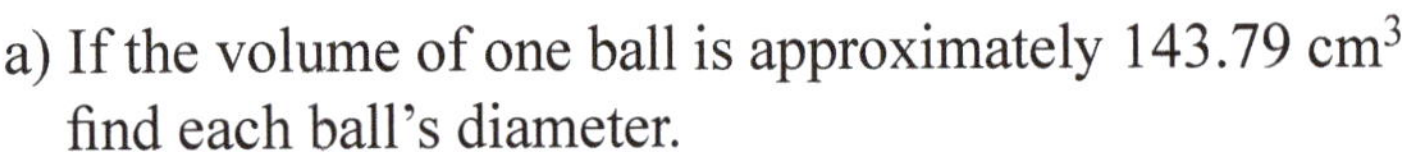

a) If the volume of one ball is approximately 143.79 cm^3 find each ball's diameter.

b) Find the height and radius of the base of the container, hence, find the percentage of air inside the container correct to two decimal places.

c) A replica of a tennis ball in bronze is melted to make a trophy in the form of a cone whose base diameter is twice its height. Find the dimensions of the trophy correct to two decimal places.

STATISTICS

The 'Australian Curriculum Mathematics' (ACM) references for this sub-strand of 'Statistics and Probability' (SP) are given below. This chapter may contain additional extension work which the author feels will be beneficial to the student.

- *Relative frequency (ACMSP 226).*
- *Quartiles and interquartile range (ACMSP 248).*
- *Boxplots (ACMSP 249).*
- *Comparing data sets (ACMSP 250).*
- *Standard deviation (ACMSP 278).*
- *Comparing means and standard deviations (ACMSP 278).*
- *The shape of a frequency distribution (ACMSP 282).*
- *Comparing different data using 'stem and leaf' plots (ACMSP 283).*

Page

Shiing - Shen Chern (1911 – 2004)

There have been so many famous and accomplished Chinese mathematicians over the centuries: Jing Fan, Shen Kuo, Chia-Chiao Liu, Chia - Tung Lee, Hua Luogeng to name just a few, and so it was a difficult choice for me to profile one particular person. Chern was a Chinese - American mathematician who made significant and fundamental contributions to 'differential geometry', and 'topology'. He has been called the 'father of modern differential geometry' and is widely regarded as one of the greatest mathematicians of the 20th century, winning numerous prestigious awards.
In 1984 he was also the co-founder of the world renowned 'Mathematics and Sciences Instititute' at the University of California, Berkeley. In honour of him, the 'Chern Medal' was established in 2011 to recognize 'an individual whose accomplishments warrant the highest level of recognition for outstanding achievement in the field of mathematics'.

RELATIVE FREQUENCY

Much of the examinable work on Statistics has been covered and explained in the Year 9 book. You are strongly advised to revise that work and make sure you have a good understanding of the following ideas: frequency distribution tables, cumulative frequency, histograms, frequency polygons, grouped frequency distribution tables, ogives etc.

We will however revise some of the Year 9 work on the pages to follow.

Besides knowing how often a particular score occurs (given by the frequency), it is also useful to know what fraction of the total results it represents.

$$\text{Relative frequency of a score} = \frac{\text{frequency of score}}{\text{total frequency}}$$

Example: Rewrite the frequency distribution table below, by adding a further column and calculating the relative frequency of each score.

Score (x)	Frequency (f)
5	9
6	14
7	12
8	5

$\sum f = 40$

Score (x)	Frequency (f)	Relative frequency
5	9	0.225
6	14	0.35
7	12	0.3
8	5	0.125

$\sum f = 40$ $\quad\sum = 1.000$

The relative frequency of 5 is $\frac{9}{40} = 0.225$

The relative frequency of 6 is $\frac{14}{40} = 0.35$ etc.

Another way of thinking about this is that the score of 6 occurred 35% of the time.

Note: A good check is that the total of the relative frequency column should add up to 1, i.e. 100%.

DOT PLOTS

As the name suggests, this graph uses dots to represent data. This is one of the simplest and more basic types of graphs, and there is usually only one horizontal number line or axis. Each dot represents one piece of data, and this is plotted directly above one of the numbers.

Example: The school computer club has 35 members, and their ages are shown below:

13, 15, 12, 13, 14, 14, 14, 16, 13, 12, 16, 15, 14, 13, 14, 15, 14

13, 14, 16, 14, 15, 17, 16, 14, 12, 14, 13, 15, 16, 12, 13, 15, 14, 17

Use a dot plot to show the ages of the students in the computer club and discuss the results.

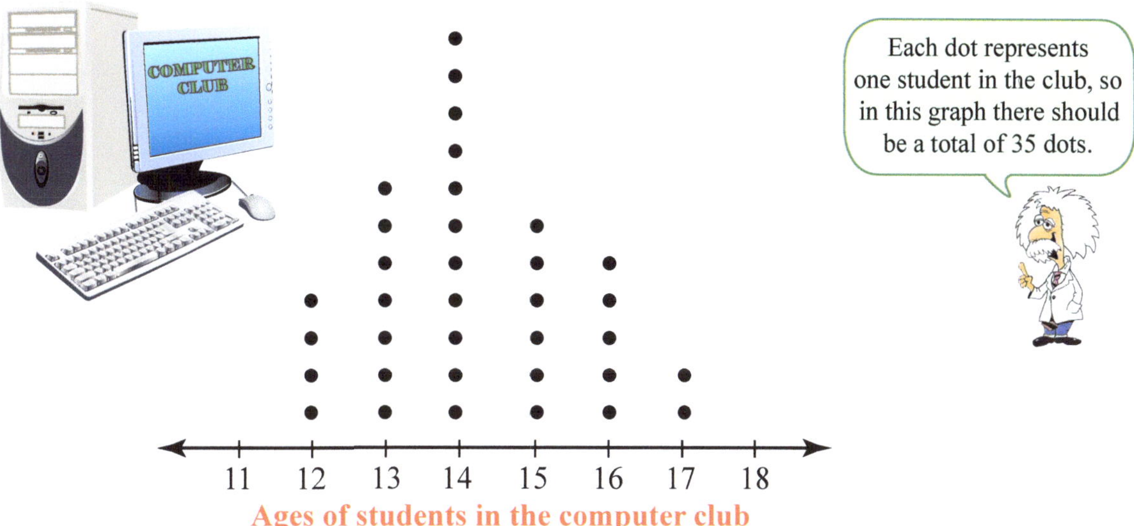

Ages of students in the computer club

Discussion: We can see immediately from the above graph the following facts:

(i) The ages of the students in the computer club range from 12 to 17 years old.

(ii) The most frequently occurring age is 14 years old, and then the numbers get smaller as the students get older. The possible reason for this is that as students get older, they have other interests and also have to spend more time studying. Maybe more of them also have part-time jobs.

(iii) The horizontal line gives us the range of ages, and we do not have to start from the number zero on the line, because the youngest age is 12 years old.

(iv) One of the disadvantages of dot plots is that it is only useful for surveys involving smaller numbers of people. If we were conducting a survey on 100 people or more, then it would be difficult and very time consuming to plot all the dots.

(v) Another disadvantage is that there is no vertical axis. So if we wanted to know how many students in the club are aged say 13, then we actually have to count the dots.

STEM AND LEAF PLOTS

A stem and leaf plot organises data by showing all the items **in order** using stems and leaves. The data is arranged according to place value. The digits containing the largest place value are recorded in the stem, and the smaller place value digits are recorded in the leaves.

A stem and leaf plot is a method used to organize statistical data. The largest place value numbers are used to form the stem, and the lesser place value numbers are used to form the leaves. The stem is usually on the left and the leaves are usually on the right.

Example 1: A class of students sat for an English exam.
Make a stem and leaf plot of their exam marks below.

54, 63, 96, 80, 62, 23, 76, 84, 78, 93, 91, 57
67, 70, 81, 83, 94, 97, 71, 85, 72, 94, 72, 92

It is advisable to first rearranged the number in increasing order.

Solution: 23, 54, 57, 62, 63, 67, 70, 71, 72, 72, 76, 78
80, 81, 83, 84, 85, 91, 92, 93, 94, 94, 96, 97

Since the data ranges from 23 to 97, the stems will range from 2 to 9. The leaves record all the place value units. The score 62 is plotted by placing the 6 in the stem column and the 2 in the leaf column.

Stems	Leaves	
2	3	This line is recording 23
5	4 7	This line is recording 54, 57
6	2 3 7	This line is recording 62, 63, 67
7	0 1 2 2 6 8	This line is recording 70, 71, 72, 72, 76, 78
8	0 1 3 4 5	This line is recording 80, 81, 83, 84, 85
9	1 2 3 4 4 6 7	This line is recording 91, 92, 93, 94, 94, 96, 97

Observations: We can see that the lowest score is 23 and the highest score is 97.
7 students had a score in the 90 — 99 range.
2 students scored the same mark of 94.
There is one score of 23 which is very different from the rest of the scores. We will explain towards the end of this chapter that such a score will be given a special classification as an 'outlier'.

FURTHER EXAMPLES

Example 2: In this example we shall compare the exam results of 2 separate classes.

Class A: 61, 68, 70, 75, 84, 86, 90, 91, 92, 94, 94
96, 99, 100
Class B: 60, 63, 70, 71, 73, 73, 73, 76, 77, 84, 85
86, 91, 92

Solution:

Class A Leaves	Stems	Class B Leaves
8 1	6	0 3
5 0	7	0 1 3 3 3 6 7
6 4	8	4 5 6
9 6 4 4 2 1 0	9	1 2
0	10	

Observations: It can be seen at quick glance that class A obtained much better results than class B.
Class B has more scores in the 70s than class A.
On the other hand, class A had 7 scores in the 90s compared to only 2 scores in the 90s for class B.

Example 3: The stem and leaf plot represents the scores of two teams in the NBA competition.
Can you work out what the following 'back to back' stem and leaf plot shows?

Leaves	Stems	Leaves
Lakers		Cavaliers
7 5 3	6	1
6 5 5 2	7	4 8
8 6 6	8	3 5
7 4 4 2	9	2 2 5 7 8 9
5	10	0 1 7
4 2	11	
3	12	3 5 5 7

Observations: We can see from the plot that the Cavaliers scored 90 points and higher much more often than the Lakers.
The Lakers scored less than 90 points in 10 games, while the Cavaliers scored less than 90 points in only 5 games.

MEASURES OF CENTRAL TENDENCY AND DISPERSION

The mean, median, mode are measures that describe something about the CENTRE of a distribution of scores. Range and interquartile range are measures of dispersion (i.e. how spread out the scores are).

Consider the following 11 scores in order to illustrate the few definitions below:

29, 12, 17, 23, 31, 35, 19, 15, 16, 23, 18

MEAN — the average of all the scores. The mean is obtained by adding up all the scores and dividing by the total number of scores.

i.e. Mean $= \dfrac{\text{sum of all scores}}{\text{total number of scores}}$

$= \dfrac{29 + 12 + 17 + 23 + 31 + 35 + 19 + 15 + 16 + 23 + 18}{11}$

$= 21.\dot{6}\dot{3}$

We could also write this as:

$\bar{x} = \dfrac{\sum x}{n}$

where $\sum$ means sum.

MEDIAN — the middle score when all the scores are arranged in order of size from the smallest to the highest. If there are an even number of scores, then it is the average of the two middle scores. Must first arrange scores in order from smallest to highest.

12, 15, 16, 17, 18, [19], 23, 23, 29, 31, 35 ⇨ Median = 19

the middle score

MODE — the score which occurs most often. The most popular score. It is the score with the highest frequency. There can be more than one mode.

∴ Mode = 23 (occurs twice)

RANGE — the highest score subtract the lowest score. The range gives us an idea of how spread out the scores are.

∴ Range = highest score – lowest score = 35 – 12 = 23

QUARTILES — Exactly one-quarter of the data falls between any two consecutive quartiles.

- Q_1 is the point below which one-quarter of the ordered data falls. (**Q_1** is also known as the **lower quartile**).
- Q_2 is the point below which half of the ordered data falls. (**Q_2** is also known as the **median**).
- Q_3 is the point below which three-quarters of the ordered data falls. (**Q_3** is also known as the **upper quartile**).

INTERQUARTILE RANGE — is the middle 50% of scores when they are arranged in order of size. It is the difference between the upper quartile (top 25% of scores) and the lower quartile (bottom 25% of scores).

12, 15, 16, 17, 18, [19], 23, 23, 29, 31, 35

$Q_1 = 16$ $Q_2 = 19$ $Q_3 = 29$

Interquartile range = 3rd quartile (Q_3) – 1st quartile (Q_1)

∴ IQR = 29 – 16 = 13

COMPARING TWO SETS OF DATA

The example below illustrates why the range and interquartile range are not completely reliable indicators of dispersion.

Example: Two cricketers scored the following runs in a sample of 8 Sheffield Shield matches:

Mark:	42	56	72	48	61	0	54	75
Steve:	28	82	33	90	22	81	52	20

Find the mean and range for each batsman. Who is the most consistent batsman?

Solution: Mark

$$\text{Mean} = \frac{42+56+72+48+61+0+54+75}{8} = \frac{408}{8} = 51$$

Range = 75 – 0 = 75

0   42   48   54   56   61   72   75

(Q_1 = 42, Q_3 = 72)

$\text{IQR} = Q_3 - Q_1 = 72 - 42 = 30$

Steve

$$\text{Mean} = \frac{28+82+33+90+22+81+52+20}{8} = \frac{408}{8} = 51$$

Range = 90 – 20 = 70

20   22   28   33   52   81   82   90

(Q_1 = 22, Q_3 = 82)

IQR = 82 – 22 = 60

Analysis: Both batsmen have the same ‘mean’ or ‘average’ of 51 runs per match. From the range, it seems like Mark’s scores are more widespread because his range is 75 compared to Steve’s range of 70.
The IQR of Mark is 30 compared to Steve’s IQR of 60, so this means the IQR is a better indicator of dispersion than range, and this shows that Mark is the most consistent batsman. In summary, both the range and the IQR are fairly basic (and sometimes unreliable) indicators of the dispersion of a set of scores.

Note: A **population** in statistics means all of the data under consideration. Sometimes all the data is too large and we need to use only a part of the population which we call a **sample**. If we used the population in the above example, then we would need to find the average scores from every shield match in which they played – and this number could be very large.

STANDARD DEVIATION

As illustrated by the example on the previous page, we need a much better system of working out the dispersion of a set of scores. A formula has been devised by Mathematicians called 'Standard Deviation' S_n, and this gives an accurate method of working out how widespread the scores are from the mean when using a sample of the population.

$$S_n = \sqrt{\frac{\sum(\bar{x} - x)^2}{n - 1}}$$

S_n = standard deviation

$\bar{x}$ = mean of scores

x = each score

n = number of scores

A large standard deviation means the scores are more spread out, while a small standard deviation means the scores are closer together on average.

Note: To find the standard deviation for a whole population (σ_n) the above formula is changed slightly to: $\sigma_n = \sqrt{\dfrac{\sum(\bar{x} - x)^2}{n}}$.

Example: From the example on the previous page, calculate the standard deviation of Mark and Steve's cricket results.

Solution: Mark

$\bar{x} = 51$

$\bar{x} - x = 51 - 42,\ 51 - 56,\ 51 - 72,\ 51 - 48,\ 51 - 61,\ 51 - 0,\ 51 - 54,\ 51 - 75$

$(\bar{x} - x)^2 = 81 \quad 25 \quad 441 \quad 9 \quad 100 \quad 2\,601 \quad 9 \quad 576$

$\sum(\bar{x} - x)^2 = 81 + 25 + 441 + 9 + 100 + 2\,601 + 9 + 576 = 3\,842$

$$\sqrt{\frac{\sum(\bar{x} - x)^2}{n - 1}} = \sqrt{\frac{3\,842}{7}} = 23.4$$

Steve

$\bar{x} = 51$

$\bar{x} - x = 51 - 28,\ 51 - 82,\ 51 - 33,\ 51 - 90,\ 51 - 22,\ 51 - 81,\ 51 - 52,\ 51 - 20$

$(\bar{x} - x)^2 = 529 \quad 961 \quad 324 \quad 1\,521 \quad 841 \quad 900 \quad 1 \quad 961$

$\sum(\bar{x} - x)^2 = 529 + 961 + 324 + 1\,521 + 841 + 900 + 1 + 961 = 6\,038$

$$\sqrt{\frac{\sum(\bar{x} - x)^2}{n - 1}} = \sqrt{\frac{6\,038}{7}} = 29.37$$

Analysis: From the above results we can see that Mark had a batting average of 51 with a standard deviation of 23.4. Steve had a batting average of 51 with a standard deviation of 29.37. Because Mark's scores are grouped more closely together (as shown by the lower standard deviation), he is a more consistent batsman than Steve.

STATISTICS AND THE CALCULATOR

Most scientific calculators have a 'Statistics Mode' which automatically calculates the mean and standard deviation of a set of scores. Firstly, make sure your calculator is switched onto **STAT** mode — the letters **STAT** will show on the display panel. This is done by pressing [MODE] 2, then pressing 1 (for 1 variable).

[=] Press this button after each score is entered. It is usually found at the bottom right hand corner.

To now obtain all your calculations, press the following:

[AC] [SHIFT] [STAT] and the number 4 (var).

You will see on the screen:

1 : n 2 : $\bar{x}$

3 : σx 4 : sx

By entering #1, you will be given the number of scores (n).

By entering #2, you will be given the mean of the scores ($\bar{x}$).

By entering #3, you will be given the standard deviation of a population (σx).

By entering #4, you will be given the standard deviation of a sample (sx).

Example 1: Consider again Mark's 8 cricket scores from the previous page:

42 56 72 48 61 0 54 75

Enter each score: 42 [=] 56 [=] 72 [=] 48 [=] 61 [=] 0 [=] 54 [=] 75

Press [AC] [SHIFT] [STAT] 4

Press [n] 8 i.e. 8 scores were entered.

Press [$\bar{x}$] 51 i.e. mean of the 8 scores.

Press [sx] 23.4 i.e. standard deviation

The calculator can also be used to calculate the mean and standard deviation from a frequency table.

Example 2:

Score (x)	Frequency (f)
4	8
5	4
6	5
7	3

In addition to the previous steps shown at the top of the page:

[SHIFT] [SET UP] (▼) 3, then 1

By pressing 1 we activate the frequency column. To enter the various frequencies along side the scores, we use (◄▲►▼) button to position the cursor.

Press [n] 20 Number of scores

Press [$\bar{x}$] 5.15 Mean of scores

Press [sx] 1.14 Standard deviation

THE SHAPE OF A FREQUENCY DISTRIBUTION

When we draw histograms, stem-and-leaf plots or dot plots it is easy to determine the overall pattern of the shape. We can draw an approximate curve around the whole shape and this can inform us if the results are symmetrical or skewed in a particular direction.

Example 1:

Stems	Leaves
3	2
4	4 7
5	5 6 9
6	3 4 6 8
7	4 5 7
8	1 3
9	5

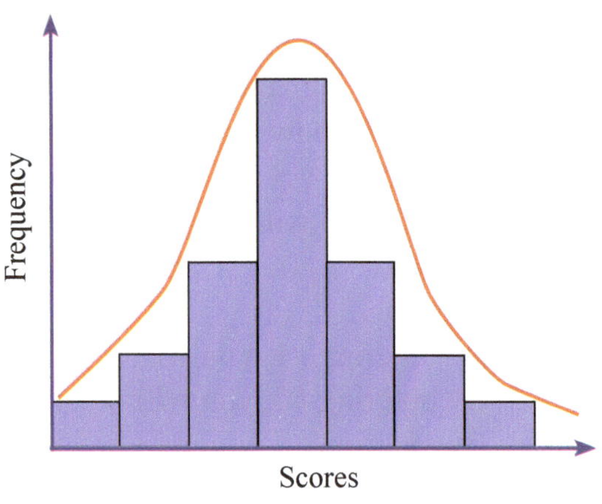

These are called 'symmetrical' distributions or 'normal' curves.

Note: They don't have to be exactly symmetrical.

If the curve is symmetrical, we often call it a 'bell shaped' curve or distribution.

Example 2:

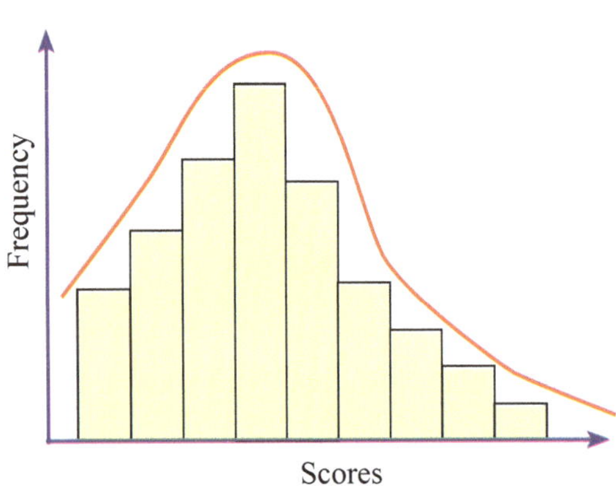

In this histogram the tail points to the right and we call this **positively skewed**.

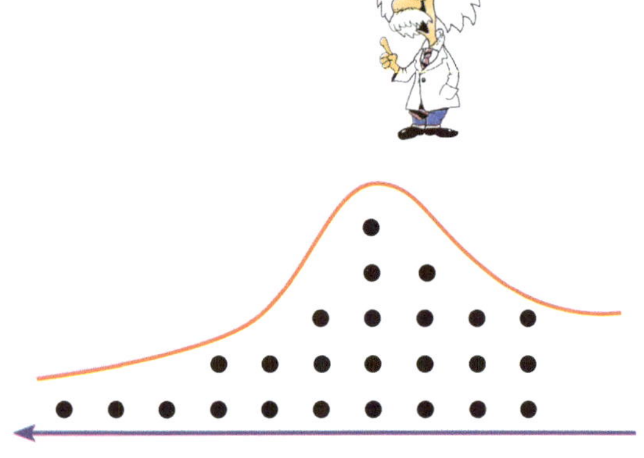

In this dot plot the tail points to the left and we call this **negatively skewed**.

Example 3:

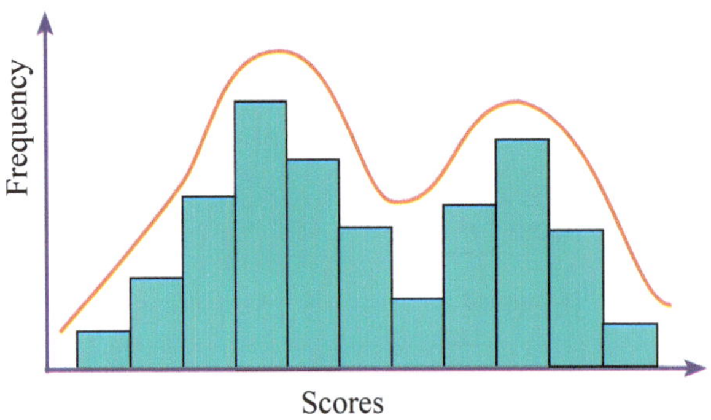

The histogram on the left has 2 peaks and we call this a **bimodal** distribution.

THE NORMAL CURVE

Consider a survey on the heights of 100 students in Year 10. Some students would be very short, some just below average, a few extremely tall — but the majority would be scattered either side of the mean or average height. If we drew a 'histogram' of the results, and then constructed a 'frequency polygon', we obtain a bell shaped or 'normal curve' as shown below.

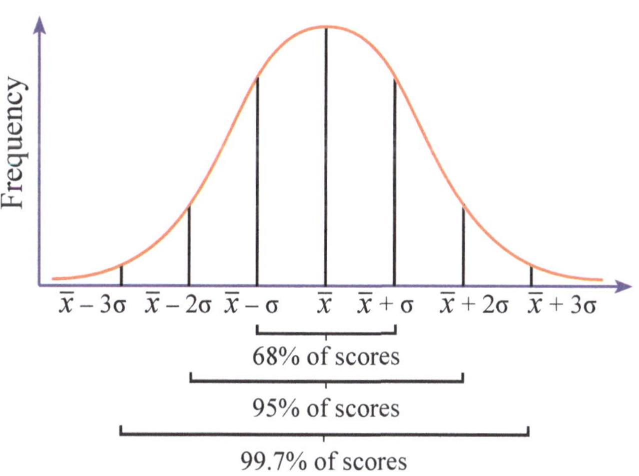

Note: This curve is only effective if we have a reasonably large set of data (i.e. ≥ 100). It doesn't work for small samples.

The standard deviation indicates how spread out the scores are. A higher standard deviation means a larger dispersion.

For all normal distributions:

68% of the scores lie within 1 standard deviation of the mean.
95% of the scores lie within 2 standard deviations of the mean.
99.7% of the scores lie within 3 standard deviations of the mean.

These are also sometimes called the '3 confidence levels'.

Example: The height of 100 students in Year 10 were measured. The mean height was found to be 170 cm and the standard deviation 6 cm.

a) What percentage have a height between 164 cm and 176 cm?
b) What percentage have a height between 158 cm and 182 cm?
c) What percentage are taller than 188 cm?

Solution: Once you have drawn up the normal curve with the mean and standard deviations, it is easy to answer the questions.

a) 68%
b) 95%
c) 99.7% have heights between 152 and 188 cm. Therefore 0.3% have heights to either side of these figures. Therefore 0.15% of students are taller than 188 cm.

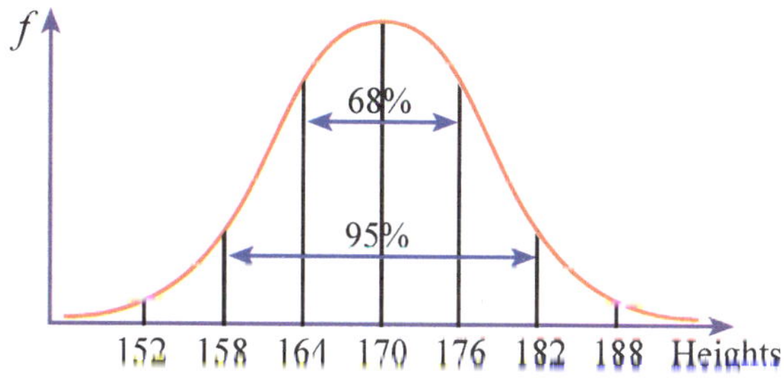

BOX PLOTS

A box and whisker plot (usually called a boxplot) is a graph that presents information from a summary of the five most important values in a data set. These important values are known as the five-number summary of the data set – the lowest and highest values, the median or the middle value and the first quartile, the middle of the lower half and the third quartile or the middle of the upper half. This type of graph is used to show the shape of the distribution, its central value, and its variability. Each section of a box plot contains 25% of the data, so its shape shows the distribution in a visual way. The difference between the first and third quartiles is called the interquartile range, IQR which represents 50% of the values in the data set.

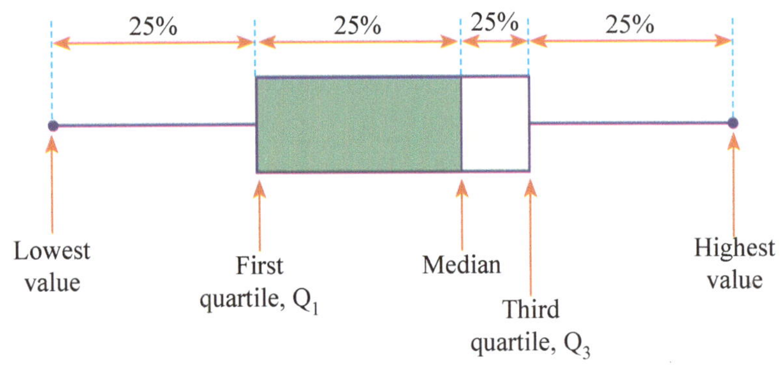

A box plot or boxplot is a method for graphically depicting a set of data through their quartiles. Boxplots also have 2 lines extending from the boxes (whiskers) indicating variability outside the upper and lower quartiles.

Example: The ages of thirty offenders who were caught during random breath testing were recorded as follows: {18, 18, 19, 19, 19, 19, 20, 20, 20, 20, 21, 21, 21, 23, 24, 25, 25, 26, 28, 30, 31, 33, 37, 41, 47, 52, 55, 58, 67, 72}.

a) List the five number summary of the ages.
b) Represent the ages on a box plot.
c) Describe the distribution of the data.

Solution: a) Lowest value: 18, First quartile (Q_1): 20, Median: $\frac{24 + 25}{2} = 24.5$,

Third quartile (Q_3): 37, Highest value: 72

b)

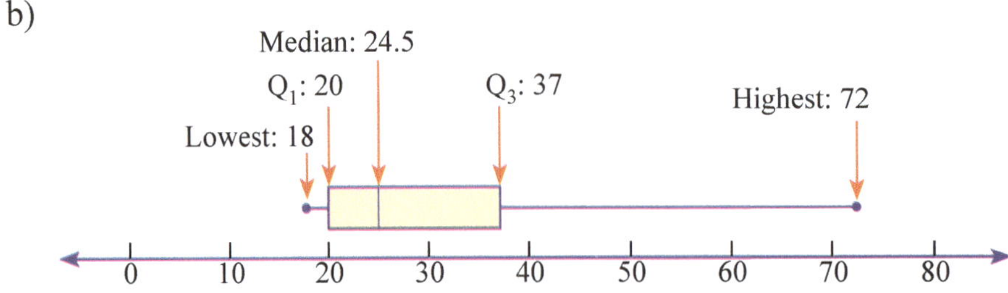

c) The trend of the data is positively skewed as its median is closer to the lower quartile.

OUTLIERS

An outlier in a set of data is a value that is considered to lie outside most of the other values in a set of data either by being much smaller or larger than the other values. For an outlier value to be judged too small it must lie 1.5 times the interquartile range below the first quartile. In a similar way, for an outlier value to be judged too large it must lie 1.5 times the interquartile range above the third quartile.

Lower outlier = $Q_1 - 1.5 \times IQR$ Higher outlier = $Q_3 + 1.5 \times IQR$

Outliers are represented with an asterisk or circle on a boxplot so that it is still included on the diagram. The whisker of the boxplot then goes to the last value that is not an outlier.

> In statistics, an outlier is a data point that differs significantly from other observations or data. An outlier may be due to variability in the measurement or it may indicate experimental errors. Hence they are sometimes excluded from the data set because they can cause serious problems in statistical analysis.

Example: a) For the data set {3, 18, 24, 25, 27, 28, 29, 32, 34, 38, 47} show that

(i) 3 is an outlier and (ii) 47 is not an outlier.

b) Show the data set as a boxplot.

Solution: a) The First quartile, $Q_1 = 24$ and the Third quartile, $Q_3 = 34$

The Interquartile Range, IQR = 34 – 24 = 10

(i) For a possible lower outlier: $Q_1 - 1.5 \times IQR = 24 - 1.5 \times 10 = 9$

As the value 3 is less than 9 then 3 is considered to be an outlier.

(ii) For a possible higher outlier: $Q_3 + 1.5 \times IQR = 34 + 1.5 \times 10 = 49$

As the value 47 is less than 49 then 47 is not considered to be an outlier.

b)

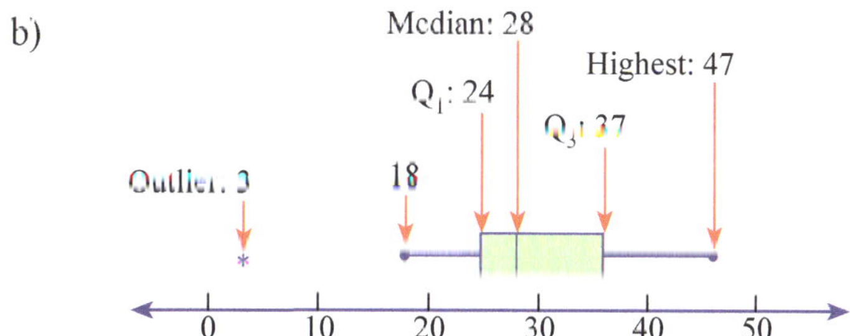

CHAPTER SUMMARY

RELATIVE FREQUENCY

Besides knowing how often a particular score occurs (given by the frequency), it is also useful to know what fraction of the total results it represents.

$$\text{Relative frequency of a score} = \frac{\text{frequency of score}}{\text{total frequency}}$$

THE INTERQUARTILE RANGE

You will already know from earlier years that the range is the difference between the highest score and the lowest score.

The interquartile range is the middle 50% of scores when they are arranged in order from the smallest to the largest. It is the difference between the upper quartile Q_3 (top 25% scores) and the lower quartile Q_1 (bottom 25% of scores).

Interquartile range (IQR) = 3rd quartile (Q_3) – 1st quartile (Q_1)

THE DIFFERENCE BETWEEN A POPULATION AND A SAMPLE

A **population** in statistics means all of the data under consideration. Sometimes all the data is too large and we need to use only a part of the population which we call a **sample.**

Example: Let us assume that we are trying to calculate various statistical data regarding the heights of students in a particular year (say year 10). Rather than considering the entire data for the whole population (which may involve over 150 students), a simpler way would be to consider the heights of a smaller sample of the students.

Note: The important difference between the two will become clearer when using the standard deviation formula shown below.

STANDARD DEVIATION

A formula has been devised by Mathematicians called the 'Standard Deviation' (S_n), and this gives an accurate method of working out how widespread the scores are from the mean.

$$S_n = \sqrt{\frac{\sum(\bar{x} - x)^2}{n - 1}}$$

A large deviation means the scores are more spread out, while a small standard deviation means the scores are closer together on average.

To find the standard deviation for a whole population (σ_n) the above formula is changed to

$$\sigma_n = \sqrt{\frac{\sum(\bar{x} - x)^2}{n}}.$$

The formula is almost identical except we use 'n' instead of '$n - 1$'.

STANDARD DEVIATION USING THE CALCULATOR

Fortunately, you don't have to memorise the formula on the previous page, or carry out the time consuming calculations which are involved. The calculator can work out the mean and standard deviation automatically, once it has been switched on to the 'Statistical mode'. Once all the data has been entered, press the following:

AC SHIFT STAT and the number 4 (var).

You will see on the screen:

1 : n	2 : $\bar{x}$
3 : σx	4 : sx

By entering #1, you will be given the number of scores (n).
By entering #2, you will be given the mean of the scores ($\bar{x}$).
By entering #3, you will be given the standard deviation of a population (σx).
By entering #4, you will be given the standard deviation of a sample (sx).

THE NORMAL DISTRIBUTION

Consider a survey on the height of 100 students in Year 10. Some students would be very short, some just below average, a few extremely tall – but the majority would be scattered either side of the mean or average height. If we draw a 'histogram' of the results, and then constructed a 'frequency polygon', we obtain a bell shaped or 'normal curve' as shown below.

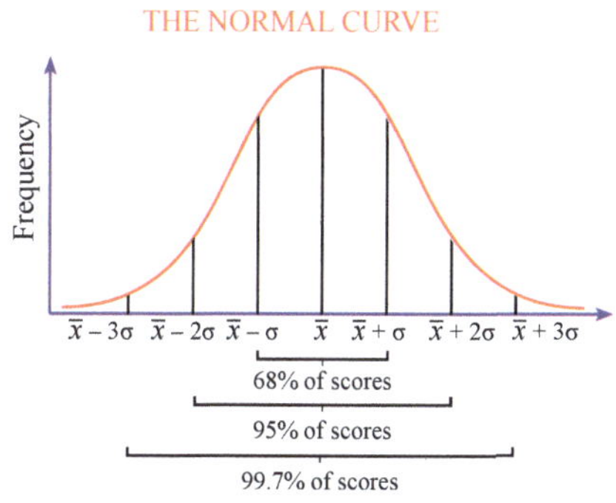

Note: This curve is only effective if we have a reasonably large set of data (i.e. ≥ 100). It doesn't work for small samples.

For all normal distributions:
68% of the scores lie within 1 standard deviation of the mean.
95% of the scores lie within 2 standard deviations of the mean.
99.7% of the scores lie within 3 standard deviations of the mean.

DIFFERENT DEVIATION STANDARDS

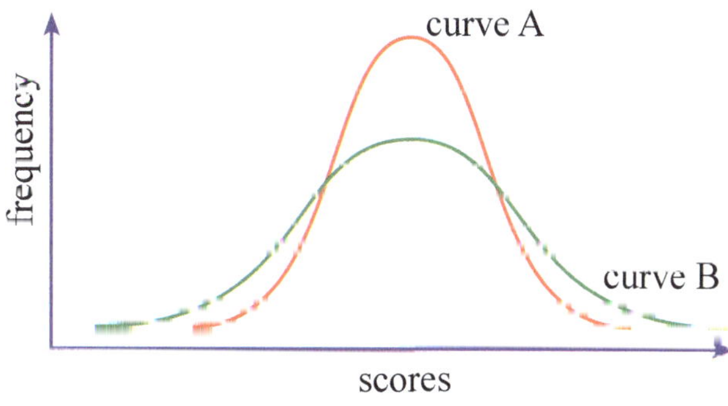

The smaller the standard deviation, the thinner and taller the bell is. The larger the standard deviation, the wider and shorter the bell is.
The diagram shows two distributions with the same mean, median and mode, but curve B has a larger standard deviation than curve A.

BOX AND WHISKERS PLOTS (OR SIMPLY A BOXPLOT)

This presents information using 5 of the most important values from a set of data. This five-number summary consists of the lowest value, the first quartile (Q_1), the median or second quartile (Q_2), the third quartile (Q_3) and the highest value.

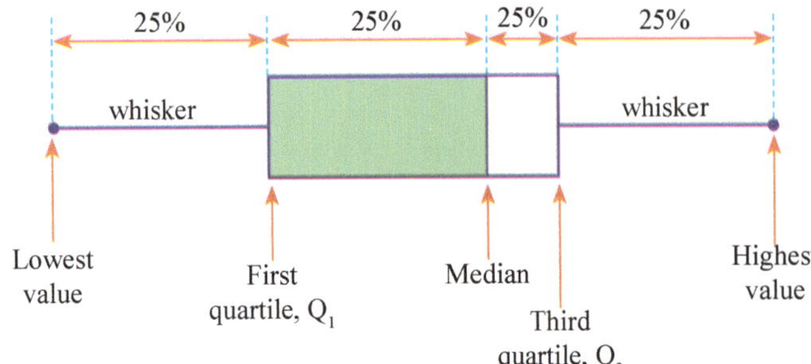

OUTLIERS

An outlier in a set of data is a value that is considered to lie outside most of the other values in the data either by being much smaller or larger than the other values. For an outlier value to be judged too small it must lie 1.5 times the interquartile range below the first quartile.
In a similar way, for an outlier value to be judged too large it must lie 1.5 times the inter quartile range above the third quartile.

Lower outlier = $Q_1 - 1.5 \times \text{IQR}$ Higher outlier = $Q_3 + 1.5 \times \text{IQR}$

Outliers are represented with an asterisk or circle on a boxplot so that it is still included on the diagram. The whisker of the boxplot then goes to the last value that is not an outlier.

SKEWED DATA

Not all graphs are symmetrical and normal (perfect bell shapes).

Positive Skew

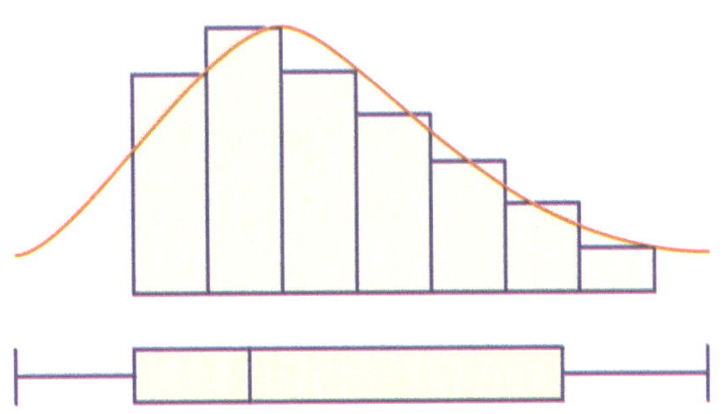

In the box plot the median is closer to the lower quartile than the upper quartile.

i.e. mode < median < mean

Negative Skew

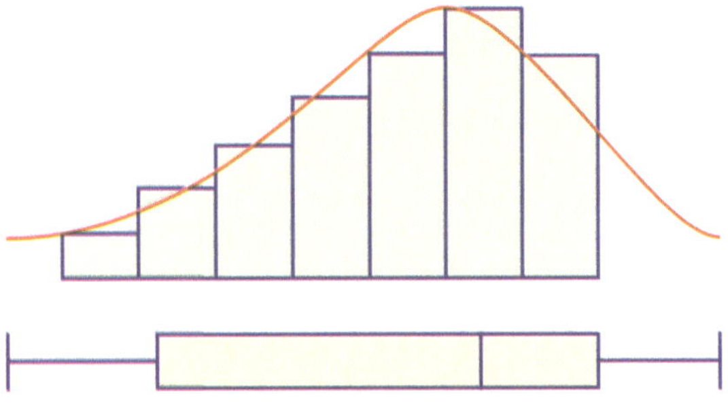

In the box plot the median is closer to the upper quartile and usually larger than the mean and less than the mode.

i.e. mean < median < mode

LEVEL 1 — STATISTICS

EASIER QUESTIONS

Note: Only turn back to page number shown if you have difficulty. | Page

Q1. The table below shows how students from Memorial High School travel to school each day. — 45, 52

Mode of transport	walk	bus	train	bicycle	motor bike	car
Students	178	136	84	124	142	75

a) What is the school population?

b) What percentage of the school population (correct to 1 d.p.)

(i) walk to school? (ii) travel by car to school?

c) A sample of 150 students are to be interviewed on how much time they spend travelling each day. How many students should be selected from the group who:

(i) walk to school? (ii) travel by car to school?

Q2. Calculate the mean, median, mode and range and standard deviation for each sample of scores: — 44, 46, 47

a)

6	3	5	6	2
8	4	6	7	3

b)

19	15	13	14	16
17	12	14	13	18
12	14	15	13	14

Q3. Calculate the mean and standard deviation S_n for the sample of scores in each table to 2 d.p.: — 47

a)

(*x*)	(*f*)
9	5
10	7
11	14
12	10
13	6

b)

(*x*)	(*f*)
15	9
16	13
17	14
18	19
19	15

Try doing Q2. and Q3. using STAT mode on your calculator. Remember to select $\boxed{sx}$ to find the standard deviation of a sample.

Q4. The temperatures in Sydney and Melbourne were recorded each day at 12 p.m. for a period of 20 consecutive days. The results were graphed on the dot plots shown below. — 41, 44

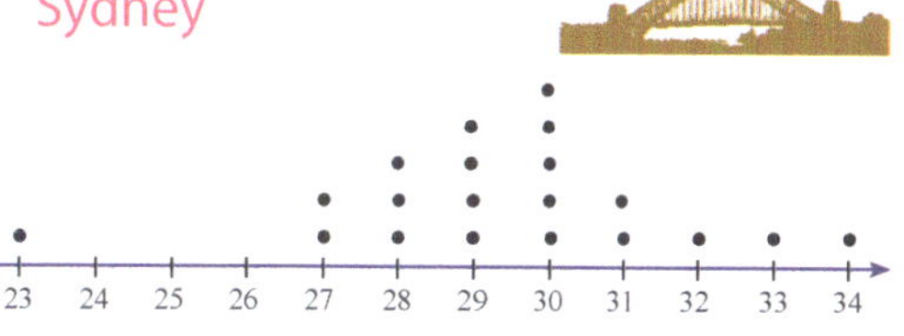

a) Find the mean, median, mode and range for each city.

b) Give reasons why you think one city has more consistent temperatures.

c) What sort of distribution does each of the cities have?

d) Calculate the 'interquartile range' for both cities.

e) Which city has the warmer temperature and give reasons for your choice.

AVERAGE QUESTIONS

Note: Only turn back to page number shown if you have difficulty. — Page

Q1. Calculate the relative frequency of each score in the tables: 40

a)

Score	frequency
10	2
11	5
12	7
13	11
14	8
15	2

(Answers to 2 d.p.)

b)

Score	frequency
40	1
50	4
60	7
70	12
80	10
90	6

(Answers to 3 d.p.)

Q2. Find the mean, range and standard deviation σ_n (correct to 1 d.p.) for the following sets of scores: 44 - 47

a) 5, 3, 9, 6, 5, 7, 6, 4, 8, 2, 4.

b) 16, 14, 12, 18, 17, 15, 16, 11, 20, 15, 19, 13.

c) 66, 53, 71, 44, 86, 76, 65, 74, 47, 31, 72, 91, 79, 58, 80, 48, 69, 54, 70, 86.

d) 65, 88, 79, 49, 57, 72, 68, 90, 63, 75, 87, 53, 69, 66, 89, 67, 69, 81, 91, 80.

Q3. Find the mean, median and standard deviation σ_n of each set of scores: 47

a)

Score	frequency
16	2
17	5
18	7
19	11
20	8

b)

Class	C.C.	f
50 - 59	54.5	6
60 - 69	64.5	7
70 - 79	74.5	12
80 - 89	84.5	7
90 - 99	94.5	3

(Answers to 2 d.p.)

Try doing Q2. and Q3. using STAT mode on your calculator. Remember to select $\boxed{\sigma_n}$ to find the standard deviation of a population. C.C. means class centre.

Q4. The mean birthweight of 100 babies was found to be 3.625 kg with a standard deviation of 250 g. Find: 49

a) the percentage of babies with a weight between 3.125 kg and 4.125 kg.

b) the percentage of babies with a weight between 3.625 kg and 3.875 kg.

c) the percentage of babies weighing less than 2.875 kg.

Q5. The number of hamburgers sold each day at a local Burger shop were recorded as follows: 50, 51

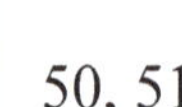

{120, 144, 165, 182, 180, 20, 150, 164, 110, 190, 187}

a) Find the five-number summary and the interquartile range of the data set.

b) Identify any value/s that could be considered as outliers.

c) Display this information as a boxplot.

Q1. The number of people booking evening reservations at two restaurants are displayed in a stem plot. The values for $\{a,b,c,d,e,f,g\}$ for Restaurants A and B respectively are: $\{0,1,2,3,4,5,6\}$ and $\{1,2,3,4,5,6,7\}$.

Stem	Leaf
a	3
b	3 4
c	5 5 8
d	0 1 6 6 7 8
e	4 7 7 8
f	5 7 9 9
g	2

Key: 1|3 = 13

a) Write the bookings of people attending each restaurant.

b) For each restaurant find the (i) mean (ii) median (iii) mode and compare each measure for each restaurant.

c) For each restaurant find the (i) range (ii) standard deviation (S_n) and compare each measure for each restaurant.

d) If the distribution for restaurant A is approximately normally distributed state the values with which 68% of the number of reservations lie using (i) the data (ii) the confidence interval of a normal distribution.

Q2. A group of people were surveyed as to the number of mangoes that each person ate in a week. The values for the relative frequency have been rounded to two decimal places. Due to a clerical error the values in the table are incomplete.

Score	Frequency	Relative frequency
0	4	
1	6	0.23
2	9	0.35
3	5	0.19
4		

a) Find the number of people who were surveyed.

b) Write the missing values into the table.

c) Find the mean and standard deviation (S_n) of the values in the table.

d) If a person was chosen at random state the probability that they selected
(i) an even number of mangoes (ii) more than one mango.

Q3. The examination scores of a number of students which is normally distributed has a mean of 40. The upper bound of the score that 68% of the people achieved was 44.

a) State the standard deviation of the scores.

b) If 48 students scored outside three standard deviations from the mean find
(i) the total number of students in the group
(ii) the number of students making scores from 32 to 48.
(iii) the lower bound of the score that 68% of the students achieved.

Q4. The ages of twenty-six first-time mothers in a local hospital are given as follows.

{25, 32, 29, 33, 31, 22, 24, 28, 38, 21, 28, 29, 30, 32, 26, 21, 22, 29, 32, 38, 25, 28, 30, 32, 33, 27}.

a) Find the five-number summary of the data set.

b) Display this Information as a boxplot.

c) Describe the distribution of the data.

Q1. The results from a Maths Test for three classes are shown on box-and-whisker plots.

a) For each class state the
(i) median (ii) lowest value (iii) highest value
(iv) first quartile (v) third quartile

b) State the interquartile range for each class and use it to rank the classes from least to most consistent.

c) The results from Class C only included the scores from five students, whereas the scores from twenty students were shown for Class A. The teacher from Class C argued that his students performed significantly better than the students in Class A. Discuss.

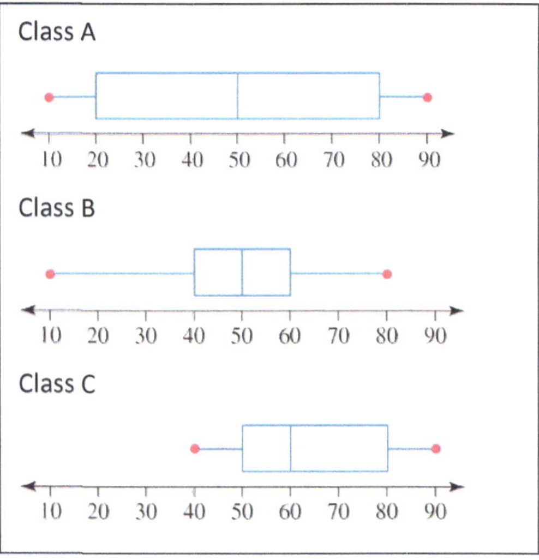

Q2. The money, rounded to the nearest whole dollar raised by twenty students for the recent Walkathon is listed as {78, 84, 61, 73, 71, 83, 87, 65, 78, 80, 60, 67, 71, 82, 84, 79, 78, 43, 65, 72}.

a) (i) List the data, from smallest to largest and hence state the lowest, highest, median, first quartile and third quartile values.
(ii) Show the amounts as a box plot and state whether the distribution is symmetrical or not.

b) Using an appropriate calculation justify why 43 is not an outlier and prove that there are no other outliers in the data set.

Q3. Statistical measures such as mean and standard deviation can either be calculated for an entire population or from a sample or part of a population. To find the standard deviation and mean for a population the data from the entire population needs to be known such as the heights for all students in a year level.

Consider the scores from a class of students in a multiple choice quiz as {15, 18, 20, 10, 11, 18, 12, 14, 15, 17, 13, 12, 8, 15, 20, 16, 17, 12, 17, 15}.

a) Find the mean and standard deviation for the population.

b) (i) Two smaller samples were selected as A: {8, 12, 13, 14, 20} and B: {15, 18, 16, 20, 20}. Find the mean and standard deviation for both samples.
(ii) Compare the mean and standard deviation for both samples and explain why they are different even though they are part of the original population.
(iii) Which sample is closer in nature to the population and describe why.

Q4. The elements in the following sets have a value/s missing.

a) Find the median and mode if the mean for the following is 5.
(i) $\{2, 3, 6, x, 8\}$ (ii) $\{4, 4, x, x, 6\}$ (iii) $\{5, 3, 2x, 6\}$

b) Find the mean if the median for the following is 5.
(i) $\{2, 3, 6, x, 14\}$ (ii) $\{2, 3, 6, x, 8, 10\}$ (iii) $\{5, x, 2x, 6\}$

PROBABILITY

The 'Australian Curriculum Mathematics' (ACM) references for this sub-strand of 'Statistics and Probability' (SP) are given below. This chapter may contain additional extension work which the author feels will be beneficial to the student.

- *Probability (ACMSP 204).*
- *Two and three step experiments (ACMSP 226).*
- *Selecting with and without replacement (ACMSP 246).*

Leonardo Da Vinci (1452 – 1519)

Da Vinci was born in Italy during the Renaissance era and he made significant contributions to an extremely wide range of disciplines and interests. Some of these included inventions, drawings, paintings, sculptures, architecture, science, music, mathematics, engineering, literature, anatomy, geology, astronomy, botany, paleontology and cartography. He is often also considered one of the greatest painters of all time even though only about 15 of his paintings have survived down the years. Two of his most famous paintings are 'Mona Lisa' and 'The Last Supper' which are each worth hundreds of millions of dollars. Although he had no formal academic training, many historians and scholars consider him to be an example of the 'The Universal Genius' and one of the most diversely talented people to have ever lived. He was a man of 'unquenchable curiosity' and 'feverishly inventive imagination'. Amongst his prolific talents he was also acclaimed as a man of great technical ingenuity. He had already conceptualised flying machines, armoured fighting vehicles, solar energy and adding machines more than 400 years before they actually came into existence in the twentieth century.

INTRODUCTION AND SAMPLE SPACES

Probability involves the study of 'chance', 'uncertainty' or 'predicting events'. It can be applied to card games, board games and many other forms of gambling: poker, black jack, bridge, backgammon, roulette wheel, horse racing, etc. It is also an important branch of Mathematics which is used a great deal in Science and Economics.

A 'sample space' is a list of all the possible outcomes in an experiment, and it is denoted by the capital letter S. If we wish to count the number of possible outcomes in the sample space, then this is denoted or shown by $n(S)$. The experiment and examples below will help you to understand the above ideas, even though some of the experiments will only be explained more fully in later pages of this chapter.

> A list of all possible outcomes is called the sample space and is denoted by the capital letter S.
> The number 'n' of outcomes in the sample space is denoted by $n(S)$.

Example: List the outcomes (S) of the following sample spaces and also determine $n(S)$.

a) Tossing one coin
$S = \{\text{head, tail}\} \Rightarrow n(S) = 2$
i.e. There is a total of 2 possible outcomes.

b) Choosing a playing card
$S = \{\text{ace of spades, king of spades, etc.}\} \Rightarrow n(S) = 52$
i.e. There is a total of 52 possible outcomes.

c) Tossing a single die
$S = \{1, 2, 3, 4, 5, 6\} \Rightarrow n(S) = 6$
i.e. There is a total of 6 possible outcomes.

d) Tossing 2 coins
Let H = head and T = tail
$S = \{\text{HH, HT, TH, TT}\} \Rightarrow n(S) = 4$
i.e. There is a total of 4 possible outcomes.

Note: A {head, tail} result can take place in 2 different ways!

e) Choosing a letter from the alphabet
$S = \{a, b, c, d, e, f..............x, y, z\} \Rightarrow n(S) = 26$
i.e. There is a total of 26 possible outcomes.

f) Triangular spinner
$S = \{\text{green, red, blue}\} \Rightarrow n(S) = 3$
i.e. There is a total of 3 possible outcomes.

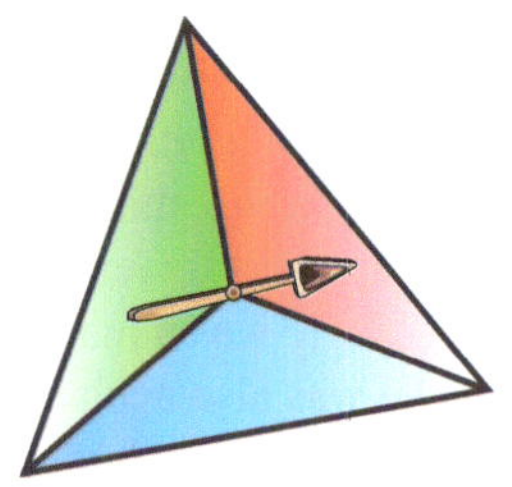

Note: The main idea to understand on this page is the 'total number of possible outcomes'.

THE PROBABILITY FORMULA

In many of the simplest experiments, we can use a common sense approach to work out probabilities. However, the questions will become more difficult later in your schooling, and you will need the important mathematical rule shown below.

$$P(E) = \frac{n(E)}{n(S)}$$

where $P(E)$ = probability of an event.
$n(E)$ = number of favourable ways the event can occur.
$n(S)$ = total number of possible outcomes in the sample space.

Example: A box contains the letter cards shown. A person picks out a card without first looking.

Find the probability of each event described below and give each answer in fraction, percentage and decimal form.

a) Picking a card with a letter B?
b) Picking a card with a letter E?
c) Picking a card with a letter A or E?
d) Picking a card with a letter M?
e) Picking a card with a letter in the first six letters of the alphabet?

Solution: There are 10 possible outcomes.

a) There is only 1 card with the letter B.
$\therefore P$ (letter B) $= \frac{1}{10}$ or 10% or 0.1

b) There are 3 cards with the letter E.
$\therefore P$ (letter E) $= \frac{3}{10}$ or 30% or 0.3

c) There are 5 cards with either the letter A or the letter E.
$\therefore P$ (letter A or E) $= \frac{5}{10} = \frac{1}{2}$ or 50% or 0.5.

d) There are NO cards (zero cards) with the letter M.
$\therefore P$ (letter M) $= \frac{0}{10} = 0$ or 0%.
This event is called **'Impossible!'**

e) All 10 cards have letters in the first six letters of the alphabet.
$\therefore P$ (A or B or C or D or E or F) $= \frac{10}{10} = 1$ or 100% or 1.0
This event is called **'Certain'**.

The total number of possible outcomes is the same as the sample space n(s) explained on the previous page.

RANGE OF PROBABILITIES

If an event is CERTAIN to occur, then its probability is 1.
If an event is IMPOSSIBLE, then its probability is 0.
Therefore it is worth remembering that all your answers should always lie between 0 and 1.

i.e.

$$0 \leq P(E) \leq 1$$

Probabilities can be expressed in fraction, decimal or percentage format, although most questions tend to use fractions.

The probability of all outcomes in the sample space must add up to 1. Therefore the probability of 'an event not occurring' is equal to '1 – probability of the event occurring'.

i.e.

$$P(\widetilde{E}) = 1 - P(E)$$

where $P(\widetilde{E})$ is the probability of the event not occurring.

Example: A bag contains 3 red, 5 green and 2 yellow marbles. What is the chance of drawing at random one marble which is

a) Red b) Green or Yellow c) Not yellow

Solutions:

Using $P(E) = \frac{n(E)}{n(S)}$

a) There is a total of 10 marbles ⇨ $n(S) = 10$
3 of these are red ⇨ $n(E) = 3$

$\therefore$ $P(\text{selecting red}) = \frac{3}{10}$

b) There are 7 marbles which are green or yellow ⇨ $n(E) = 7$

$\therefore$ $P(\text{selecting green or yellow}) = \frac{7}{10}$

c) $\therefore$ $P(\text{selecting yellow}) = \frac{2}{10} = \frac{1}{5}$

$\therefore$ $P(\text{not selecting yellow}) = 1 - \frac{1}{5} = \frac{4}{5}$

Using $P(\widetilde{E}) = 1 - P(E)$

NORMAL PACK OF PLAYING CARDS

There are many different card games, and all of these games involve different chances. Often you will be required to answer questions about a pack of cards, so it is important that you have a good understanding of all the different cards in a pack. On this page we are going to investigate a normal pack of 52 playing cards, and also we are going to assume that all jokers have been removed.

The pack is split up equally into 4 main suits called spades, hearts, diamonds and clubs. Each suit consists of 13 cards, and there are 2 black suits (spades and clubs) and 2 red suits (hearts and diamonds).

Spades	♠	black	Ace, King, Queen, Jack, 10, 9, 8, 7, 6, 5, 4, 3, 2	(13 cards)
Hearts	♥	red	Ace, King, Queen, Jack, 10, 9, 8, 7, 6, 5, 4, 3, 2	(13 cards)
Diamonds	♦	red	Ace, King, Queen, Jack, 10, 9, 8, 7, 6, 5, 4, 3, 2	(13 cards)
Clubs	♣	black	Ace, King, Queen, Jack, 10, 9, 8, 7, 6, 5, 4, 3, 2	(13 cards)

TOTAL = 52 cards

The highest card in each suit is the ace, then comes the king, then the queen, etc. all the way down to the lowest card in each suit, which is the two. So each pack contains 4 different aces, 4 different kings, 4 different queens, etc.

> There are 52 cards in a normal pack. They are equally split up into 4 major suits called spades, hearts, diamonds and clubs. Two suits are black (spades and clubs) and two suits are red (hearts and diamonds).

Example: Consider a normal pack of playing cards. If I shuffle the cards, and then pick one out of the pack, work out the following chances.

a) What is the probability of picking the ace of clubs?
b) What is the probability of picking any heart card?
c) What is the probability of picking any red card?
d) What is the probability of picking any ace?

Solutions: a) The ace of clubs is one special, unique card out of a total of 52 cards.

Therefore the probability of picking it is '1 out of 52' or $\frac{1}{52}$.

b) 13 of the cards in the pack are hearts.

Therefore the probability of picking a heart is '13 out of 52' or $\frac{13}{52}$.

$\frac{13}{52}$ simplifies down to $\frac{1}{4}$.

c) The hearts and the diamonds are all red cards.

Exactly half the pack (26 cards) are red cards.

Therefore the probability of picking a red card is $\frac{26}{52}$ or $\frac{1}{2}$.

d) There are 4 different aces in the pack.

Therefore the probability of selecting an ace is $\frac{4}{52}$.

$\frac{4}{52}$ is a fraction which can simplify to $\frac{4 \div 4}{52 \div 4} = \frac{1}{13}$.

TOSSING TWO COINS

Jenny tossed 2 coins forty times, and recorded the results in the table on the right. Each time she tossed a (head, head) she would write a single stroke on that line in the table. Similarly, if she tossed a (head or tail), she would write a single stroke on that line. This is called the 'tally column', and it is very useful for recording results in an experiment. On every fifth stroke we put a line through like this 𝍸. It is now so easy to add up the tally column, and write the sum of each line in the total column.

Result	Tally	Total
Head, Head	𝍸 IIII	9
Head or Tail (in any order)	𝍸 𝍸 𝍸 𝍸	20
Tail, Tail	𝍸 𝍸 I	11

You will see from the table in Jenny's experiment that she tossed a (head, head) 9 times, and she tossed a (head or tail) 20 times, and she tossed a (tail, tail) 11 times. Let us try to answer why she got nearly double the amount of (head or tail) results.

Explanation:

If we labelled the coins, and called them coin 1 and coin 2, let us look at all the outcomes we can get. Jenny only wrote down 3 outcomes on her tally table, but we can now see on the right hand side that there are 4 possible outcomes. A head on coin 1 and a tail on coin 2 is a different outcome to a tail on coin 1 and a head on coin 2.

Coin 1	Coin 2
Head	Head
Head	Tail
Tail	Head
Tail	Tail

The probability of (head, head) is one chance out of four or $\frac{1}{4}$.

The probability of (tail, tail) is one chance out of four or $\frac{1}{4}$.

But (head, tail) or (tail, head) is two chances out of four or $\frac{2}{4}$ or $\frac{1}{2}$.

Therefore if we tossed 2 coins 40 times, we would expect

(head, head) to come up $40 \ \frac{1}{4} = 10$ times

(tail, tail) to come up $40 \times \frac{1}{4} = 10$ times

(tail, head) or (head, tail) $40 \times \frac{1}{2} = 20$ times

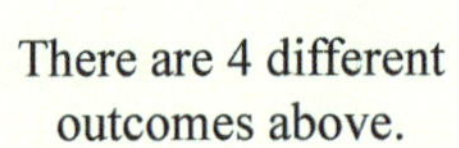

Therefore the practical experiment that Jenny carried out was very close to the results that one would expect. She obtained 9 (head, head), 11 (tail, tail) and 20 (tail or head) combinations.

Challenge Activity: If you toss 3 coins, can you list all the possible outcomes? I will give you a hint by saying that there are 8 different possible outcomes.

TOSSING A PAIR OF DICE

When we toss 2 dice, it is very similar to tossing 2 coins.
As we explained on the previous page, a head on coin 1 and a tail on coin 2 is a different outcome from a tail on coin 1 and a head on coin 2.

If you understand that idea, then it will also be easy for you to understand that a 3 on die 1 and a 5 on die 2 is a different outcome from a 5 on die 1 and a 3 on die 2. In the diagram on the top right, we have die 1 in white and die 2 in grey shading.

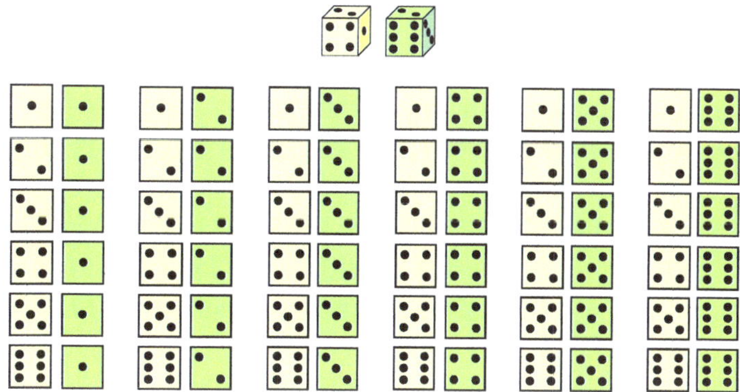

When you toss a pair of normal 6-sided dice, there is a total of 36 different outcomes.

Example 1: What is the chance of getting a total (sum) of 4 when rolling 2 dice?

There are 3 ways of getting a total of 4

 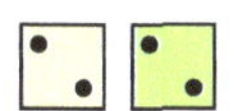 

There are 36 different possible outcomes.

Therefore the chance (or probability) of rolling a total of 4 is $\frac{3}{36}$ or $\frac{1}{12}$.

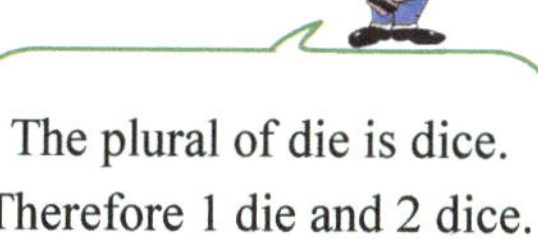

The plural of die is dice.
Therefore 1 die and 2 dice.

Example 2: If you tossed 2 dice 360 times, how many times would you expect a total of 4 to appear?

We have already worked out in Example 1 that the chance of rolling a total of 4 is $\frac{3}{36}$ or $\frac{1}{12}$.

Therefore, we would expect a total of 4 to come up about one twelfth $(\frac{1}{12})$ of the time.

$\therefore \quad \frac{1}{12}$ of 360 throws = 30 times.

Example 3: If we roll 2 dice, which total (sum) has the most chance of occurring?
The total of 7 has the most chance because there are 6 ways of obtaining this sum.

  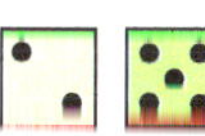  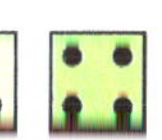 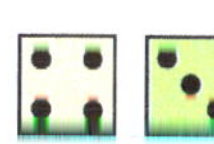 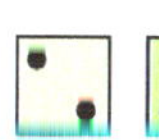  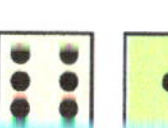 

The probability of rolling a total of 7 is $\frac{6}{36}$ or $\frac{1}{6}$

OBSERVED AND EXPECTED FREQUENCIES

If we toss a coin 10 times, in theory, what would we expect the perfect result to be?
The chance of a head is equal to one half or 50%, and the chance of a tail is also equal to one half or 50%. So we would expect heads to turn up half the number of throws (or trials), and tails to turn up half the number of throws (or trials).
Therefore we would **EXPECT** 5 heads and 5 tails during the course of 10 throws.

Experiment 1: When Steve tried this experiment practically, he got 8 heads and 2 tails as shown in the table on the right. These are called the **OBSERVED** results.

Result	Total
Heads	8
Tails	2

How can you explain the difference between the observed results and the expected results?

Explanation: The problem is that Steve only did a very small number of trials (only 10 throws). As you can see the expected frequency was 50% heads and 50% tails, but the actual observed frequency was 80% heads and 20% tails.

The observed results will become much closer to the expected results, as the number of trials increases, or gets larger.

A larger number of trials means at least 100 trials.

Experiment 2: Steve decided to test out the above Mathematical rule or hypothesis, and so he did the experiment again, and this time he tossed the coin 100 times and noted down the results.

Result	Total
Heads	56
Tails	44

On this occasion he got 56 heads and 44 tails. There is still a difference of 8 compared to the expected results.

However, if we convert these observed results to percentages, they are very much closer to the expected results than in Experiment 1. Heads occur 56% of the time and tails occur 44% of the time, and these are very much closer to 50% than the experiment involving only 10 trials. With only 10 trials, heads occurred 80% of the time and tails occurred 20% of the time !!!

So the rule or hypothesis above really does work. When doing any experiment it is better to have as many trials as possible, because this will give you a much better and more reliable result.

Activity: If you have the time to do 1 000 trials (toss the coin 1 000 times) your observed results will get even closer to the expected results. Try this experiment yourself.

TREE DIAGRAMS

In some questions (such as a pack of 52 playing cards) it is relatively easy to determine the number of favourable events, and also the total number of possible events in the sample space. However in most Probability questions it can be quite difficult to determine all the possible events, and this is when a systematic method is very useful – if not essential! By using a TREE DIAGRAM, all the possible events are clearly listed at the end of the branches, and it then becomes an easy matter to determine the favourable events, and hence answer the questions.

Example: The numbers 1, 2, 3 and 4 are written on separate cards. One card is drawn at random to give the first number of a 2 digit number (i.e. tens digit). A second card is then drawn from the remaining 3 cards to give the second number (i.e. units number).

a) What is the probability that the number formed will be divisible by 3?
b) What is the probability that the number formed will be larger than 31?

Solution:

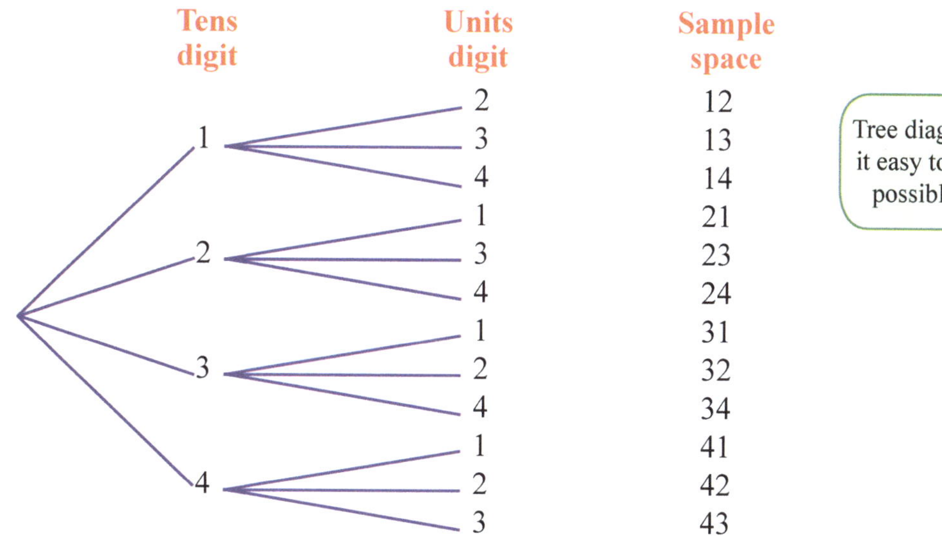

Tree diagrams make it easy to list all the possible events!

a) It can be clearly seen from the tree diagram that the total number of possible events in the sample space $n(S) = 12$. The numbers which are divisible by 3 are 12, 21, 24 and 42.

Using $P(\text{number divisible by } 3) = \dfrac{n(E)}{n(S)}$

$$= \frac{4}{12} = \frac{1}{3}$$

b) From the tree diagram there are 5 numbers which are larger than 31, i.e. 32, 34, 41, 42, and 43.
Therefore number of favourable events $n(E) = 5$.

Using $P(\text{number larger than } 31) = \dfrac{n(E)}{n(S)}$

$$= \frac{5}{12}$$

Note: Without the aid of a TREE DIAGRAM it would have been easy to get in a muddle and miss some of the possible events.

TOSSING THREE COINS

These questions, or variations of them, also tend to occur regularly in tests.

When tossing ONE COIN, there are 2 equally possible outcomes.

Sample space = {H, T}

When tossing TWO COINS, it is still easy to list the 4 different possible outcomes.

Sample space = {HH, HT, TH, TT}

Note that a head on coin one and a tail on coin two is a different outcome to a tail on coin one and a head on coin two.

When tossing THREE COINS, it is best to use a TREE DIAGRAM to work out all the possible outcomes.

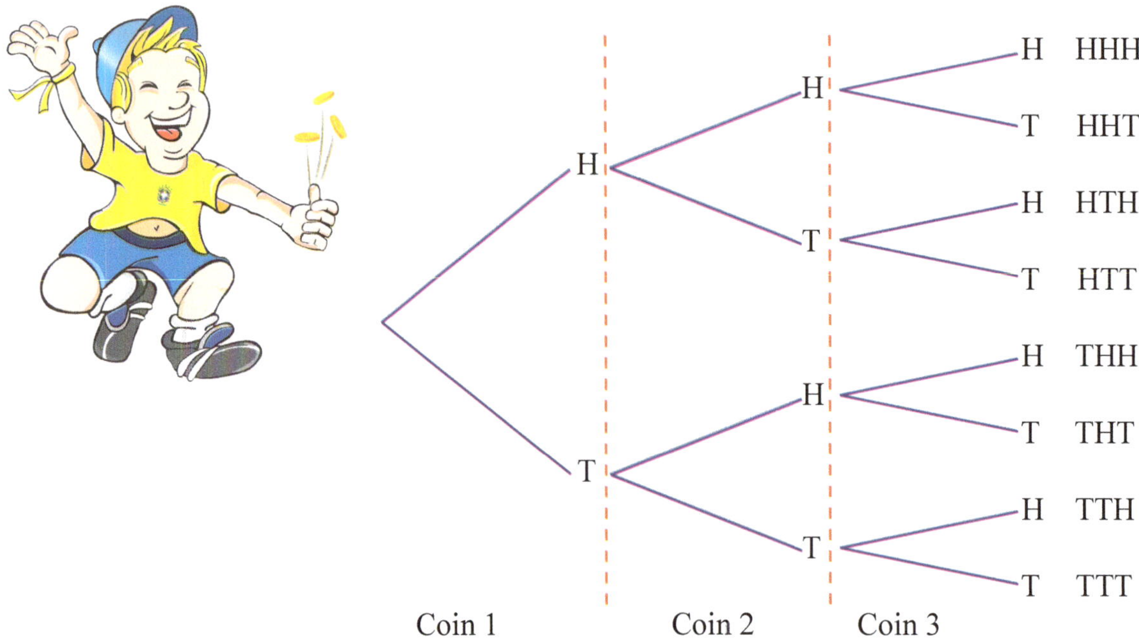

It can be clearly seen from the TREE DIAGRAM that there are 8 possible outcomes in the sample space. In this experiment, each of the 8 events are equally likely to occur.

Therefore $P(\text{head, head, head}) = \frac{1}{8}$

Note: When tossing FOUR COINS there would be 16 different possible outcomes. And for FIVE COINS there would be 32 possible outcomes, etc.

COMMON QUESTIONS

With the help of the theory shown on the opposite page, work through these important examples and make sure that you clearly understand each step.

Example 1: If two coins are tossed, what is the probability of obtaining a head and tail combination?

Sample space = {HH, HT, TH, TT)

Therefore $P(\text{HT or TH}) = \frac{2}{4} = \frac{1}{2}$

Example 2: If three coins are tossed, what is the probability of obtaining 2 heads and a tail in any order?

This event can occur in 3 ways ⇨ {HHT, HTH, THH}

From previous page, sample space = 8

$\therefore \quad P(\text{2 heads and 1 tail}) = \frac{3}{8}$

Example 3: In three child families, what is the probability of having 3 sons?

This is a variation of tossing a coin three times.

There are 8 possible outcomes, and there is only one way to obtain {son, son, son}.

$\therefore \quad P(\text{3 sons}) = \frac{1}{8}$

Example 4: If three coins are tossed, what is the probability of obtaining AT LEAST 1 HEAD?

This idea of 'AT LEAST' is a popular question and therefore important to understand. One could list the 7 outcomes which contain at least 1 head. But this is long-winded and time consuming. You will notice that the only branch that doesn't contain at least one head is {T, T, T}.

$\therefore \quad P(\text{at least one head}) = 1 - P(\text{all tails})$

$\therefore \quad P(\text{at least one head}) = 1 - \frac{1}{8}$

$= \frac{7}{8}$

Note: All probabilities down the different branches must add up to a total of 1.

REPLACEMENT PROBLEMS

In these types of questions, a TREE DIAGRAM will make the problem very much easier and quicker to solve, because they provide you with the complete SAMPLE SPACE.

> The probability of a particular outcome is obtained by MULTIPLYING the probabilities down the branches. This is called the PRODUCT RULE.

Example: A jar contains 3 green and 7 blue marbles. Two marbles are drawn out of the jar, but the first marble is replaced before drawing out the second one.
a) What is the probability of selecting 2 blue marbles?
b) What is the probability of selecting 2 marbles of a different colour?

Solution: Because the first marble is replaced, the probabilities of selecting either colour on the second draw remain exactly the same.

Let $P(\text{G})$ = Probability of drawing a Green marble
Let $P(\text{B})$ = Probability of drawing a Blue marble

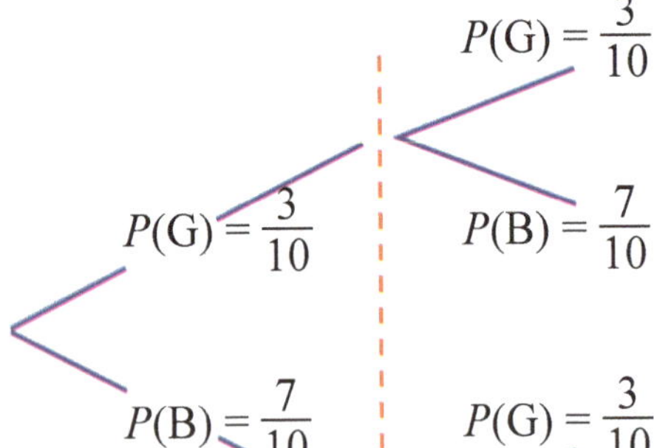

Draw 1: $P(\text{G}) = \frac{3}{10}$, $P(\text{B}) = \frac{7}{10}$

Draw 2: $P(\text{G}) = \frac{3}{10}$, $P(\text{B}) = \frac{7}{10}$, $P(\text{G}) = \frac{3}{10}$, $P(\text{B}) = \frac{7}{10}$

$P(\text{GG}) = \frac{3}{10} \times \frac{3}{10} = \frac{9}{100}$

$P(\text{GB}) = \frac{3}{10} \times \frac{7}{10} = \frac{21}{100}$

$P(\text{BG}) = \frac{7}{10} \times \frac{3}{10} = \frac{21}{100}$

$P(\text{BB}) = \frac{7}{10} \times \frac{7}{10} = \frac{49}{100}$

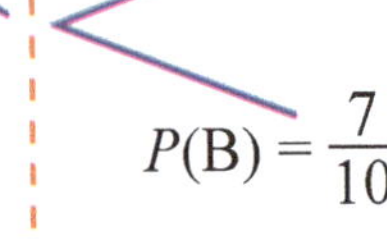

Draw 1 Draw 2

The product rule is used to determine the probabilities at the end of the branches.

a) $P(\text{BB}) = \frac{49}{100}$ (read directly off the bottom branch of the diagram)

b) $P(\text{BG}) \text{ or } P(\text{GB}) = \frac{21}{100} + \frac{21}{100}$

$= \frac{42}{100}$

$= \frac{21}{50}$

Probabilities are added down the branches.

NON REPLACEMENT PROBLEMS

In some questions, it is stated that replacement doesn't occur after the first draw. The probabilities on the second draw then change, depending on the outcome of the first draw.
This idea is best explained by following the example below.

Example: A jar contains 3 green and 7 blue marbles. Two marbles are drawn out of the jar. but the first marble is not replaced before drawing out the second one.
a) What is the probability of selecting 2 blue marbles ?
b) What is the probability of selecting 2 marbles of a different colour.

Solutions: Because the first marble is not replaced, the probabilities of selecting a particular colour on the second draw change, because there are only 9 marbles to choose from.

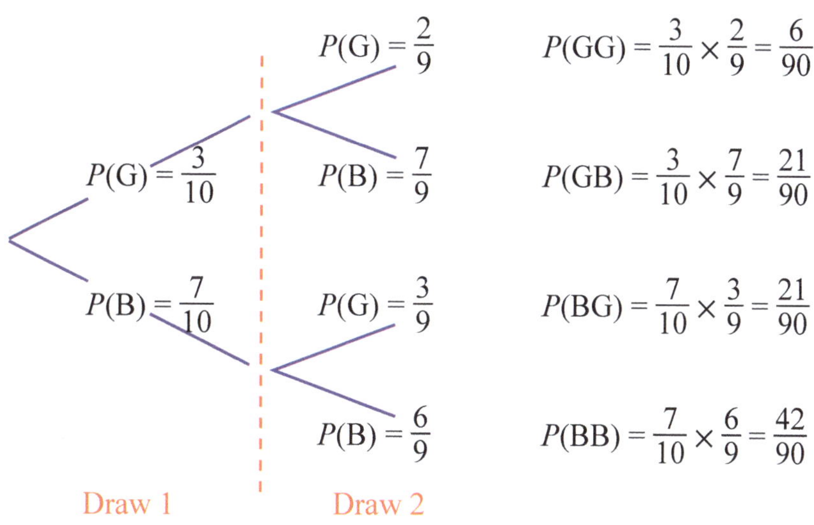

a) $P(BB) = \frac{42}{90}$ (read directly off the bottom branch of the diagram)

b) $P(BG) \text{ or } P(GB) = \frac{21}{90} + \frac{21}{90}$

$= \frac{42}{90}$

$= \frac{7}{15}$

Note: Compare the answers in this example to the ones on the opposite page where replacement took place!

DIFFERENT COMBINATIONS

This section of work will become increasingly important in later years of schooling, but it is useful to have a basic knowledge of it now. It is best explained by studying the examples below.

Example 1: I wish to cycle from town A to town B, and then from town B to Town C. There are 2 ways of getting from town A to Town B which we will call road ① and road ②. There are 3 ways of getting from Town B to Town C which we will call road ③, road ④ and road ⑤. How many different ways can I cycle from Town A to Town C?

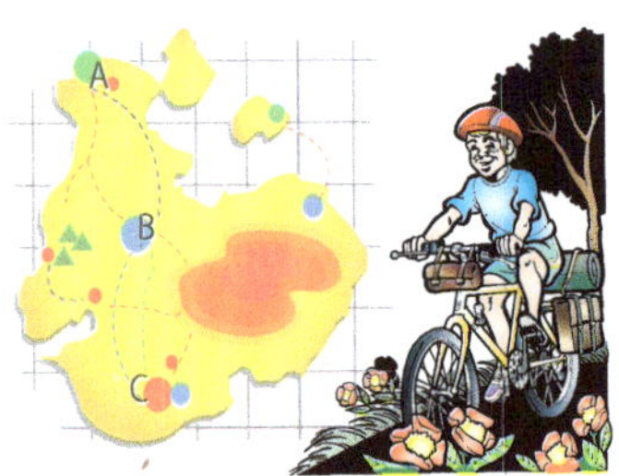

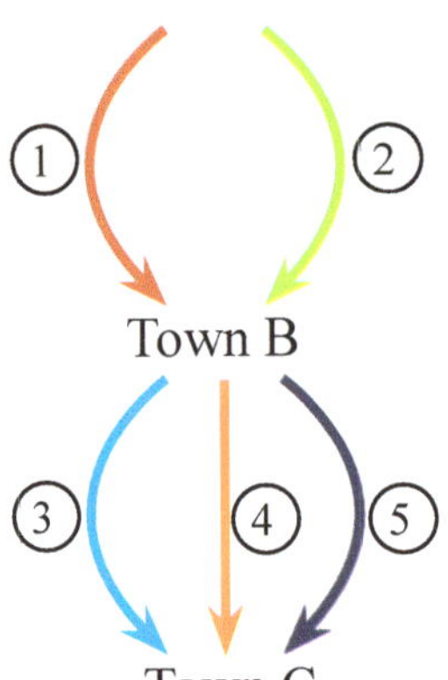

Solution: ① ⟶ ③ ① ⟶ ④ ① ⟶ ⑤

② ⟶ ③ ② ⟶ ④ ② ⟶ ⑤

OR 13 14 15 23 24 25

There are 6 different ways I can travel from Town A to Town C.

> Combinations means how many different ways or arrangements that we can have.

Example 2: I have 4 different T-shirts (blue, red, green and brown) and 2 different pairs of jeans (black and white). How many different ways can I wear these? (or how many different combinations can I wear?)

Solution:
black jeans + blue T-shirt
black jeans + red T-shirt
black jeans + green T-shirt
black jeans + brown T-shirt

white jeans + blue T-shirt
white jeans + red T-shirt
white jeans + green T-shirt
white jeans + brown T-shirt

There are 8 different combinations that I can wear.

MULTIPLYING OUTCOMES

Although tree diagrams are very useful for working out all the outcomes in smaller sample spaces (e.g. tossing 3 coins), they can also be very time consuming for larger sample spaces (e.g. tossing two dice). In some questions, we only need to know the total number of possible outcomes, rather than what each outcome is. In these types of problems, we can find this out by using a simple multiplication technique.

Example 1: 6 cards labelled 1,2,3,4,5,6 are placed in a hat. 3 cards are chosen without replacing any of them, and these are placed face up next to each other.

a) How many 3 digit numbers are possible?

b) What is the probability of the number being 345?

c) What is the probability of the number being less than 300?

Solution: a) There are 6 ways of selecting the first card. There are now only 5 cards left in the hat, so there are 5 ways of selecting the second card. Similarly there are only 4 ways of selecting the third card.

Therefore total number of selections possible $= 6 \times 5 \times 4$
$= 120$

b) There is only one way of obtaining 345 ⇨ $n(E) = 1$
There are 120 selections possible ⇨ $n(S) = 120$

Using $P(E) = \dfrac{n(E)}{n(S)}$

$\therefore \quad P(345) = \dfrac{1}{120}$

c) For the number to be less than 300, the first digit drawn must either be 1 or 2.

$$\therefore \quad P(\text{number} < 300) = \frac{2}{6} = \frac{1}{3}$$

Example 2: In an eight horse race, how many different trifectas are possible? Also, what is the probability of making the correct bet, assuming each trifecta is equally likely? (A trifecta is a term used in horse racing whereby you must pick the first, second and third horse of a race in that exact order.)

Solution: First position can be won by any of the 8 horses.
Second position can be filled by any of the remaining 7 horses.
Third position can be filled by any of the remaining 6 horses.

$\therefore$ Number of trifectas $= 8 \times 7 \times 6 = 336$

Probability of picking the winning trifecta $= \dfrac{1}{336}$

CHAPTER SUMMARY

THE BASIC LAW

There is really only one important formula which is used continually throughout this topic:

$$P(E) = \frac{n(E)}{n(S)}$$

where $P(E)$ = probability of an event
$n(E)$ = number of favourable ways an event can occur
$n(S)$ = total number of possible outcomes in the sample space or experiment

RANGE OF PROBABILITIES

If an event is **CERTAIN** to occur, then its probability is 1.
If an event is **IMPOSSIBLE**, then its probability is 0.
Therefore it is worth remembering that all your answers should always lie between 0 and 1.

i.e. $0 \le P(E) \le 1$

Probabilities can be expressed in fraction, decimal or percentage format, although most questions tend to use fractions.
The probability of all outcomes in the sample space must add up to 1. Therefore the probability of 'an event not occurring' is equal to '1 – probability of the event occurring'.

i.e. $P(\tilde{E}) = 1 - P(E)$ where $P(\tilde{E})$ is the probability of the event not occurring.

TOSSING TWO COINS

When tossing **ONE COIN**, there are 2 equally possible outcomes.
Sample space = {H, T}
When tossing **TWO COINS**, it is still easy to list the 4 different possible outcomes.
Sample space = {HH, HT, TH, TT}
Note that a head on coin one and a tail on coin two is a different outcome to a tail on coin one and a head on coin two.

PROBABILITIES INVOLVING DICE

When tossing one die, it should be fairly obvious that there are 6 possible outcomes.
When tossing two dice, there are 36 possible outcomes. When answering questions, it is best to list all possible outcomes as a set of ordered pairs.
A 3 on the first die and 5 on the second die is denoted by (3, 5). This is a different outcome to a 5 on the first die and 3 on the second die, which is denoted by (5, 3).

i.e. (1, 1) (1, 2) (1, 3) (1, 4) (1, 5) (1, 6)
(2, 1) (2, 2) (2, 3) (2, 4)............etc.
........etc.

Remember there are 36 possible, equally likely outcomes.

TREE DIAGRAMS

In many probability questions it can be quite difficult to determine all the possible events, and this is when a systematic method is very useful – if not essential! By using a TREE DIAGRAM, all the possible events are clearly listed at the end of the branches, and it then becomes an easy matter to determine the favourable events, and hence answer the questions.

For example, when tossing **THREE COINS**, it is best to use a tree diagram to work out all the possible outcomes.

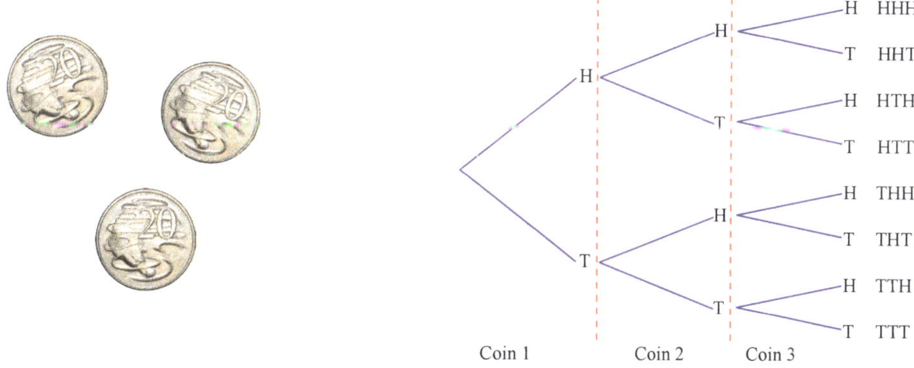

It can be clearly seen from the tree diagram that there are 8 possible outcomes in the sample space. In this experiment, each of the 8 events are equally likely to occur.

REPLACEMENT PROBLEMS

In some questions, items are replaced after they have been chosen so that the sample space remains the same for subsequent selections. In these types of questions, a tree diagram will make the problem very much easier and quicker to solve, because it provides you with the complete sample space.

Note: The probability of a particular outcome is obtained by **MULTIPLYING** the probabilities across the branches. This is called the **PRODUCT RULE**.

NON REPLACEMENT PROBLEMS

In some questions, items are not replaced before subsequent selection. This reduces the sample space each time an item is chosen. The probabilities on the branches of the second draw then change, depending on the outcome of the first draw.

COMBINATIONS AND MULTIPLYING OUTCOMES

Combinations means how many different ways or arrangements are possible in a given situation. In some questions, tree diagrams become too cumbersome to use and difficult to draw because there are simply too many outcomes. We only need to know the total number of outcomes, rather than what every outcome there is. When this happens we can use a multiplication technique.

LEVEL 1 — PROBABILITY

EASIER QUESTIONS

Note: Only turn back to page number shown if you have difficulty.

	Page
Q1. The numbers 1 to 30 are written on cards. If one card is chosen at random, what is the probability that the number on the card will be: a) even b) divisible by 6? c) less than 17? d) a factor of 36? e) one with 4 as a digit? f) greater than 12?	61
Q2. A fruitbowl contains 3 apples, 5 oranges and 2 peaches. If a piece of fruit is selected at random from the bowl, what is the probability that the fruit is: a) a peach? b) an orange or an apple? c) not an apple?	61
Q3. In a family of 4 children what is the probability of: a) 3 daughters and 1 son? b) 4 sons? c) at least one daughter?	68
Q4. Three coins are tossed together. What is the probability of: a) 1 head and 2 tails? b) 2 heads and 1 tail? c) no tails? d) at least 1 head?	68
Q5. A pair of dice is thrown. What is the probability of throwing: a) a total of 9? b) a total less than 6? c) a total greater than 7? d) a total score that is odd? e) at least one 6? f) a double 3?	65
Q6. Two cards are drawn at random from a pack of playing cards (i.e. one card is chosen and then the second is chosen without the first card being replaced). What is the probability of: a) drawing a heart for the first card? b) both cards being diamonds? c) both cards being red? d) both cards being of the same suit? e) both cards being kings?	63,71
Q7. The letters A, B, C and D are written on cards and placed in a box. A card is drawn out, the letter on it is noted down, and the card is placed back in the box. If this is done twice what is the probability of: a) drawing the same letter both times? b) drawing different letters each time? c) not getting the letter B?	70
Q8. Twenty coloured balls are placed inside a jar. The 4 different colours are yellow, blue, red and green. Firstly, count how many balls of each colour there are. Then work out the chance (using PERCENTAGE) of picking out the following colours: a) a yellow ball b) a blue ball c) a red ball d) a green ball e) How many blue balls have to be added in order to have an even chance of being picked? f) How many red balls have to be added in order to have a 60% chance of being picked? g) How many blue balls have to be add in order to have a 70% chance of being picked?	61

AVERAGE QUESTIONS

Note: Only turn back to page number shown if you have difficulty. | Page

Q1. A hundred students from Grade 6 are surveyed about the drinks they like or do not like. 50 of them are boys and 50 are girls. The results of the survey are shown in the two-way table above. If one person is chosen at random, calculate the following probabilities And write each of your answers as a percentage. 61

	Like Milk	Don't like Milk	Like Juice	Don't like Juice
Boys	22	28	29	21
Girls	37	13	32	18

a) What is the probability that the person is a girl who likes milk?
b) What is the probability that the person is a boy who does not like milk?
c) What is the probability that the person likes juice?
d) What is the probability that the person is a boy who doesn't like juice?

Q2. A jar contains 40 coloured marbles. Chang picks up a marble at random, notes down the colour, and then returns the marble to the jar. She patiently repeats this process two hundred times. From the results in the table above, make a calculated guess of how many marbles of each colour there are in the jar. Give a reason for your answers. 66

Colour	Total	Percentage
Red	122	61%
Blue	56	28%
Yellow	22	11%

Q3. A two digit number is formed from the digits 5, 6, 7 and 8. If each digit can only be used once, what is the probability of the number being: 70

a) even? b) divisible by 5? c) less than 78?

Q4. Three husbands and their wives (□□, △△, ○○) are sitting in a row on the park bench watching the ducks feeding in the nearby lake. 72

a) If the husbands wish to sit next to their wives and couple □□ wishes to sit at one end nearer the lake, how many different seating arrangements are possible?
b) If the 3 husbands wish to sit together, and the 3 wives wish to sit next to each other NEAREST to the lake, how many different seating arrangements are possible?

Q5. A box contains five red balls, three blue balls and two white balls. A ball is chosen at random from the box and then replaced before a second ball is withdrawn from the box. 70, 71

a) Find the probability that (i) both balls had the same colour (ii) both balls had different colours (iii) At least one of the balls was either blue or red.

b) Balls were chosen from a box containing the same number of coloured ball as before; however, when the first ball was chosen it was not replaced before the second ball was chosen. Find the probability that (i) both balls had the same colour (ii) both balls had different colours (iii) At least one of the balls was either blue or red.
Given that a red ball was chosen first find the probability that a (iv) blue ball (v) white ball (vi) red ball was chosen next.

EXTENSION QUESTIONS

Q1. A game is played where a fair dice is rolled, a fair coin is tossed and a card from a full playing card deck is selected at random.
Find the probability of obtaining:
a) a number less than 4 with the dice, a tail on the coin toss and a card showing a diamond.
b) at least a 5 on the dice, a head on the coin and a picture card or ace from the deck of cards.
c) a prime number on the dice, a head or tail on the coin and a red card.

Q2. A circular spinner is to be made so that when it is spun the probability of obtaining the number 4 is somewhere between 2/5 and 3/4. Find the angle range for the sector that the value of 4 should take.

Q3. A dice is biased so that the probabilities that the numbers showing uppermost is shown in the following table where x is a value between 0 and 1 .

Number	1	2	3	4	5	6
Probability	0.1	0.05	0.2	0.15	x	0.25

a) Find the value of x.
b) Which number is expected to be rolled the most. Explain your answer.
c) List the numbers in increasing order of likelihood and predict the expected number of times each number will show when the dice is rolled 1 000 times.
d) When William rolled the dice 120 times he expected the number 2 to show uppermost 6 times.
(i) Explain why he expected the number 2 to occur 6 times.
(ii) William suspected that something was wrong with the dice as when he rolled the dice 120 times the number 2 did not show at all. Do you agree with William, discuss.

Q4. A spinner with the values 2, 4, 6, 8 and 10 is made where the probability of each number showing is proportional to its value.
a) Complete the table to show the probability for each number occurring on the spinner.

Number	2	4	6	8	10
Probability					

b) If the spinner was spun twice find the probability
(i) that two 4's were spun
(ii) both numbers were the same
(iii) both numbers were divisible by 4
(iv) one number was less than 5

Q5. Margaret buys three tickets in a raffle in which 500 tickets are sold.
What is the probability of her winning?
a) 1st prize
b) 2nd and 3rd prize?
c) 1st and 3rd prize?
d) no prize?
e) all 3 prizes?

Q6. In the Melbourne Cup horse race, the final field is limited to 24 horses. If all the horses have an equal chance of winning, what is the probability of choosing the correct trifecta?

GRAPHING LINES AND CURVES

The 'Australian Curriculum Mathematics' (ACM) references for this sub-strand of 'Number and Algebra' (NA) are given below. This chapter may contain additional extension work which the author feels will be beneficial to the student.

- *Solving simultaneous equations graphically (ACMNA 237).*
- *Identifying different graphs (ACMNA 239).*
- *The circle with centre at (0, 0) and centre at (p, q) (ACMNA 239).*
- *The axis of symmetry and vertex of a parabola (ACMNA 267).*
- *The hyperbola (ACMNA 267).*
- *The cubic curve (NSW).*

Albert Einstein (1879 – 1955)

Einstein was a German born theoretical physicist, mathematician and philosopher who developed the theory of relativity, one of the two cornerstones and pillars of modern Physics. Although he contributed to many areas of Physics, he is most famous for his mass-energy formula $E = mc^2$, which is often dubbed the 'the world's most famous equation'. He received the Nobel Prize for Physics in 1921 for his discovery of the laws relating to the 'photo electric effect', a major step forward in the development of quantum theory. He showed signs of his genius in many different disciplines at a young age and, besides being a very talented musician, he had also mastered many academic subjects, including 'differential and integral' calculus by the age of fourteen.

GRAPHING STRAIGHT LINES

There are 4 different ways of expressing a relationship between the two variables x and y.

1. ALGEBRAIC

   This is written as an equation with an 'equal sign'. It shows in algebra form the connection between x and y.

Example: $y = 2x + 1$

This means: y must always equal twice the value of x plus the number 1 added on.
In this particular relationship, if x has the value of 10, then y must have the value of twice ten plus one, i.e. $y = 21$

2. TABLE FORM

   A table consisting of at least three x values is written down. The y values are then obtained by substituting the three x values into the algebraic equation.

x	0	1	2
y	1	3	5

A table of values for $y = 2x + 1$

These first 3 pages should be simple revision of work explained in previous years.

When $x = 0$, $y = 2 \times 0 + 1 = 1$
$x = 1$, $y = 2 \times 1 + 1 = 3$
$x = 2$, $y = 2 \times 2 + 1 = 5$

The x values chosen are usually 0, 1 and 2 because they make the calculations easier. However, any 3 numbers could have been chosen.

3. ORDERED PAIRS

   This is done by simply changing the 3 pairs of numbers from the table into ordered pair form. From the table above, the ordered pairs are (0,1) (1,3) and (2,5).

4. GRAPH

   The 3 ordered pairs are now plotted on the number plane. A straight line is then drawn through the points. The line shouldn't stop at the end of the 2 outermost points. It must extend beyond them, with arrows on as shown in the diagram.

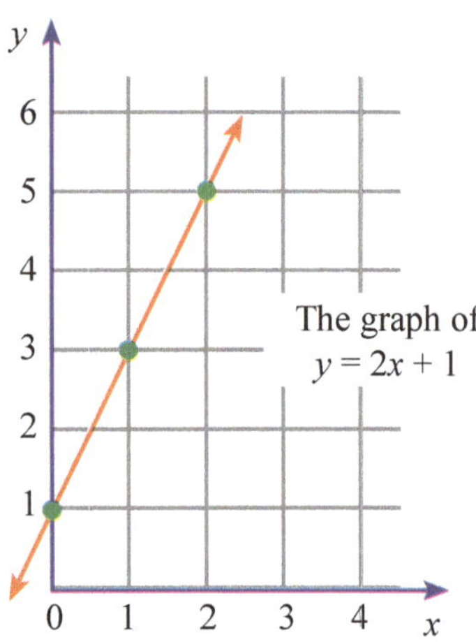

Example: For the relation $y = 3x - 1$

a) Construct a table of values

b) Write down the corresponding ordered pairs

c) Graph the relation on the number plane.

Solution: a) $y = 3x - 1$

x	0	1	2
y	–1	2	5

When $x = 0$, $y = 3 \times 0 - 1 = -1$

$x = 1$, $y = 3 \times 1 - 1 = 2$

$x = 2$, $y = 3 \times 2 - 1 = 5$

b) The above table corresponds to the following 3 sets of ordered pairs:

(0, –1) (1, 2) (2, 5)

c) Simply plot the 3 sets of ordered pairs on the number plane.
Then draw a straight line through these 3 points as shown below.

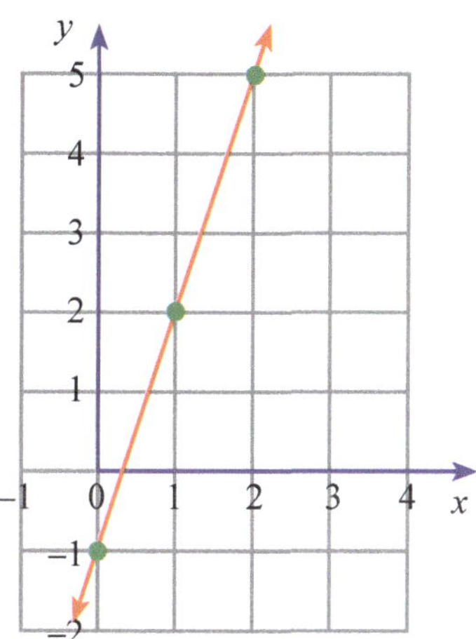

A graph of the relation

$y = 3x - 1$

The positive gradient of 3 means that the line slopes upwards to the right.

You will recall from Year 9 that a straight line can have the format $y = mx + b$ where m = gradient and b = y intercept.

If we compare $y = 3x - 1$ to $y = mx + b$, we can see immediately that the gradient $m = 3$ and the y intercept $b = -1$.

Note: If the 3 points are not in line, then a mistake has been made and you should then carefully check all your working. That is the reason why 3 points are usually plotted instead of the minimum requirement for a line of 2 points. The third point acts as a check point.

FURTHER EXAMPLES

It is not necessary to go through every step shown in the examples in the previous section. These steps were done so that you could fully understand the 4 main ways of expressing a relation between 'y' and 'x'.

However, as you gain confidence, it should be a simple process to draw up a table of values, and then plot these directly onto the number plane.

Example 1: Graph y = 3 – x on the number plane.

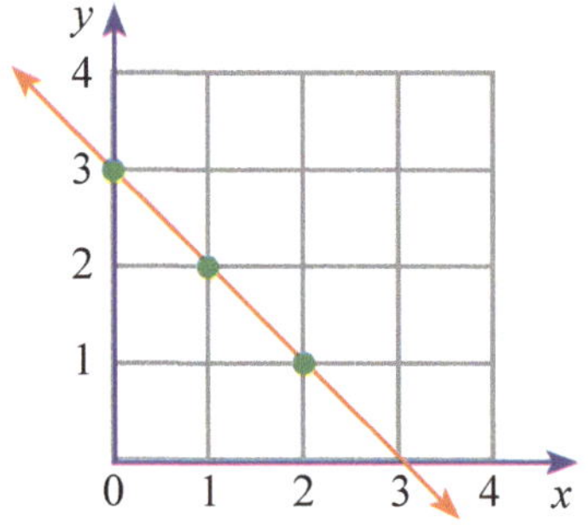

Solutions: When $x = 0$, $y = 3 - 0 = 3$
$x = 1$, $y = 3 - 1 = 2$
$x = 2$, $y = 3 - 2 = 1$

i.e.

x	0	1	2
y	3	2	1

If we compare $y = -x + 3$ (or $y = 3 - x$) to $y = mx + b$, we can see immediately that the gradient $m = -1$ and that y intercept $b = +3$. Also the negative gradient means the line slopes upwards to the left.

Example 2: Does the point (2, –1) lie on the line $y = 2x - 5$?

Solutions: We can work this problem out without having to plot the line and point on a graph. Simply substitute $x = 2$ and $y = -1$ into the equation of the given line $y = 2x - 5$.

$\therefore\ -1 = 2 \times 2 - 5$ TRUE!

The left hand side equals the right hand side.
Therefore the point does lie on the line.

Note: If a given point SATISFIES (holds true) the equation of a line, then the point lies on the line.
If a given point doesn't SATISFY the equation of a line, then the point doesn’t lie on die line.

SKETCHING A LINE USING THE INTERCEPTS

Drawing up a table of values is time consuming, and in the more senior years of schooling we usually only only want to make a quick sketch of the line.

To graph a straight line only requires two definite points. Therefore the fastest way of graphing or sketching a straight line is to find out where it cuts the x and y axis. This is called the INTERCEPT METHOD. We use the idea that the equation $y = 0$ is another way of describing the x-axis, and the equation $x = 0$ is another way of describing the y-axis.

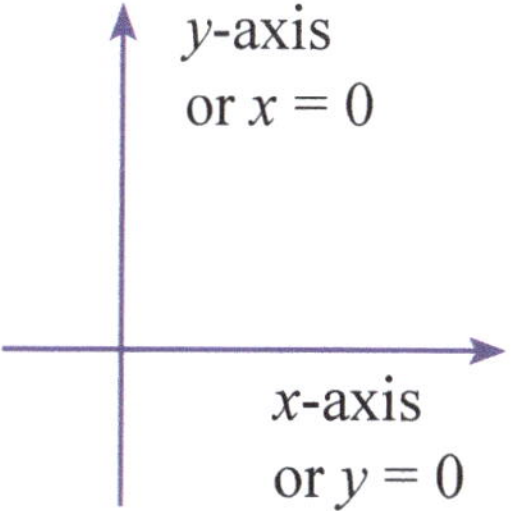

> To find the y intercept, let $x = 0$ and solve the resulting equation.
>
> To find the x intercept, let $y = 0$ and solve the resulting equation.

Example 1: Sketch $3x - 2y = 6$

Let $x = 0 \quad \therefore \; -2y = 6 \Rightarrow y = -3$

i.e. y intercept $(0, -3)$

Let $y = 0 \quad \therefore \; 3x = 6 \Rightarrow x = 2$

i.e. x intercept $(2, 0)$

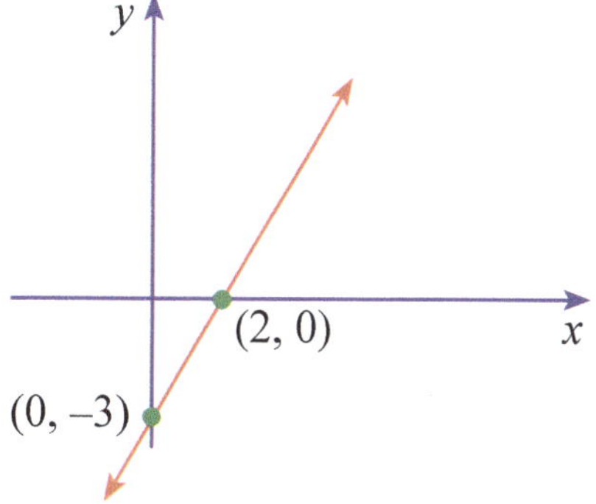

Example 2: Sketch $y = -3x + 8$

Let $x = 0 \Rightarrow y = +8$

i.e. y intercept $(0, 8)$

Let $y = 0 \quad \therefore \; 0 = -3x + 8$

$\therefore \; 3x = 8 \Rightarrow x = 2\frac{2}{3}$

i.e. x intercept $\left(2\frac{2}{3}, 0\right)$

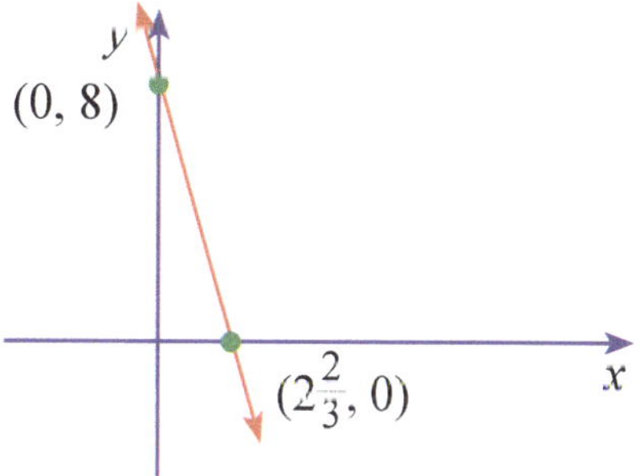

GRAPHING A PARABOLA

To recognise the equation of a PARABOLA, you will find that 'y' is always a function of 'x^2'.

Examples: $y = x^2 - 5x + 6$, $y = 9 - x^2$, $y = 3x^2 + 6x$, $y = 2x^2 + 6x - 3$

The GENERAL EQUATION of a parabola is:

$$y = ax^2 + bx + c$$

Where a, b, c are numerical constants and $a \neq 0$.

The graph of $y = ax^2 + bx + c$ will have a MINIMUM if $a > 0$ (i.e. positive)

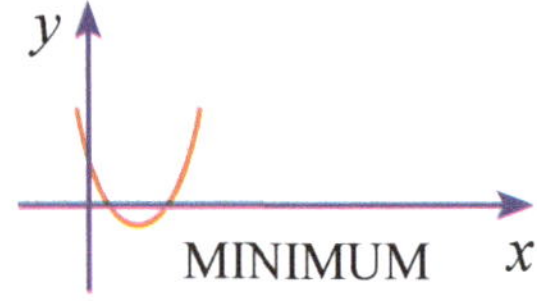

The graph of $y = ax^2 + bx + c$ will have a MAXIMUM if $a < 0$ (i.e. negative).

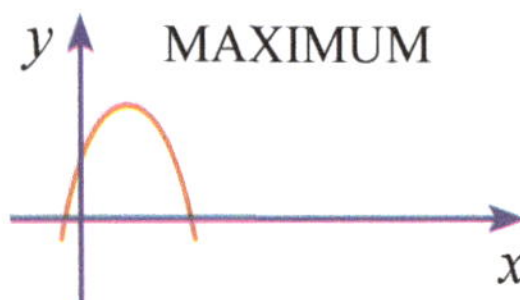

The points at which the parabola cuts the x-axis are particularly important. They are called the ROOTS of the equation. The simplest parabola is $y = x^2$, but there are an infinite number of parabolic graphs depending on the values of a, b and c in $y = ax^2 + bx + c$.

The most basic way of graphing a parabola is by using a table of values.

Example: Using a table of values, graph the function $y = x^2 - 2x - 8$ for $-3 \le x \le 5$

Solution: Substitute x values from -3 to $+5$ into $y = x^2 - 2x - 8$.

x	–3	–2	–1	0	1	2	3	4	5
y	7	0	–5	–8	–9	–8	–5	0	7

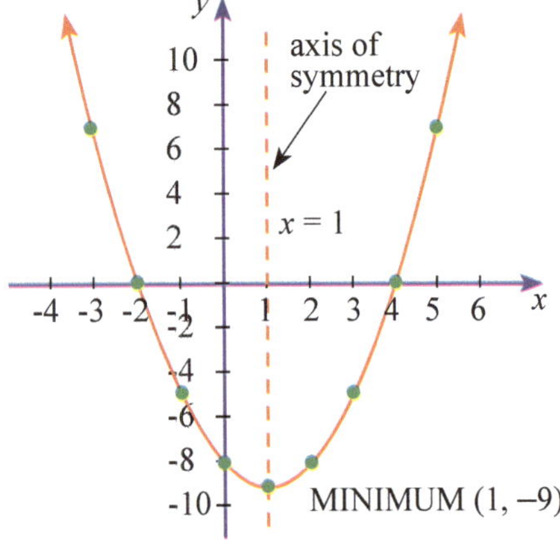

Note: (i) The minimum has coordinates (1, –9)

(ii) The ROOTS of the equation are $x = -2$ and $x = 4$.

(iii) One half of the parabola reflects onto the other half in a line called the AXIS OF SYMMETRY. The axis of symmetry in this example is $x = 1$.

GRAPHING REGIONS

This process involves shading a certain region on the number plane which satisfies one of the following inequality signs: ($>, \geq, <, \leq$).

If the inequation involves either '$\geq$' or '$\leq$', then a SOLID LINE is drawn.

If the inequation involves either '$>$' or '$<$' then a BROKEN LINE is drawn.

The examples below are easier to understand, because they involve only horizontal lines, (y = constant), or vertical lines (x = constant).

Example 1: Shade in the region $y \geq 2$.

Draw the line $y = 2$ and then shade in the region above it as shown.

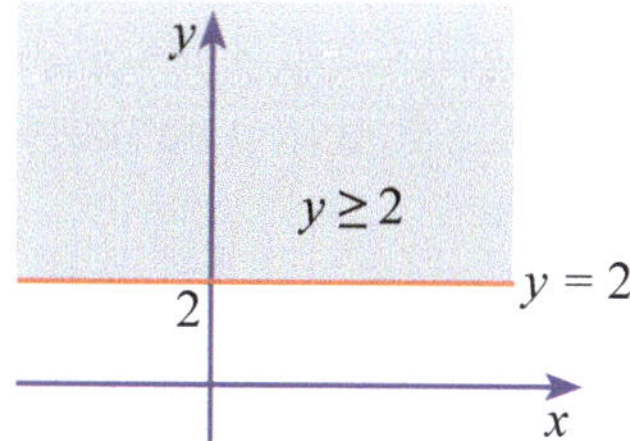

Example 2: Shade in the region $x < 3$.

Draw the broken line $x = 3$, and then shade in the region to the left of it as shown.

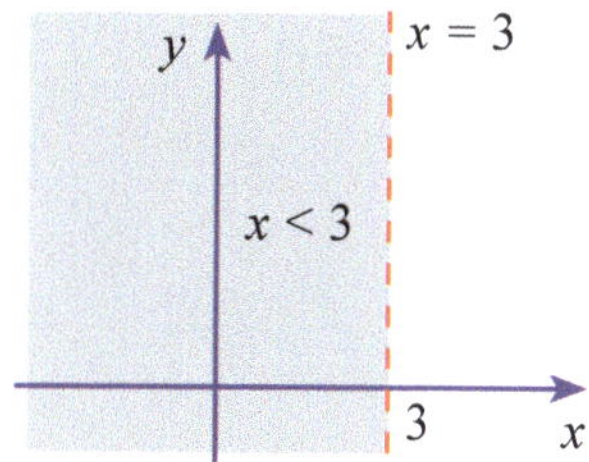

Example 3: Shade in the region satisfied by:

$\{(x, y) : x < 3\} \cap \{(x, y) : y \geq 2\}$

$\cap$ is a Maths symbol representing INTERSECTION. The region shaded must now satisfy both inequalities at the same time, as shown in the diagram.

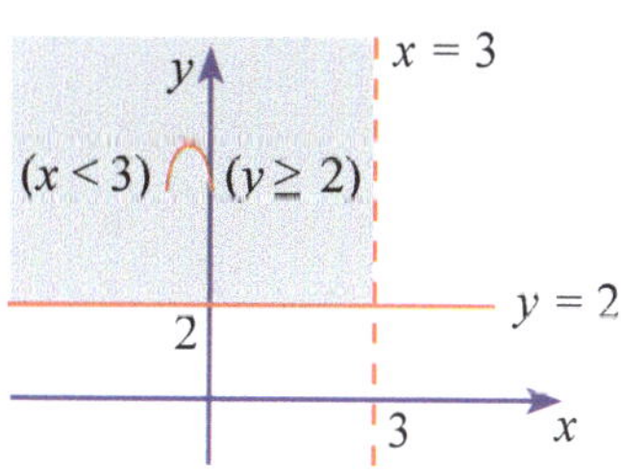

On the previous page, it was explained how to graph more simple regions involving inequalities of lines which were either vertical or horizontal. The steps outlined below relate to the graphing of any region: sloping lines, parabolas, circles, etc.

STEPS:

(i) First sketch the relation as shown in the intercept method.
(ii) Choose a suitable TEST POINT and determine whether the inequality holds.
(iii) Shade in the required region.

Note: Unless the function actually passes through the origin, the TEST POINT (0,0) is the simplest to choose.

Example 1: Shade the region satisfied by $3x - 2y \geq 6$

Step (i) Sketch $3x - 2y = 6$
let $x = 0 \Rightarrow y = -3$
let $y = 0 \Rightarrow x = 2$

Step (ii) Choose test point (0, 0)
$3 \times 0 - 2 \times 0 \ngeq 6$
The test point (0, 0) doesn't satisfy $3x - 2y \geq 6$

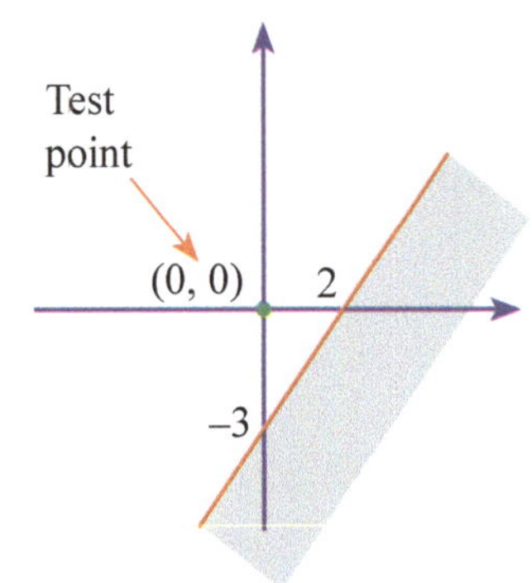

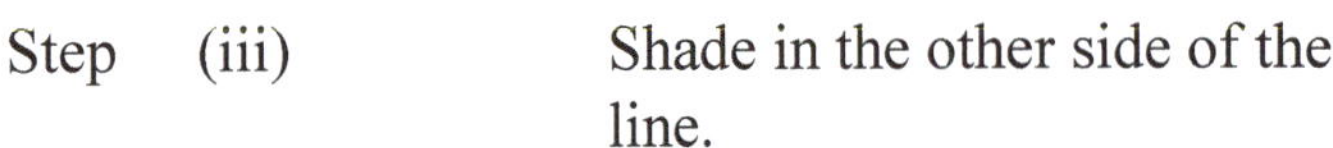

Step (iii) Shade in the other side of the line.

Example 2: Shade the region satisfied by $y < x^2$

Step (i) Sketch $y = x^2$ using a broken line.
This is a basic parabola shape.

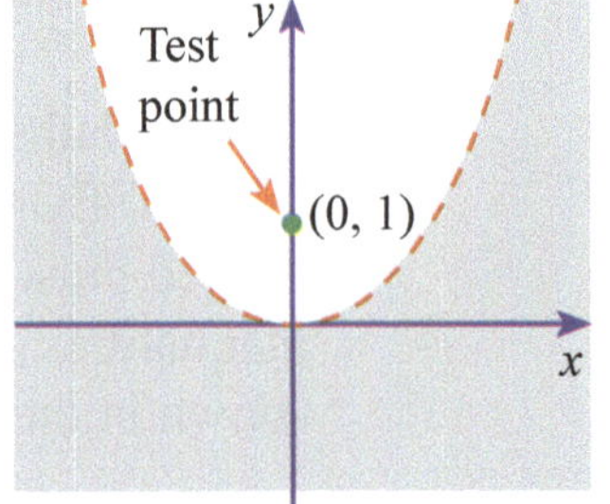

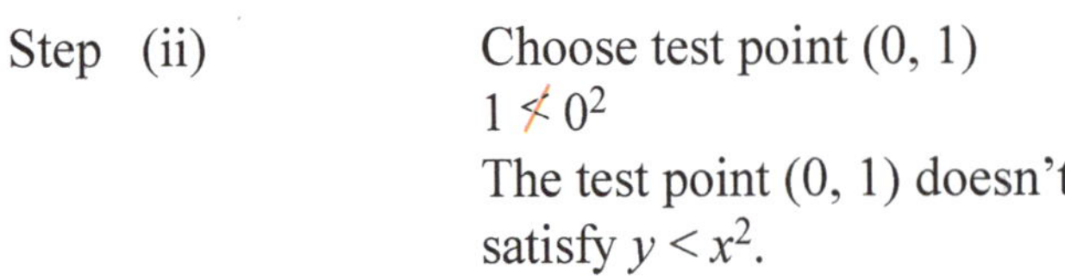

Step (ii) Choose test point (0, 1)
$1 \nless 0^2$
The test point (0, 1) doesn't satisfy $y < x^2$.

Step (iii) Shade in the other side of the parabola using a broken line.

THE CUBIC CURVE

An equation involving 'x' raised to the power of 3 is called a CUBIC.

i.e. $y = ax^3 + bx^2 + cx + d$

Where a, b, c, d are numerical constants and $a \neq 0$.

There are faster and more efficient methods of sketching the main features of a cubic curve which you may cover in the optional topic called POLYNOMIALS. However, at this stage it is important to understand the basic shape patterns, and most questions will require you to graph the curve by drawing up a table of values.

The simplest equation is given by $y = x^3$ and its curve is shown on the right.

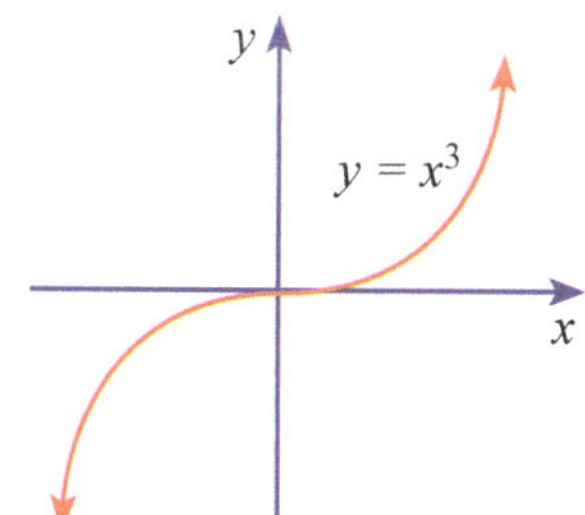

However most cubic curves have one of the 2 shapes shown below:

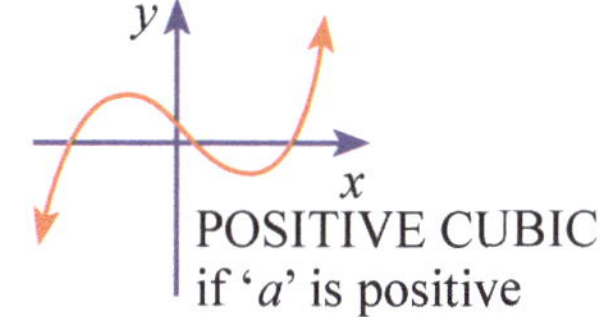

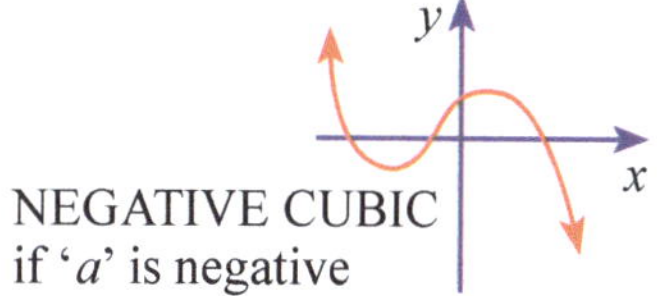

Example: If $y = x(x-3)(x+2)$ construct a table of values for x between -3 and $+4$. and graph the curve.

Solution:

x	-3	-2	-1	0	1	2	3	4
y	-18	0	4	0	-6	-8	0	24

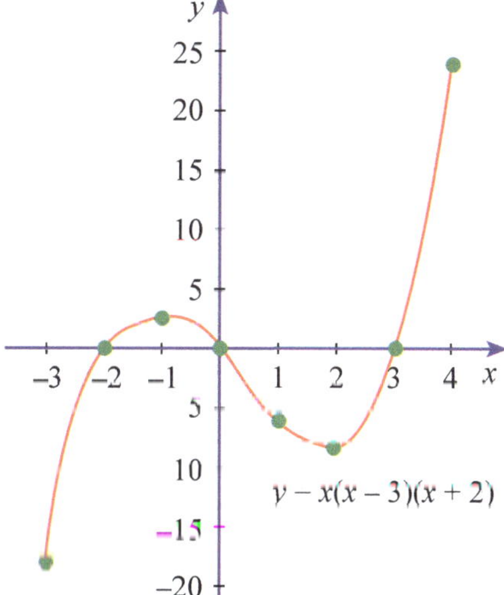

Note: $y = x(x-3)(x+2)$ when expanded out equals $x^3 - x^2 - 6x$ which is a positive cubic.

THE HYPERBOLA

The basic hyperbolic equation has the following pattern:

$$xy = c$$

or $$y = \frac{c}{x}$$

where c is some constant which could either be positive or negative.

Some examples include: $xy = 3$ or $y = \frac{3}{x}$

$xy = -5$ or $y = \frac{-5}{x}$

There are two basic shapes depending on whether c is positive or negative.

If c is positive

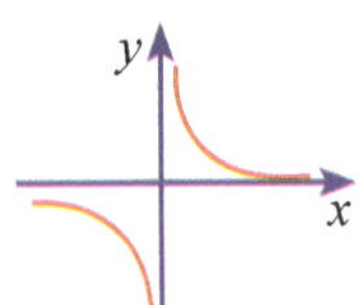

If c is negative

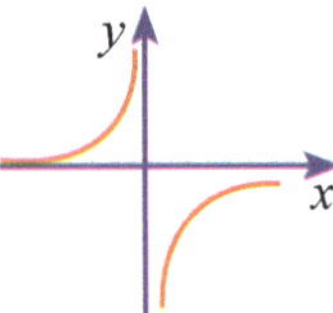

As you can see, hyperbolic equations have graphs which are always split up into two (or more) separate curves as shown.

Example: Draw up a table of ordered pairs for $y = \frac{1}{x}$, where $-4 \leq x \leq 4$.

Graph the resulting ordered pairs on a number plane and draw a smooth curve through them.

Solution:

x	-4	-2	-1	$-\frac{1}{2}$	$-\frac{1}{4}$	$\frac{1}{4}$	$\frac{1}{2}$	1	2	4
y	$-\frac{1}{4}$	$-\frac{1}{2}$	-1	-2	-4	4	2	1	$\frac{1}{2}$	$\frac{1}{4}$

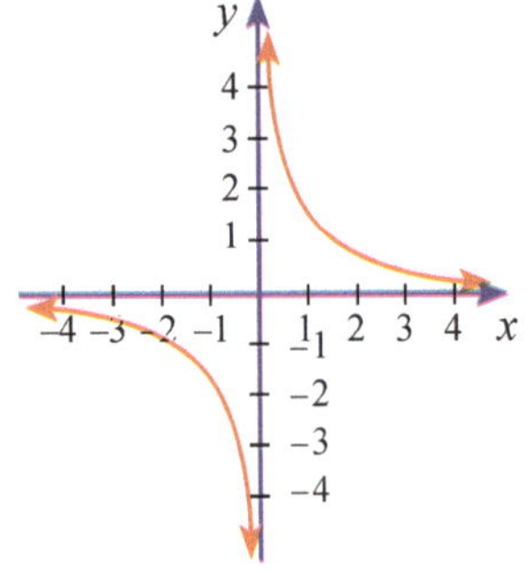

Note: You cannot substitute $x = 0$ into $y = \frac{1}{x}$. One can never divide by zero in Mathematics. Therefore there is a discontinuity or break in the curve at $x = 0$. Hyperbolic graphs never involve one smooth continuous curve because there is always at least one break somewhere.

THE CIRCLE

A 'LOCUS' is a 'set of points which satisfies a given condition'.

A circle is defined as 'the locus of points which is equidistant from a fixed point'.

To determine the equation of a circle of centre (p, q) and radius (r), use:

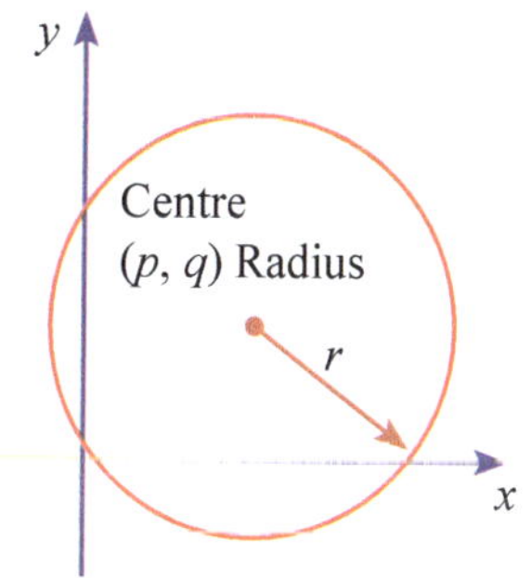

$$(x - p)^2 + (y - q)^2 = r^2$$

If the centre of the circle is at the origin (0, 0) then this equation simplifies to:

$$x^2 + y^2 = r^2$$

Example 1: Find the equation of the circle with centre $(3, -4)$ and radius 5.

Using $(x - p)^2 + (y - q)^2 = r^2$

$(x - 3)^2 + (y + 4)^2 = 25$

It is normally left in this form, so there is no need to expand out the brackets.

Example 2: a) For the circle $(x + 3)^2 + (y - 2)^2 = 16$, find the centre and radius.

b) Hence graph the region satisfied by $(x + 3)^2 + (y - 2)^2 < 16$.

Solution: a) This requires the reverse procedure to the previous example.

Comparing $(x + 3)^2 + (y - 2)^2 = 16$ to

$(x - p)^2 + (y - q)^2 = r^2$

we have: centre $(-3, 2)$ and radius = 4.

b) Choose a test point $(-3, 2)$.

$(-3 + 3)^2 + (2 - 2)^2 < 16$

The test point chosen does satisfy the inequality.

Therefore shade the region inside the circle as shown.

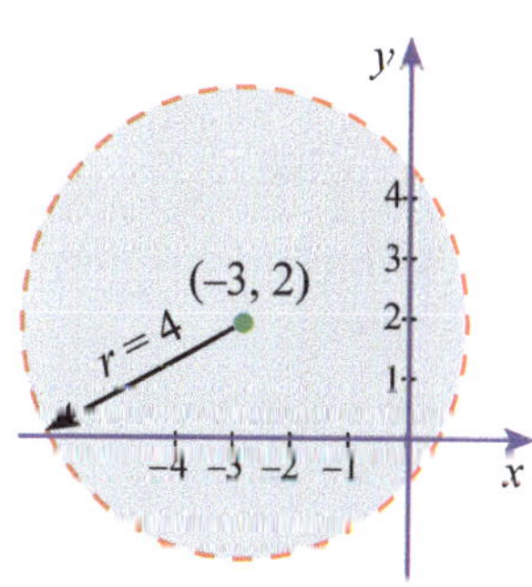

SOLVING SIMULTANEOUS EQUATIONS GRAPHICALLY

Two equations are said to be solved SIMULTANEOUSLY, when you find an ordered pair (x, y) which **satisfies both equations**. This can be done graphically by drawing graphs of the two equations, and finding their point of intersection.

Example: Solve the following pair of simultaneous equations graphically;

$$2x + y = 5 \quad \text{(i)}$$
$$y = x + 2 \quad \text{(ii)}$$

Solution: It is first necessary to draw up a table of values for each equation. and then plot these ordered pairs on the number plane.

$2x + y = 5$

x	0	1	2
y	5	3	1

$y = x + 2$

x	0	1	2
y	2	3	4

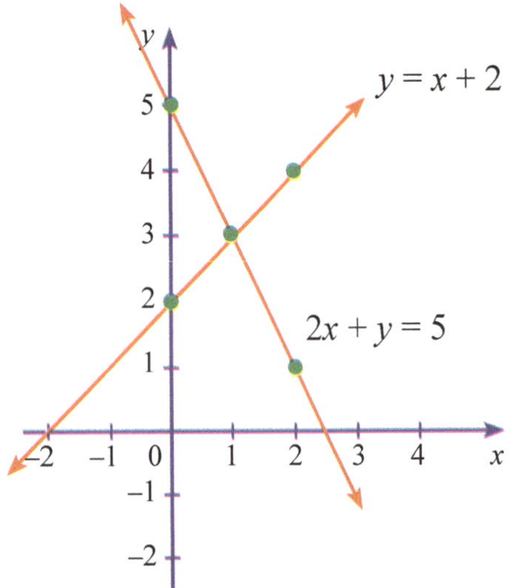

From the graph, the point of intersection is (1, 3).

Therefore the solution which satisfies both equations simultaneously are $x = 1$ and $y = 3$.

CHECKING THE ANSWERS: Once again, we can easily check the solution by substituting $x = 1$ and $y = 3$ back into the original equations.

$$2x + y = 5$$
$$\therefore \quad 2 \times 1 + 3 = 5$$
TRUE

Also

$$y = x + 2$$
$$\therefore \quad 3 = 1 + 2$$
TRUE

As you can sec the solution $x = 1$ and $y = 3$ SATISFIES BOTH EQUATIONS SIMULTANEOUSLY.

CHAPTER SUMMARY

SKETCHING STRAIGHT LINES

You will already have learned from earlier years methods of graphing a straight line.

1. Draw up a table of values using 3 pairs of x and y coordinates.
2. Use $y = mx + b$ where m = gradient and $b = y$ intercept.
3. The 'intercept method'.

We only need a minimum of 2 points to determine a line.

To find the y intercept, let $x = 0$ and solve the resulting equation.
To find the x intercept, let $y = 0$ and solve the resulting equation

FORMAT OF THE PARABOLA

To recognise the equation of a parabola, you will find that 'y' is always a function of 'x^2'.
The GENERAL EQUATION of a parabola is:
$y = ax^2 + bx + c$ where a, b and c are numerical constants and $a \neq 0$

The graph of $y = ax^2 + bx + c$ will have a MINIMUM if $a > 0$ (i.e. positive)

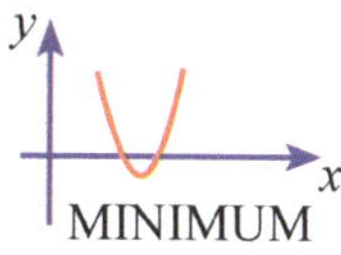

The graph of $y = ax^2 + bx + c$ will have a MAXIMUM if $a < 0$ (i.e. negative)

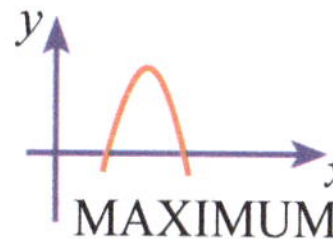

The points at which the parabola cuts the x-axis are particularly important. They are called the ROOTS of the equation. The simplest parabola is $y = x^2$, but there are an infinite number of parabolic graphs depending on the values of a, b and c in $y = ax^2 + bx + c$.

GRAPHING REGIONS

This process involves shading in a certain region on the number plane which satisfies one of the following inequality signs: $\{> \geq \leq <\}$.

If the inequation involves either '$\geq$' or '$\leq$', then a **SOLID LINE** is drawn.

If the equation involves '$>$' or '$<$', then a **BROKEN LINE** is drawn.

The steps outlined below relate to the graphing of any region - horizontal and vertical lines, sloping lines, parabolas, circles, etc.

Step 1: First sketch the relation as shown in the intercept method.

Step 2: Choose a suitable test point and determine whether the inequality holds.

Step 3: Shade in the required region.

The test point (0, 0) is the simplest and best to choose – unless the function happens to pass through the origin.

STRAIGHT LINE GRAPHS

$y = mx + b$

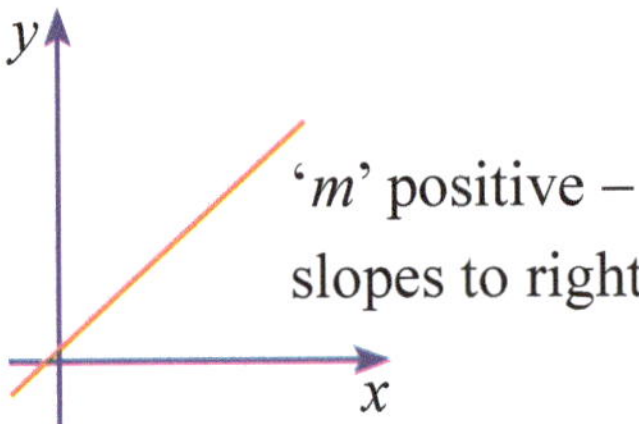

'm' negative –
slopes to left

THE PARABOLA

$y = ax^2 + bx + c$

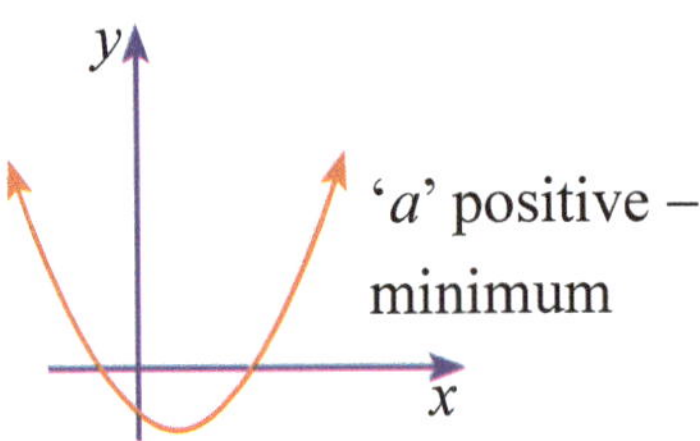

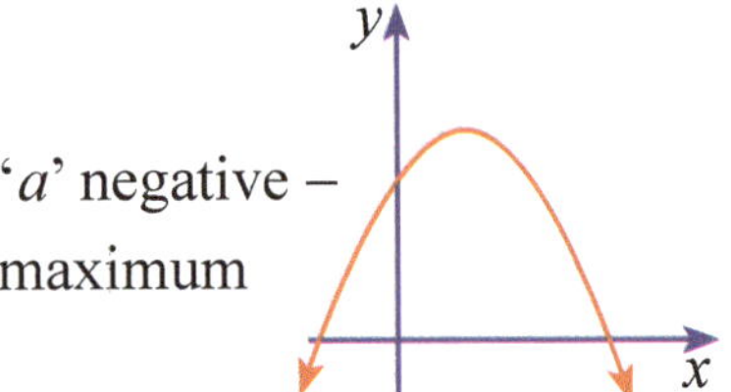

THE CUBIC

$y = ax^3 + bx^2 + cx + d$

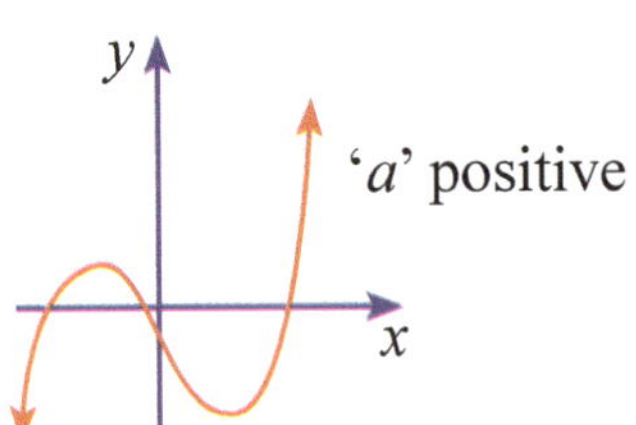

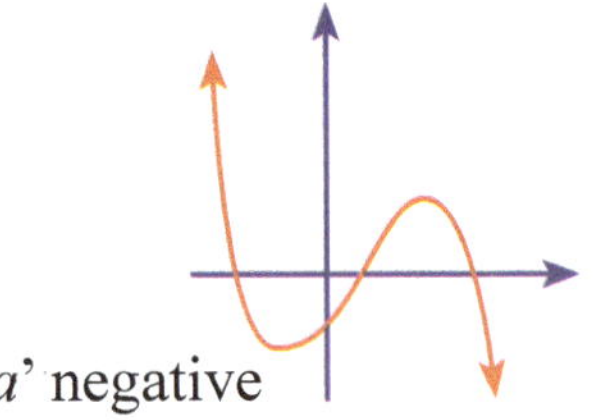

THE HYPERBOLA

$y = \frac{c}{x}$ or $y = \frac{-c}{x}$

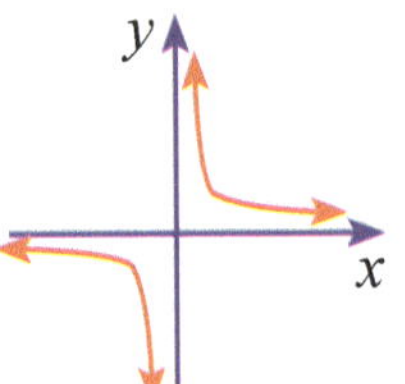

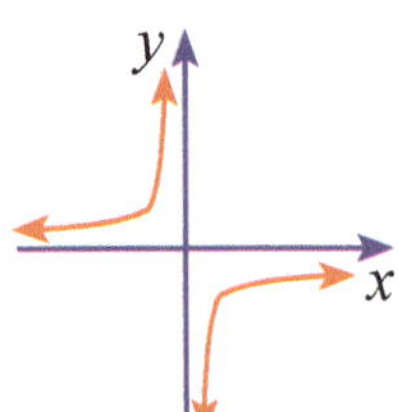

THE EXPONENTIAL

$y = a^x$ or $y = a^{-x}$

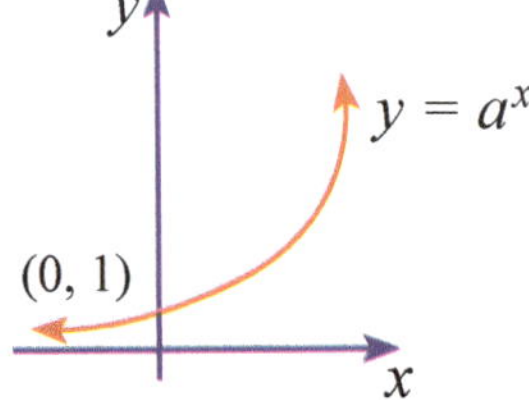

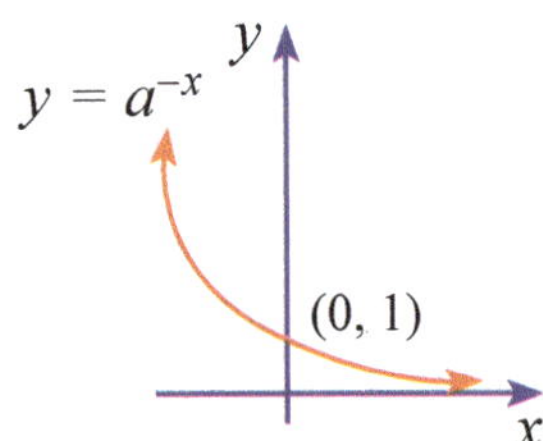

Note: The exponential curve has not been explained in the chapter, but it is useful to know its basic shape.

THE CIRCLE

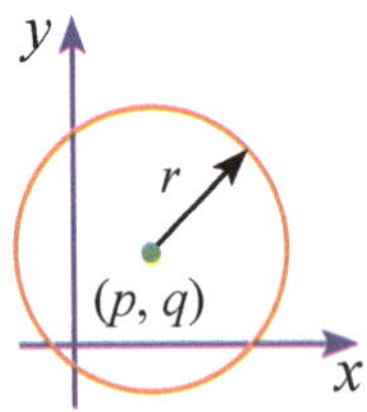

$(x - p)^2 + (y - q)^2 = r^2$

Centre (p, q), Radius 'r'

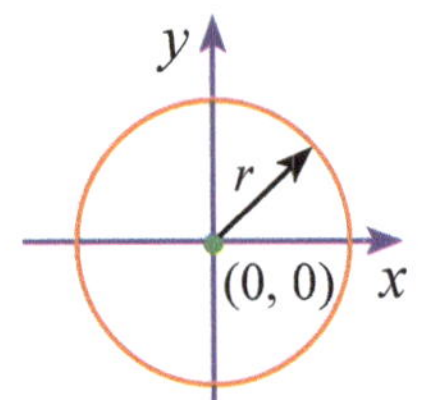

$x^2 + y^2 = r^2$

Centre origin (0, 0), Radius 'r'

LEVEL 1 — GRAPHING LINES AND CURVES

EASIER QUESTIONS

Note: Only turn back to page number shown if you have difficulty. — Page

Q1. On a number plane draw the solution set of each equation: 80 - 82

a) $y = 3x - 4$

x	0	1	2
y			

b) $y = -2x + 1$

x	0	1	2
y			

c) $x + y = 5$

x	0	1	2
y			

Q2. Complete the following tables and then graph the parabolas on the same number plane: 84

a) $y = x^2$

x	−4	−3	−2	−1	0	1	2	3	4
y									

b) $y = x^2 + 3$

x	−4	−3	−2	−1	0	1	2	3	4
y									

c) $y = 2x^2$

x	−4	−3	−2	−1	0	1	2	3	4
y									

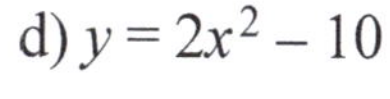

d) $y = 2x^2 - 10$

x	−4	−3	−2	−1	0	1	2	3	4
y									

Q3. Complete the table of values below and use this to graph $y = 2x^3$ for values of x from −2 to 2: 87

x	−2	−1	−0.8	−0.6	−0.4	−0.2	0	0.2	0.4	0.6	0.8	1	2
y													

Q4. Complete the table of values for $y = \frac{3}{x}$, and hence graph the curve for values of x from −4 to 4: 88

x	−4	−3	−2	−1	−0.5	0	0.5	1	2	3	4
y											

Q5. By drawing up a table of values for each straight line, for values of x from 0 to 4, find the point of intersection of each pair of lines: 90

a) $y = x + 3$
$y = 4x$

b) $y = 2x - 1$
$y = x + 2$

c) $y = 4x - 5$
$y = -2x + 1$

Q6. Give the equation for each of the following circles: 89

a) Centre (0, 0), radius = 5 units
b) Centre (0, 0), radius = $\sqrt{7}$ units
c) Centre (3, −5), radius = 4 units
d) Centre (−2, −1), radius = $\sqrt{10}$ units
e) Centre (−4, 0), radius = 2 units
f) Centre (0, 6), radius = $\sqrt{3}$ units

LEVEL 2 — GRAPHING LINES AND CURVES

AVERAGE QUESTIONS

Note: Only turn back to page number shown if you have difficulty. — Page

Q1. Sketch the following lines using the intercept method: 83

a) $y = 2x - 6$ b) $4x + 2y = 10$ c) $x - y - 3 = 0$

Q2. Complete the table of values and sketch the results on a number plane: 84

a) $y = x^2 + 4x + 4$

x	–4	–3	–2	–1	0	1	2	3
y								

b) $y = x^2 - 2x - 3$

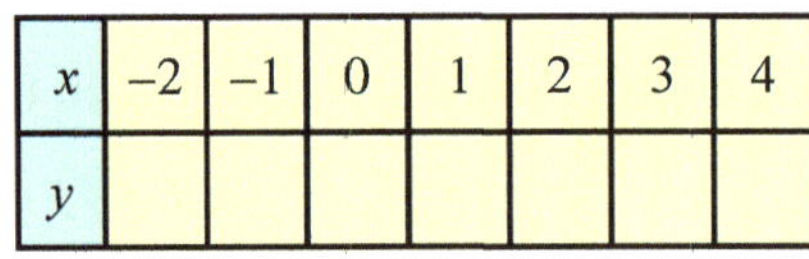

x	–2	–1	0	1	2	3	4
y							

Q3. Find the axis of symmetry and hence the coordinates of the maximum or minimum for each of these parabolas: 84

a) $y = x^2 + 2x + 1$ b) $y = x^2 - 4x + 5$ c) $y = 2x^2 + 6x - 20$

Q4. Shade the regions satisfied by the following conditions: 85, 86

a) $y \geq x$ b) $y < 3 + x$ c) $x - y > 2$

d) $2x + y \leq 4$ e) $y > 2x^2$ f) $y \leq -x^2$

Q5. Sketch the following cubic curves by first completing a table of values for each: 87

a) $y = x(x - 1)(x + 2)$ for $-3 \leq x \leq 2$

b) $y = (x + 1)(x - 1)(x - 3)$ for $-2 \leq x \leq 4$

c) $y = -2x^3$ for $-3 \leq x \leq 3$

Q6. 90

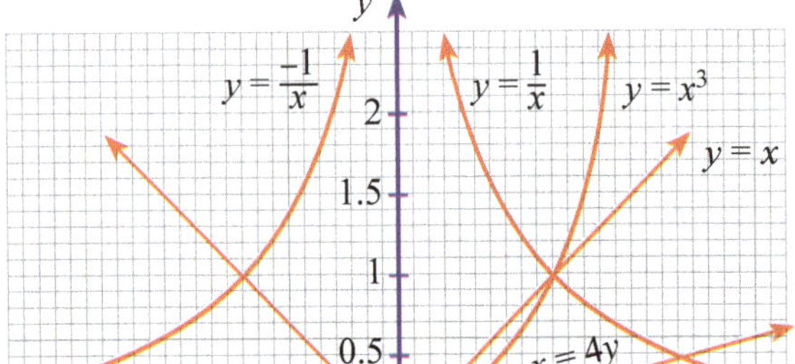

Use the diagram to solve these simultaneous equations:

a) $y = -x$, $x = 4y$ b) $y = \frac{1}{x}$, $x = 4y$

c) $y = x^3$, $y = -x$ d) $y = -\frac{1}{x}$, $y = -x$

e) $y = \frac{1}{x}$, $y = x^3$ f) $y = x^3$, $x = 4y$

g) $y = \frac{1}{x}$, $y = x$ h) $y = x^3$, $y = x$

LEVEL 3 — GRAPHING LINES AND CURVES

HARDER QUESTIONS

Q1. For the linear relations below:

a) (i) Find the values of a, b and c.
(ii) State the equation of the line.
(iii) Show both graphically and using algebra that $(1\frac{1}{2}, 7\frac{1}{2})$ does not lie on the line.

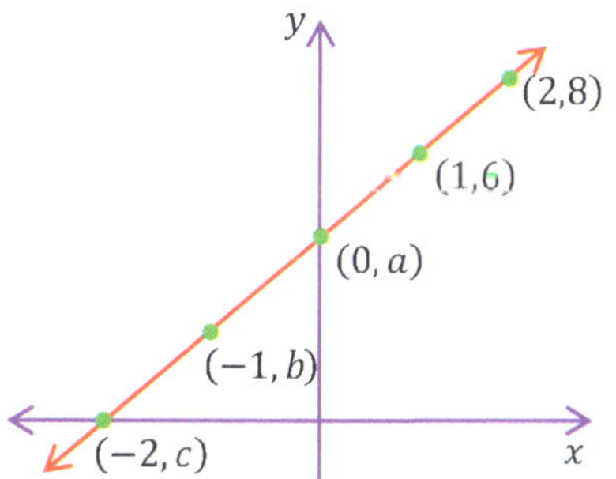

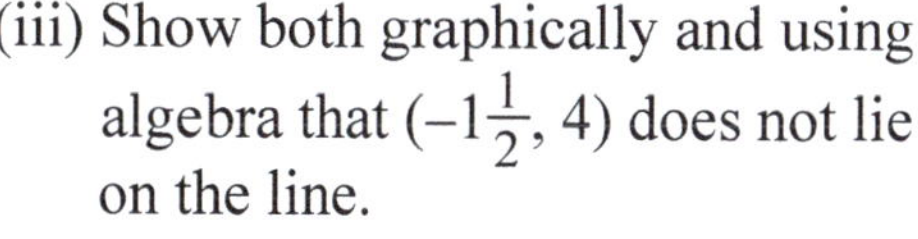

b) (i) Find the values of a, b and c.
(ii) State the equation of the line.
(iii) Show both graphically and using algebra that $(-1\frac{1}{2}, 4)$ does not lie on the line.

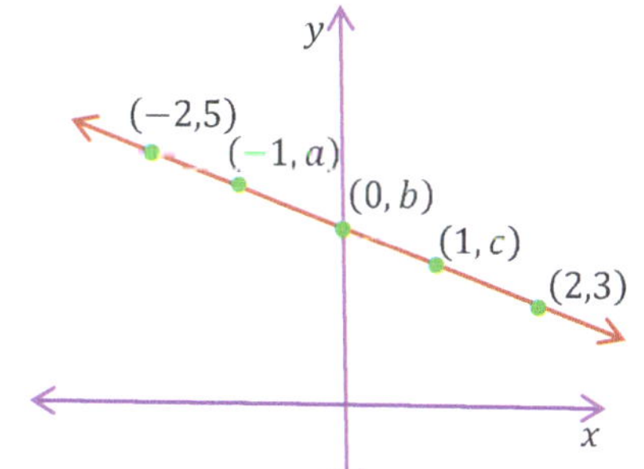

Q2. Two lines are shown on the set of axes. The equations of the lines are: $y = \frac{1}{2}x + 3$ and $y = -\frac{1}{2}x + 5$.

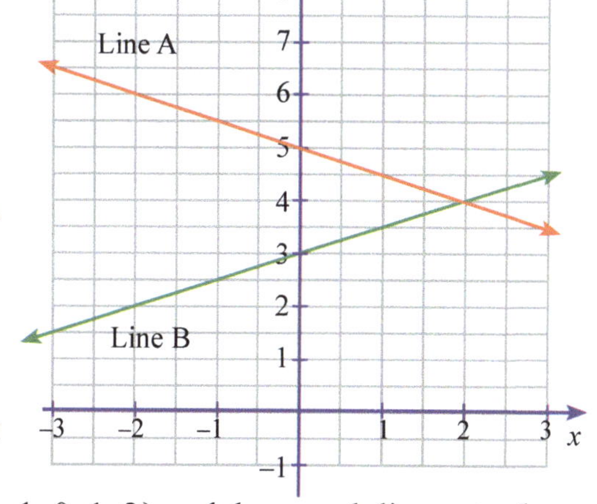

a) (i) Using the gradient and y-intercept explain how you are able to match each line with its equation.
(ii) Using algebra show that the point at which the lines intersect is (2, 4).

b) (i) Using algebra find the value of a so that lines with equations, $y = ax + 1$ and $y = -ax + 7$ both contain the point (2, 4) and state the equation for each line.
(ii) For each line make a table of values for $x = \{-2, -1, 0, 1, 2\}$ and draw each line onto the set of axes.

Q3. The graph of $y = 2x^2 + 4x - 16$ is shown on the set of axes.

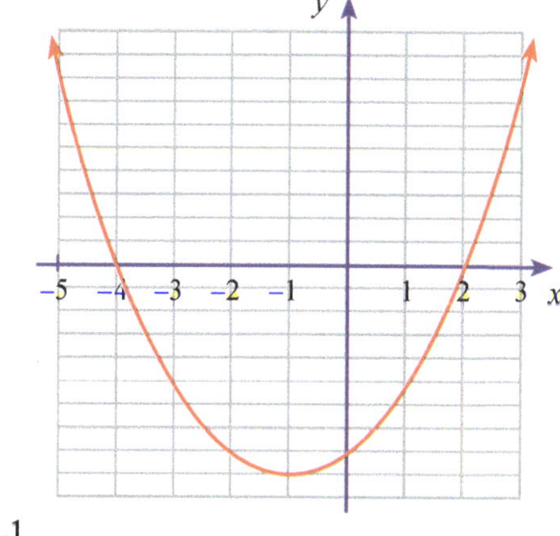

a) By expanding the brackets show that $2x^2 + 4x - 16 = 2(x - 2)(x + 4)$

b) Make a table of values for $x = \{-5, -4, -3, -2, -1, 0, 1, 2, 3\}$ and label the points onto the graph in coordinate form.

c) Explain what the x-intercepts for a graph are and use the table of values to explain why the x-intercepts occur for $x = -4$ and $x = 2$.

d) (i) Explain why the minimum value for y occurs for $x = -1$.
(ii) The x-value for the minimum point is halfway between the x-intercepts of the graph. Show that $x = -1$ is halfway between the x-intercepts of the graph.

e) (i) For the line with equation $y = 6x - 12$ make a table of values for $x = \{-1, 0, 1, 2, 3\}$.
(ii) Draw the line onto the graph and state the coordinates of the points of intersection between the parabola and the straight line.

LEVEL 4 — GRAPHING LINES AND CURVES

EXTENSION QUESTIONS

Q1. The graph of $y = x^3 - 2x^2 - 8x$ is shown on the set of axes.

a) By expanding the brackets show that $x^3 - 2x^2 - 8x = x(x + 2)(x - 4)$.

b) Make a table of values for $x = \{-3, -2, -1, 0, 1, 2, 3, 4, 5\}$ and label the points onto the graph in coordinate form.

c) Explain how the factorised expression for the cubic can be used to find the x-intercepts.

d) (i) For the line with equation $y = 3x - 12$ make a table of values for $x = \{-3, -2, -1, 0, 1, 2, 3, 4, 5\}$

(ii) Draw the line onto the graph and state the coordinates of the points of intersection between the cubic and the straight line.

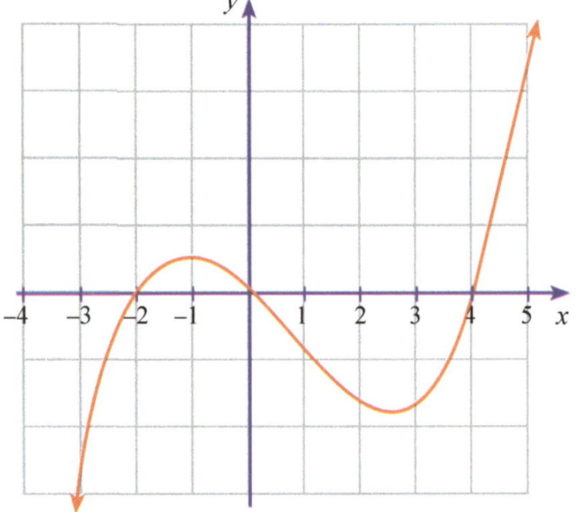

Q2. The graph of $y = \frac{1}{x-2} + 4$ is shown on the set of axes.

a) Make a table of values for $x = \{-2, -1, 0, 1, 2, 3, 4, 5, 6\}$ and label the points that exist onto the graph in coordinate form.

b) (i) Explain why the point for $x = 2$ doesn't exist.

(ii) Explain why the point for $y = 4$ doesn't exist.

c) The orange lines with equations $x = 2$ and $y = 4$ are shown on the graph.

Each line is called an asymptote. Explain what an asymptote is in relation to the graph of the hyperbola.

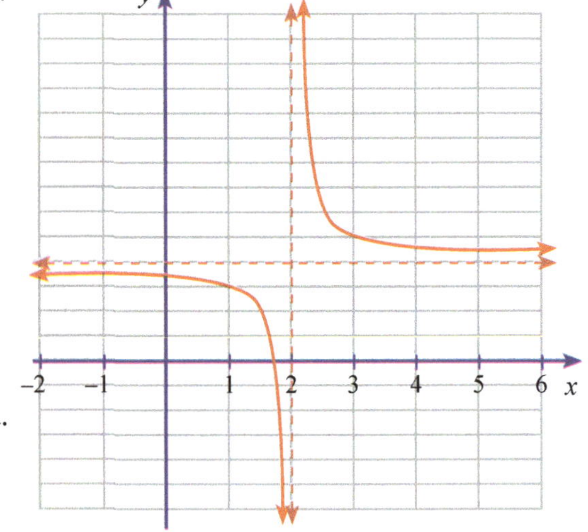

Q3. a) For the following make a table of values for $x = \{-4, -3, -2, -1, 0, 1, 2, 3, 4\}$.

(i) $y = \frac{12}{x}$ (ii) $y = 6x + 6$

(iii) $y = x^2 + 7x + 4$

b) Use the table of values to find the points of intersection for all three curves.

c) Plot each curve onto the set of axes showing the points of intersection of the curves in coordinate form.

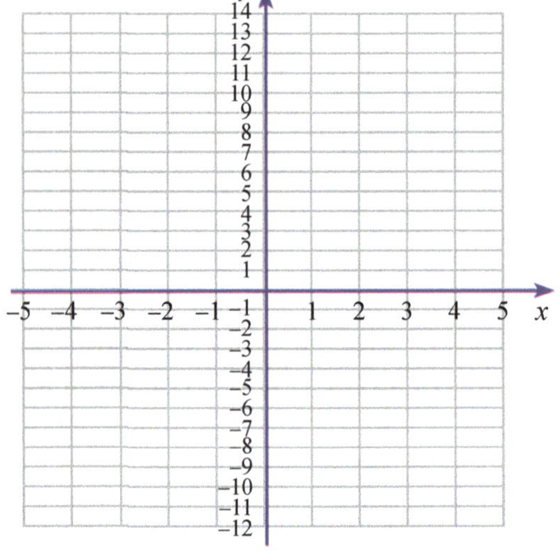

CIRCLE GEOMETRY

The 'Australian Curriculum Mathematics' (ACM) references for this sub-strand of 'Measurement and Geometry' (MG) are given below. This chapter may contain additional extension work which the author feels will be beneficial to the student.

- *Parts of a circle (ACMMG 272).*
- *Angle properties of circles (ACMMG 272).*
- *Chord properties of circles (ACMMG 272).*
- *Tangent properties of circles (NSW).*
- *Proofs using circle theorems (NSW).*

Note: The author feels that it is not necessary to provide a summary for this relatively short chapter.

Emmy Noether (1882 – 1935)

There have been many accomplished and famous women mathematicians (Sofia Kovalevskaya, Sophie Germain etc.), particularly in the 20th century with the emancipation of women. Noether was a German mathematician who made many important contributions to the development of abstract algebra. She also proved a famous theorem in mathematical physics known as Noether's theorem which explains the connection between symmetry and the laws of conservation. Albert Einstein and other leading scholars have described her as the most important woman in the history of mathematics. As one of the leading mathematicians of her time she developed the theory of rings, fields and algebras. She was born to a Jewish family and initially planned to become a teacher of French and English after having passed the required exams, but instead decided to further her studies and research in mathematics. At that time women were excluded from academic positions, so she worked at the Mathematics Institute of Erlangen for 7 years without pay after which she became a leading member of the prestigious Gottingen university maths department. In 1933, Germany's Nazi government dismissed all Jews from University positions and so she left and took up a lecturing position in the USA. In 1935, she underwent surgery for an ovarian cyst and, despite signs of recovery, died 4 days later at the relatively young age of only 53.

IMPORTANT PARTS OF A CIRCLE

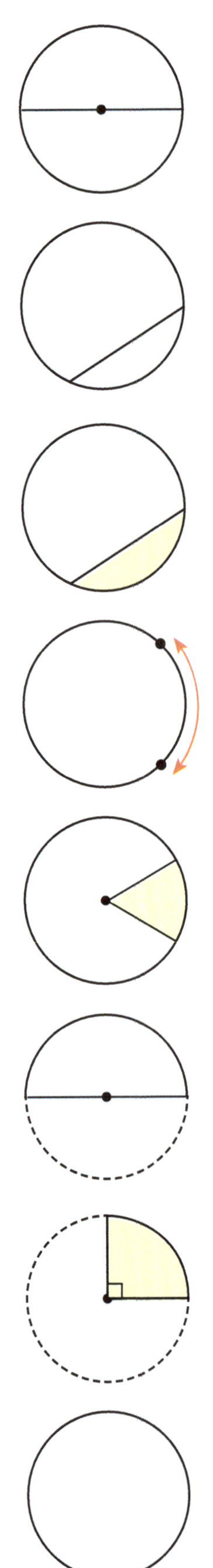

DIAMETER is any chord drawn through the centre of the circle.
RADIUS = diameter ÷ 2

CHORD is any straight line touching the circumference of the circle in two places.

SEGMENT is an area cut off by a chord. Minor segment is shaded yellow in the diagram. The major segment is shaded blue.

ARC is the length between two points on the circumference of a circle.

SECTOR is the area bounded by two radii and the circumference.

SEMICIRCLE is half a circle. The straight edge will always be the diameter.

QUADRANT is one quarter of a circle. It is a sector which has a 90° angle at the centre.

TANGENT is a straight line which touches the circle on the circumference at one point only.

ANGLE THEOREMS

There are several 'theorems' or 'laws' relating to circles which you should know. This page explains the 4 main angle properties of a circle. The proof of most of these theorems will be covered in the level 3 assignment.

1.

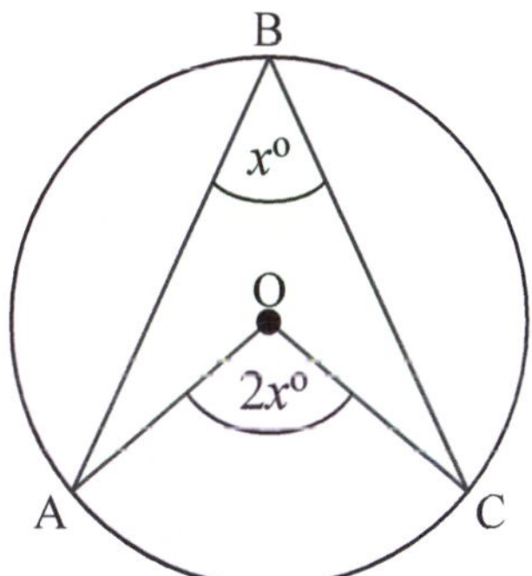

> The angle at the centre of the circle is double the angle at the circumference standing on the same arc.

i.e. $\angle AOC = 2 \times \angle ABC$
Both angles are drawn from the same arc AC.

2.

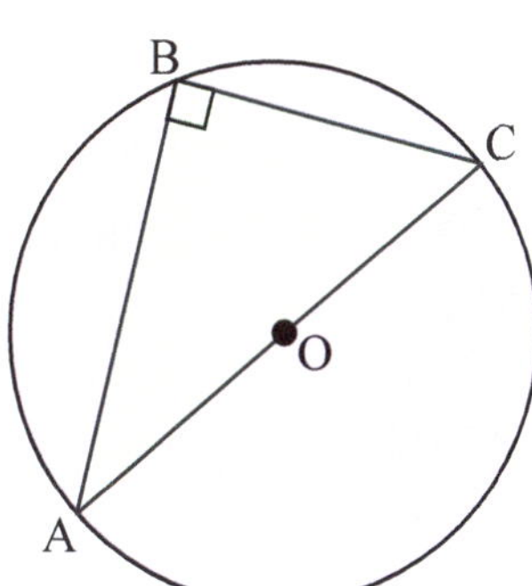

> The angle in a semicircle is a right angle or 90°.

This follows automatically from the previous theorem.
$\angle AOC = 2 \times \angle ABC$
Since $\angle AOC = 180°$ then $\angle ABC = 90°$.

3.

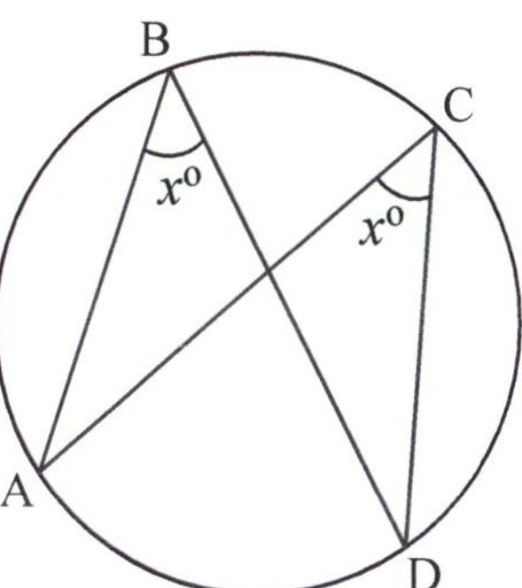

> Angles at the circumference of the circle standing on the same arc are equal.

i.e. $\angle ABD = \angle ACD$
They are both drawn from the same arc AD.

4.

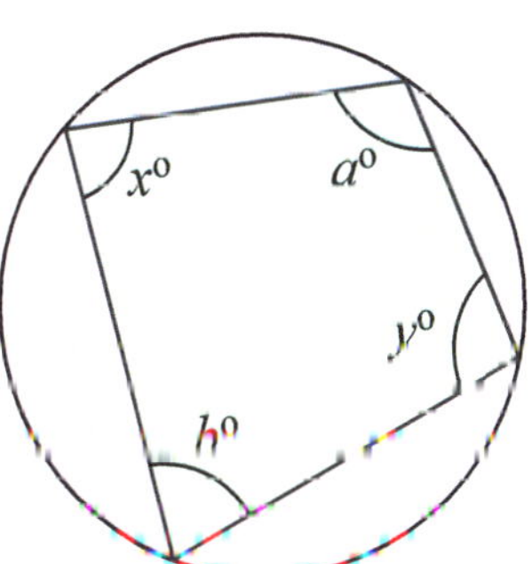

If the vertices (or corners) of a 4-sided figure (quadrilateral) lie on the circumference of a circle, then it is called a CYCLIC QUADRILATERAL.

> The opposite angles of a cyclic quadrilateral are supplementary.

In the diagram: $x^o + y^o = 180^o$
and $a^o + b^o = 180^o$

EXAMPLES

Find the values of the unknown pronumerals in the examples below:

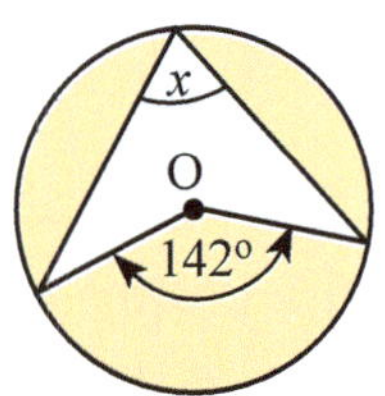

From theorem 1, x is half the angle at the centre.
$x = 71°$

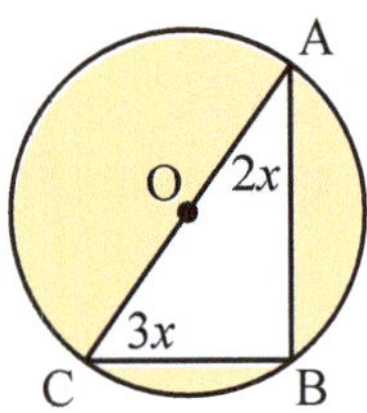

From theorem 2, $\angle ABC = 90°$.

$\therefore \quad 2x + 3x + 90 = 180$ Angle sum of $\triangle = 180°$

$\therefore \quad 5x = 90°$

$\therefore \quad x = 18°$

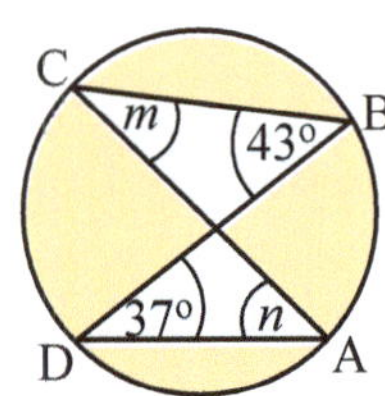

From theorem 3, $\angle ADB$ and $\angle ACB$ have been drawn from the same arc.

$\therefore \quad m = 37°$

Also, $\angle CBD$ and $\angle CAD$ have been drawn from the same arc.

$\therefore \quad n = 43°$

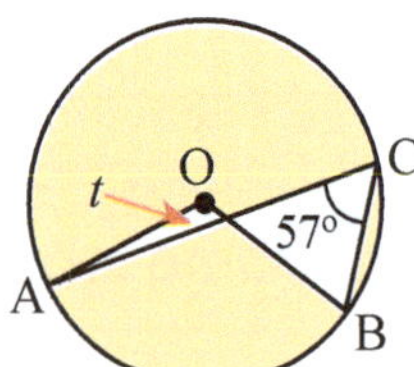

From theorem 1, the angle at the centre is double the angle at the circumference.

$\therefore \quad t = 114°$

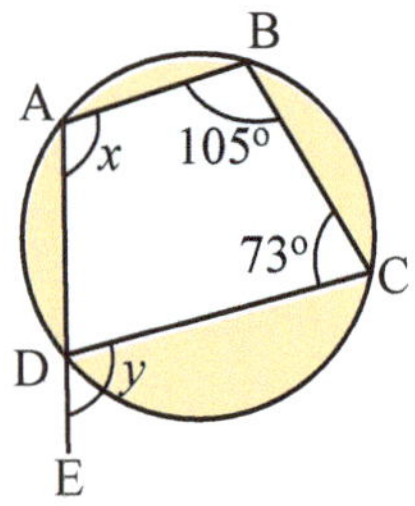

From theorem 4, opposite angles of a cyclic quadrilateral are supplementary.

$\therefore \quad x + 73° = 180°$

$\therefore \quad x = 107°$

Also $\angle ADC = 75°$

$\therefore \quad y = 180° - 75°$

$= 105°$

Because $\angle$'s on a straight line are supplementary

$\triangle OQR$ is isosceles because OQ = OR (same radius)

$\therefore \quad \angle OQR = \angle ORQ$

$\therefore \quad n = 40°$

From theorem 2, $\angle PQR = 90°$ (angle in a semicircle)

$\therefore \quad \angle OQP = 50°$

But $\triangle PQO$ is also isosceles because OQ = OP (same radius).

$\therefore \quad m = 50°$.

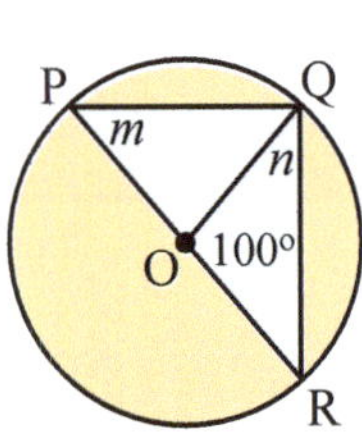

CHORD THEOREMS

The 4 properties are listed below, because they are very closely connected.

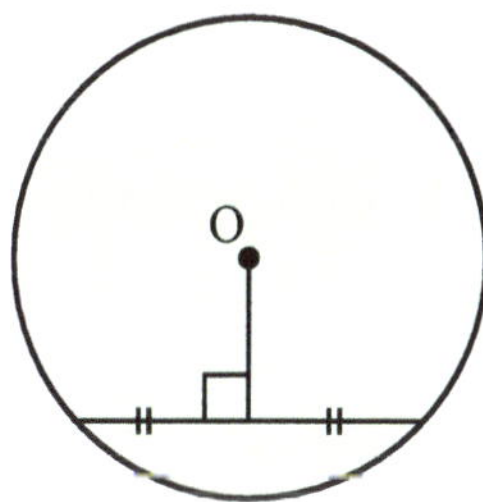

(i) The line from the centre of the circle to the midpoint of a chord is perpendicular to the chord.

(ii) The line from the centre perpendicular to the chord, bisects the chord.

(iii) The bisector of a chord passes through the centre.

(iv) Equal chords are the same distance from the centre, and subtend equal angles at the centre.

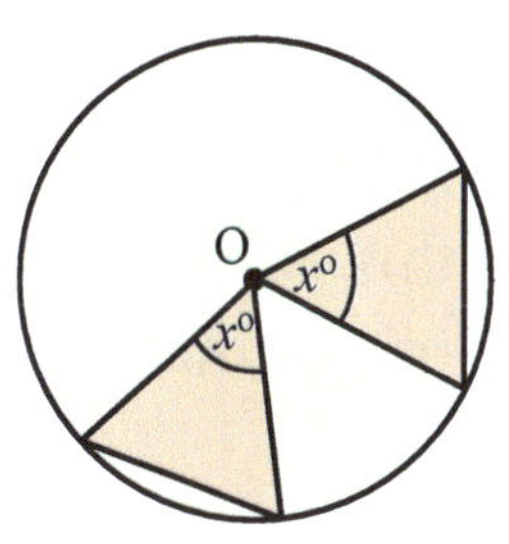

Theorems (ii) and (iii) are really the reverse of theorem (i).

Example: Find the length of the chord AC in the diagram.
Use the theorem of Pythagoras to calculate AB.

$OA^2 = OB^2 + AB^2$

$\therefore \quad 5^2 = 3^2 + AB^2$

$\therefore \quad 16 = AB^2$

$\therefore \quad AB = 4$ units

From Theorem (ii) above, it is known that OB must bisect AC.

$\therefore \quad AC = 8$ units

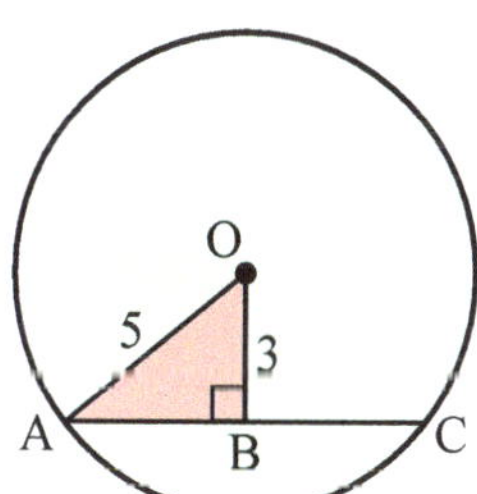

TANGENT THEOREMS

Any line that touches a circle at just one point is called a TANGENT to that circle. There are two properties relating to tangents.

1.

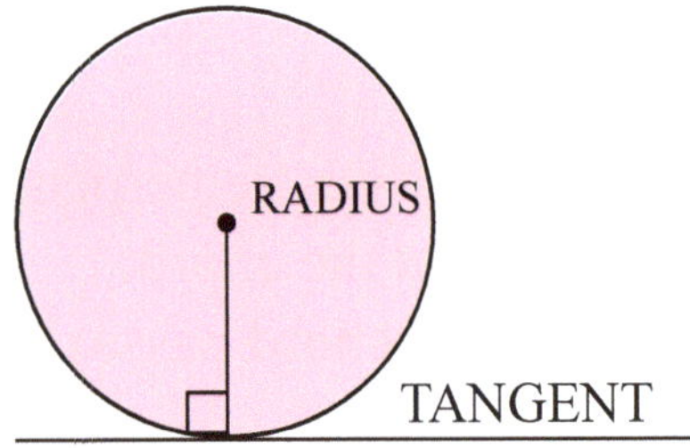

A tangent to a circle is perpendicular to the radius drawn at the point of contact.

2.

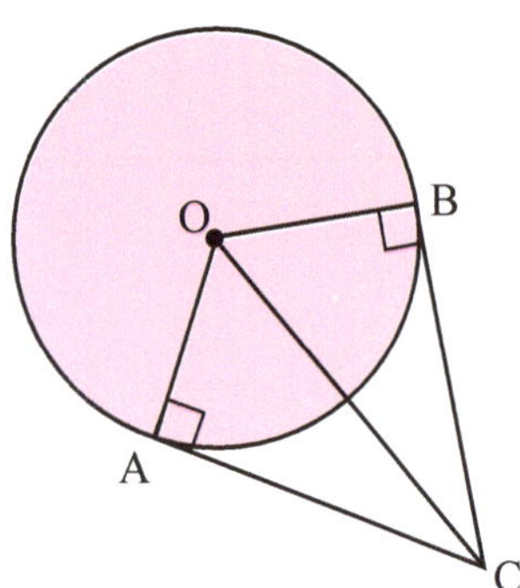

Two tangents drawn to a circle from some outside point are equal in length. The line joining this point to the centre is an axis of symmetry.

i.e. CA = CB in the diagram.
OC is an axis of symmetry.
ΔOAC is congruent to ΔOBC.

Example: If AB is a tangent to the circle at A, then
(i) Calculate the size of x.
(ii) Calculate the length of OB.

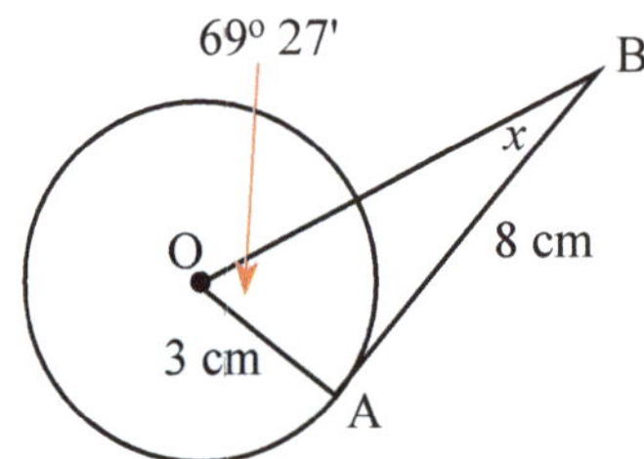

Solution: (i) Because AB is a tangent, then from Theorem 1 above, $\angle$OAB must be 90°. The angle sum of a triangle is 180°.

$\therefore \quad 69^\circ 27' + 90^\circ + x = 180^\circ$
$\therefore \quad x = 20^\circ 33'$

(ii) Because it is a right-angled triangle, Pythagoras' Theorem can be applied.

$OB^2 = 3^2 + 8^2$
$\therefore \quad OB^2 = 73$
$\therefore \quad OB = \sqrt{73}$

MISCELLANEOUS EXAMPLES

Find the values of the unknown pronumerals in each question, and give reasons for your answers.

(i)

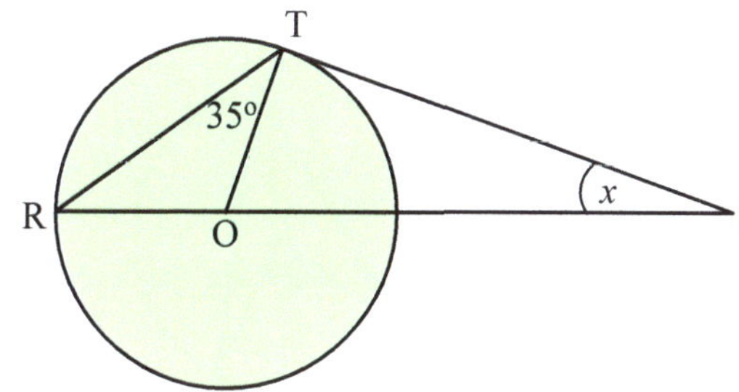

$T\hat{R}O = 35°$ (Isosceles Δ, OT = OR)

$O\hat{T}P = 90°$ (Tangent is perpendicular to radius)

In ΔRTP, angle sum = 180°

$\therefore \ x = 180° - 35° - 125°$

$= 20°$

(ii)

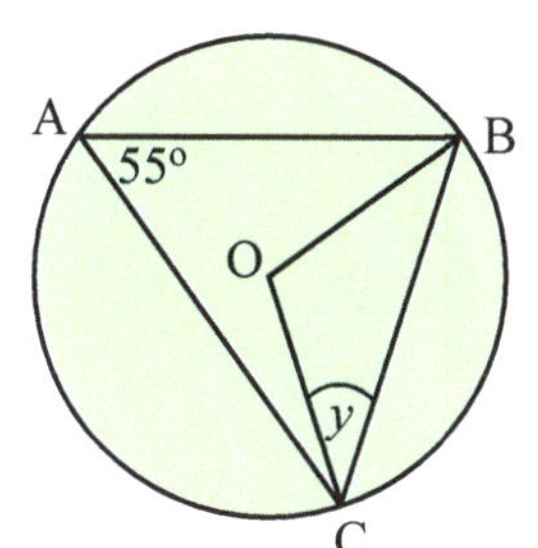

$C\hat{O}B = 110°$ (Angle at centre is twice angle at circumference)

ΔOBC is isosceles with OB = OC

$\therefore \ y = (180° - 110°) \div 2$

$= 35°$

(iii)

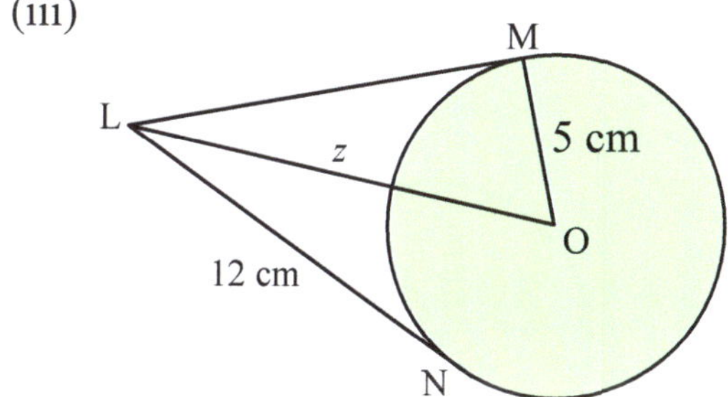

LM = LN = 12 cm (Tangents from external points are equal)

ΔLMO has a right angle at M (Tangent is $\perp$ to radius)

$\therefore \ z^2 = 12^2 + 5^2$ Pythagoras

$\therefore \ z^2 = 169$

$\therefore \ z^2 = \sqrt{169} = 13$ cm

(iv) A chord of length 12 cm has a perpendicular distance of 4 cm from the centre of a circle. Find the radius of the circle.

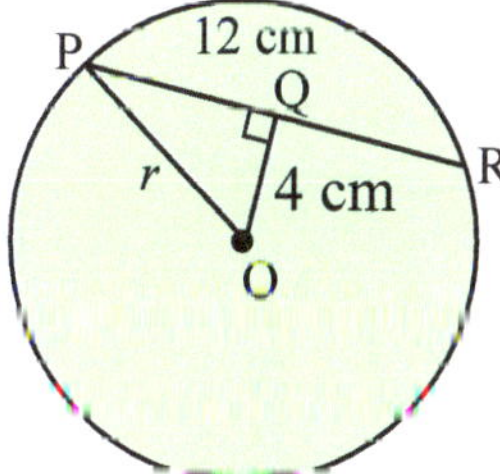

PQ = 6 cm (OQ $\perp$ PR $\therefore$ PQ = QR)

$\therefore \ r^2 = 4^2 + 6^2$ Pythagoras' Theorem

$\therefore \ r^2 = 16 + 36$

$\therefore \ r = \sqrt{52}$ cm or $2\sqrt{13}$ cm

LEVEL 1 — CIRCLE GEOMETRY

EASIER QUESTIONS

Note: Only turn back to page number shown if you have difficulty.

	Page
Q1. Find the value of the pronumeral in each case:	99, 100
Q2.	101
Q3. Find the value of the pronumerals in each case:	102, 103

Q1. Find the value of the pronumeral in each case:

a)

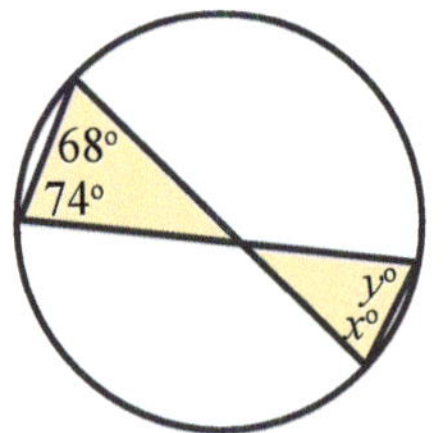

b)

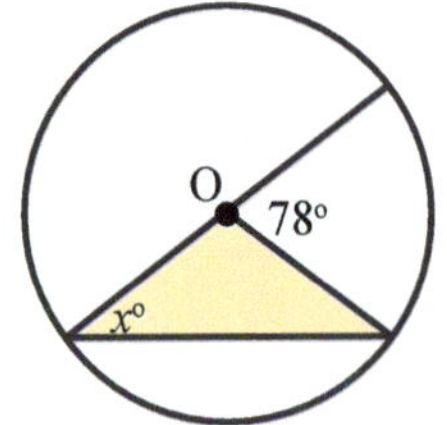

c)

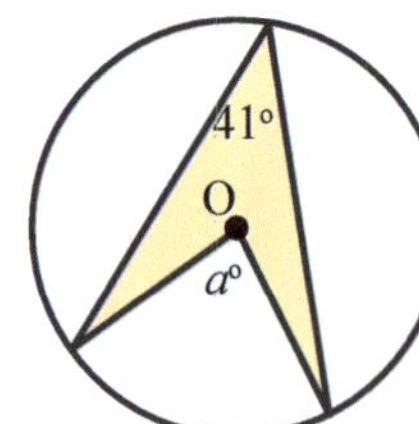

d)

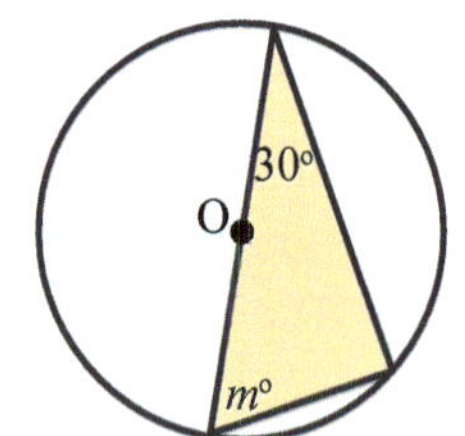

e)

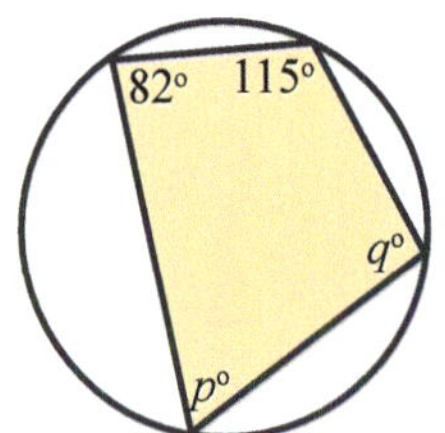

f)

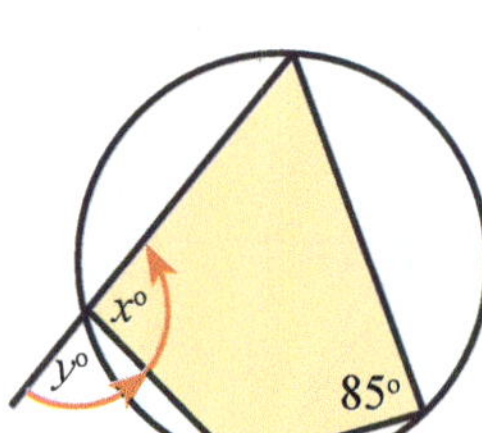

g)

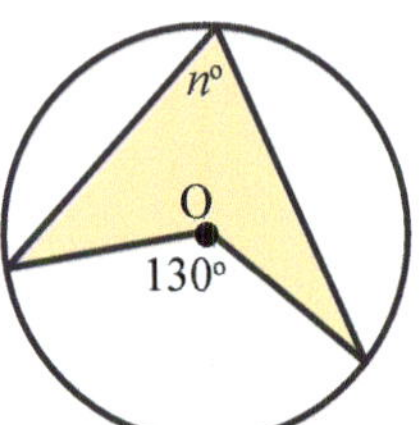

h)

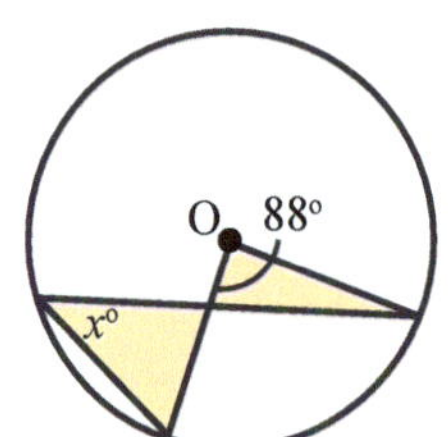

i)

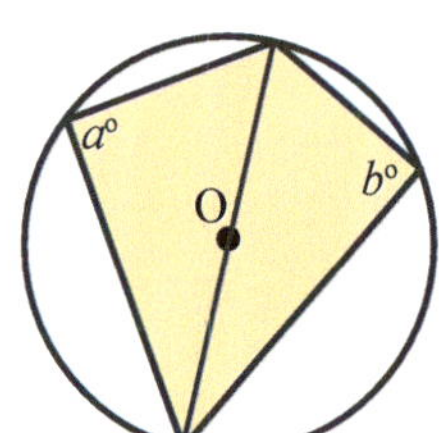

Q2. a)

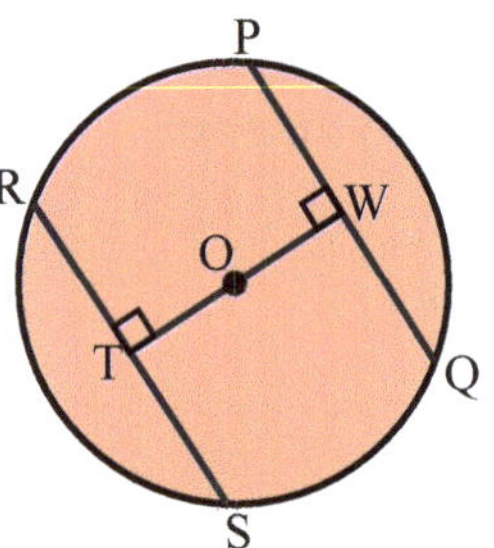

OT = OW, PQ = 15 cm
Find RS.

b)

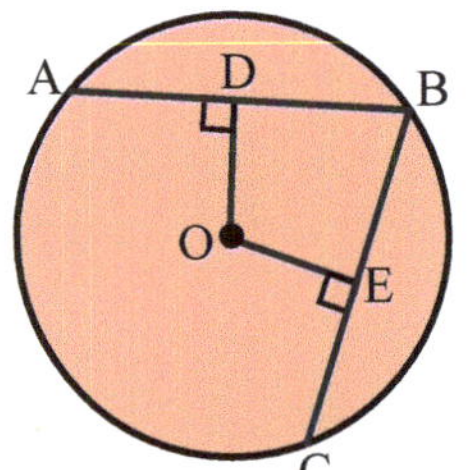

AB = BC, OD = 3.2 cm
Find OE.

c)

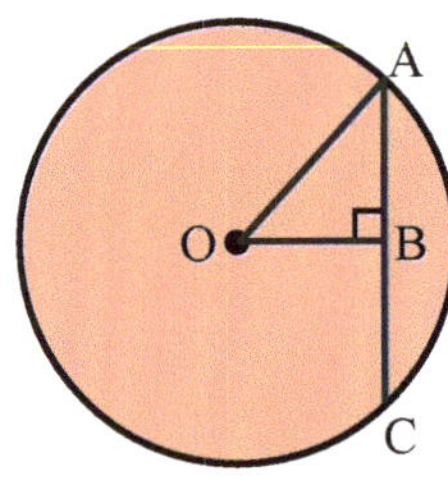

AC = 24 cm, OB = 5 cm
Find OA.

Q3. Find the value of the pronumerals in each case:

a)

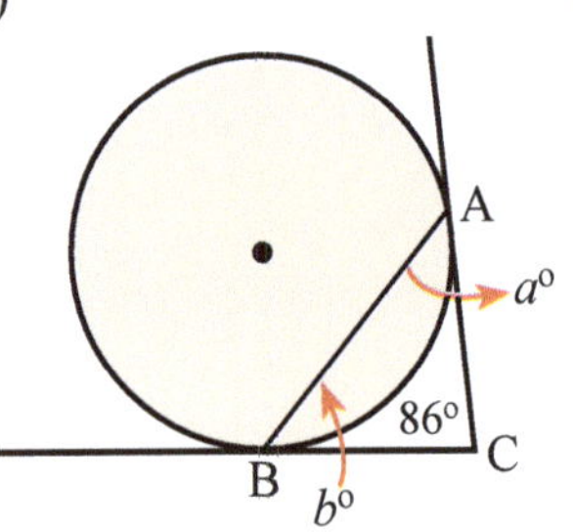

b)

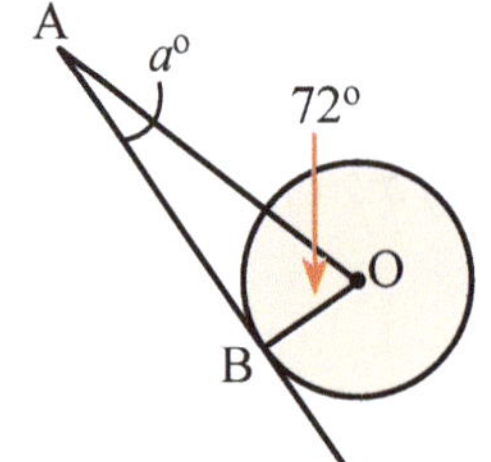

c)

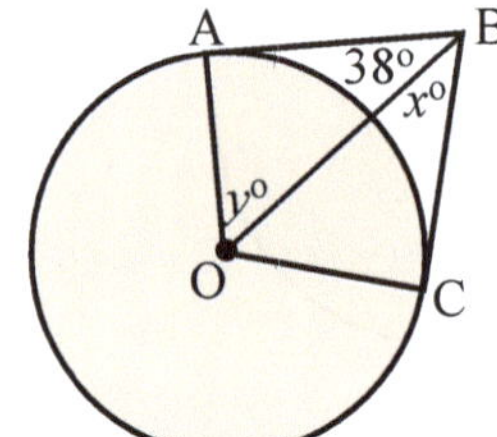

 AVERAGE QUESTIONS

Note: Only turn back to page number shown if you have difficulty. | Page

Q1. Find the value of the pronumerals giving reasons for your answer. 99, 100

a)

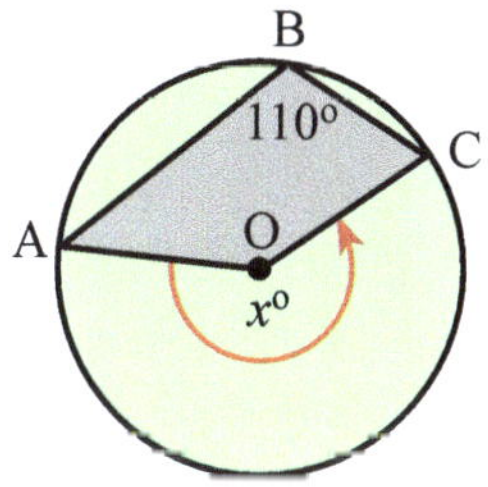

b)

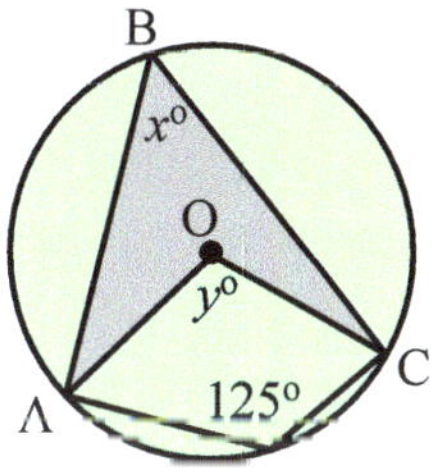

c)

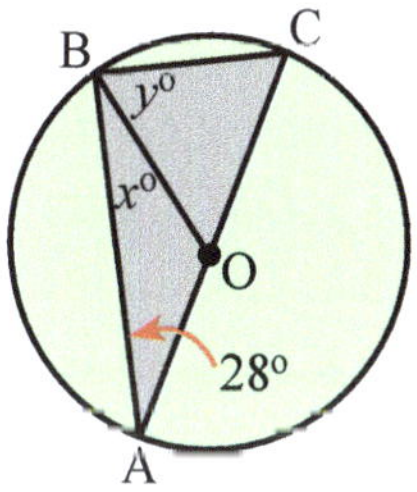

d)

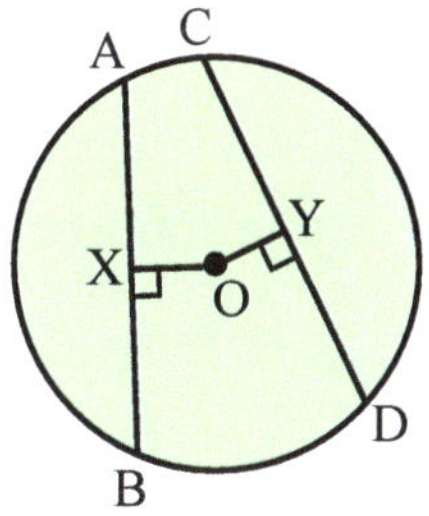

CD = 14 cm, AX = 7 cm,
OX = 3 cm.
Find OY.

e)

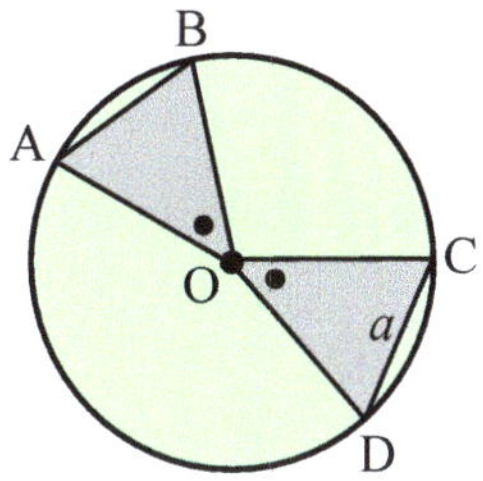

∠AOB = ∠DOC
AB = 2.3 cm
Find DC.

f)

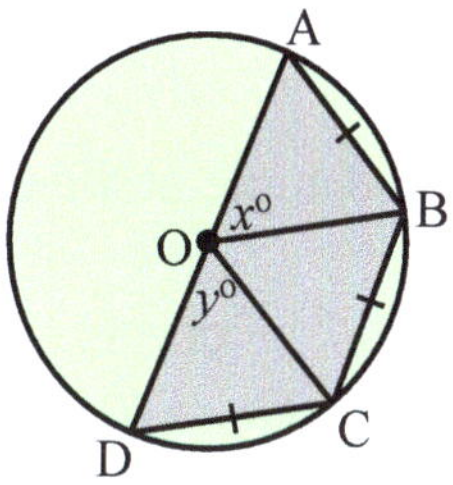

AB = BC = CD
Find $x°$ and $y°$.

101

g)

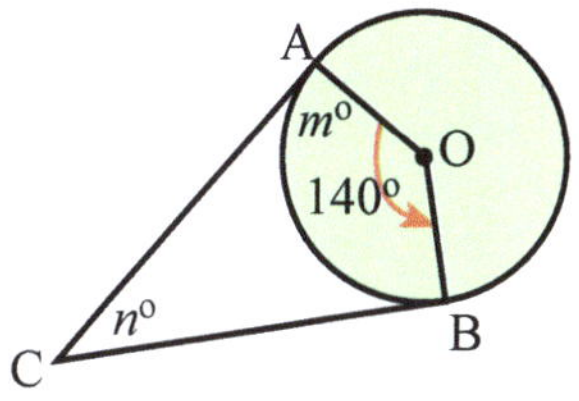

h)

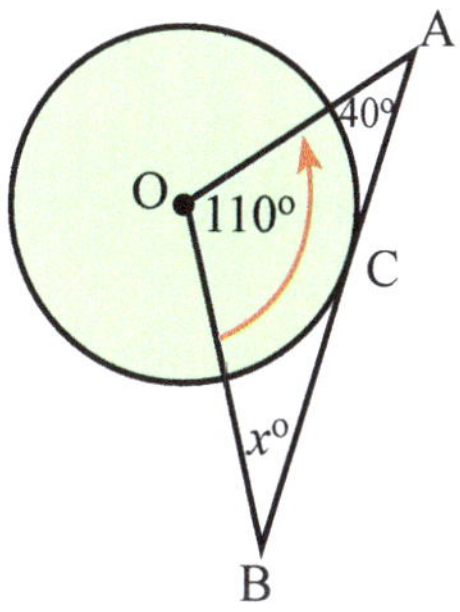

i)

102, 103

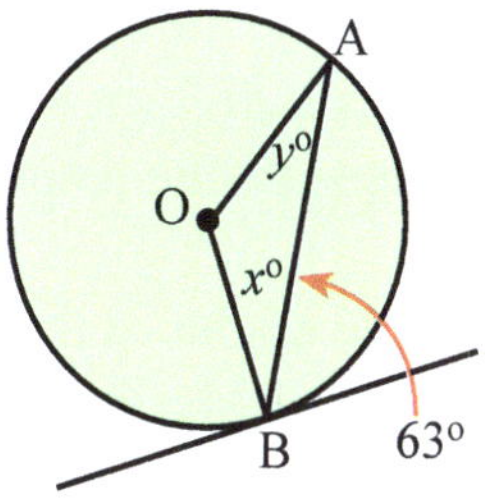

j)

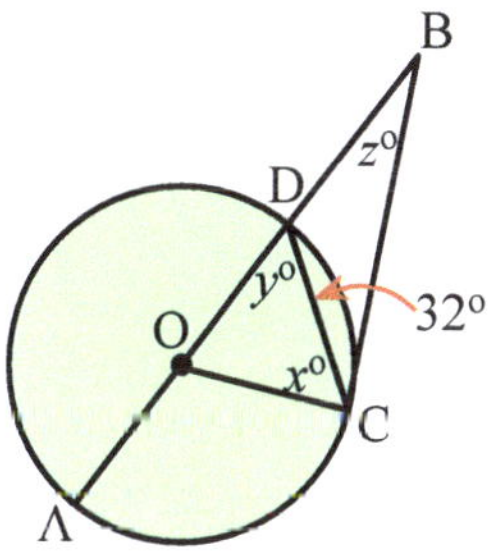

k)

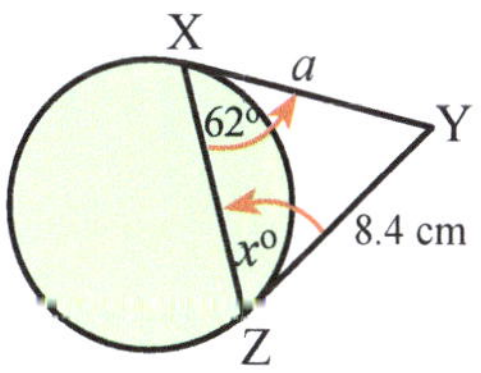

l)

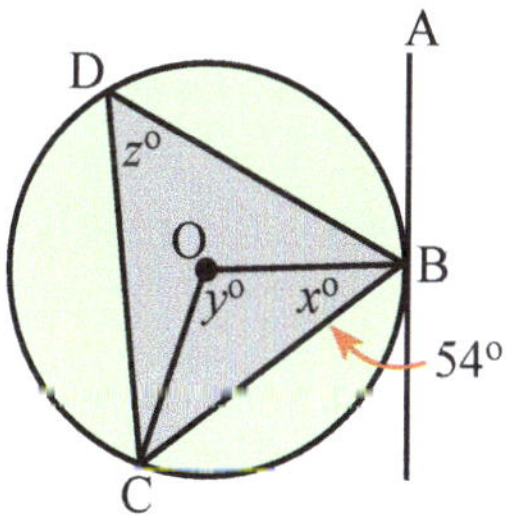

m)

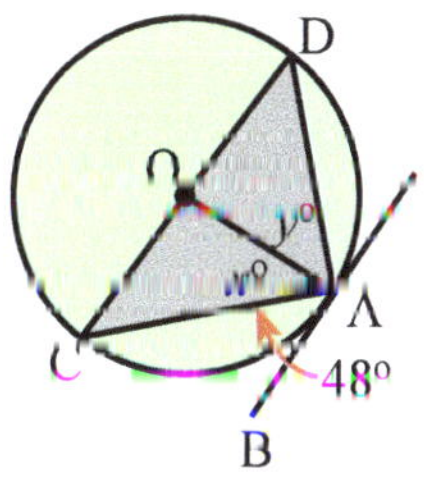

n)

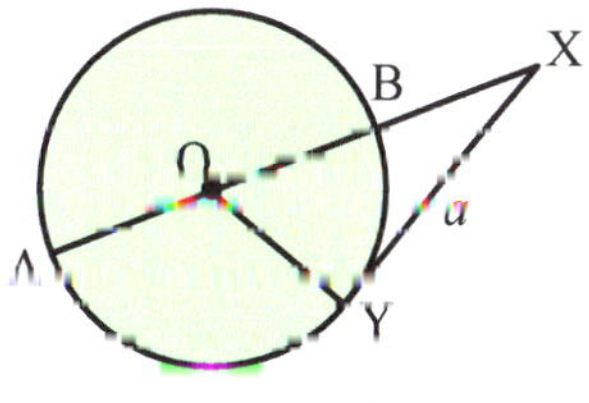

AB = 10 cm, BX = 5 cm
OY = 5 cm. Find XY. (2 d.p.)

o)

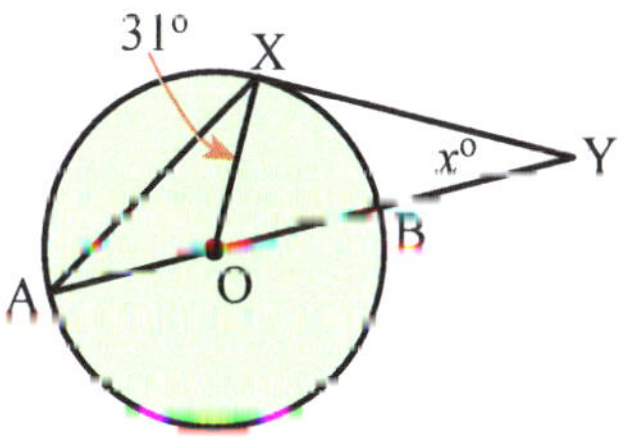

EXTENSION QUESTIONS

Q1. Proof that the angle subtended by a semi-circle is a right-angle.
Let A, B and C be points on a circle, centre O.
$\angle AOB = \alpha$, $\angle BOC = \beta$

a) (i) Explain why $\angle OAB = \angle OBA$
 (ii) Express $\angle OBA$ in terms of α

b) (i) Explain why $\angle OCB = \angle OBC$
 (ii) Express $\angle OBC$ in terms of β

c) (i) Express $\angle ABC$ in terms of α and β
 (ii) Express $\angle AOC$ in terms of α and β
 (iii) Express $\angle ABC$ in terms of $\angle AOC$
 (iv) Hence, state the result of the angle theorem.

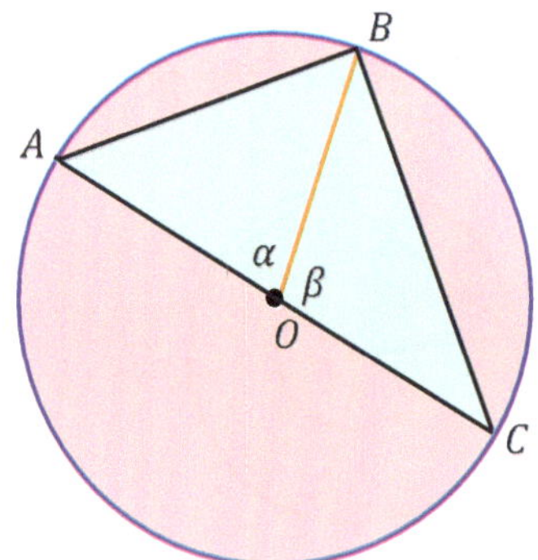

Q2. Proof that the angle at the centre of a circle is double the angle at the circumference standing on the same arc.
Let A, B and C be points on a circle, centre O.
$\angle AOB = \alpha$, $\angle BOC = \beta$

a) (i) Explain why $\angle OAB = \angle OBA$
 (ii) Express $\angle OBA$ in terms of α

b) (i) Explain why $\angle OCB = \angle OBC$
 (ii) Express $\angle OBC$ in terms of β

c) (i) Express $\angle ABC$ in terms of α and β
 (ii) Express the minor angle, $\angle AOC$ in terms of α and β
 (iii) Express $\angle AOC$ in terms of $\angle ABC$
 (iv) Hence, state the result of the angle theorem.

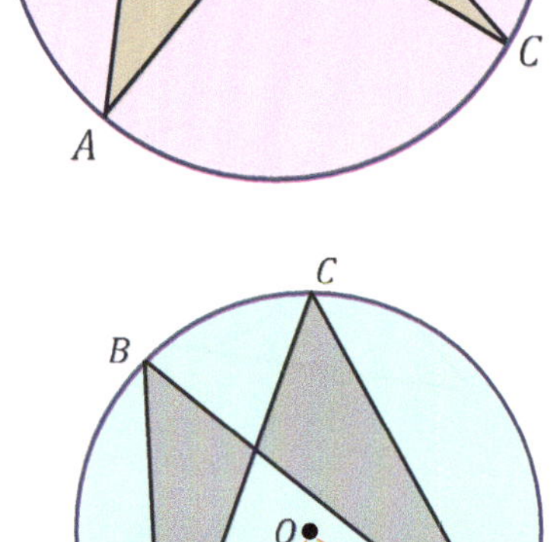

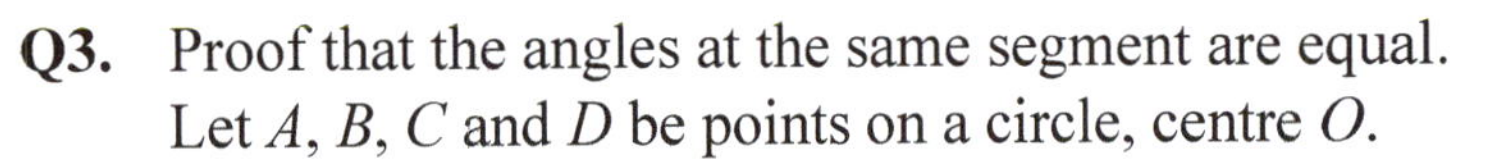

Q3. Proof that the angles at the same segment are equal.
Let A, B, C and D be points on a circle, centre O.
Lines ABD and ACD have the points A and D in common.
Let $\angle AOD = 2\alpha$

a) Express (i) $\angle ABD$ in terms of α
 (ii) $\angle ACD$ in terms of α

b) Express $\angle ABD$ in terms of $\angle ACD$.

c) Hence, state the result of the angle theorem.

Q4. Proof that the sum of the opposite angles in a quadrilateral is 180°.
Let A, B, C and D be points on a circle, centre O.
Let major angle: $\angle AOC = \alpha$
Let minor angle: $\angle AOC = \beta$

a) Express $\angle ABC$ in terms of α

b) Express $\angle ADC$ in terms of β

c) Express $\angle ABC + \angle ADC$ in terms of α and β.

d) Given that $\alpha + \beta = 360°$, show that $\angle ABC + \angle ADC = 180°$

e) Hence, state the result of the angle theorem.

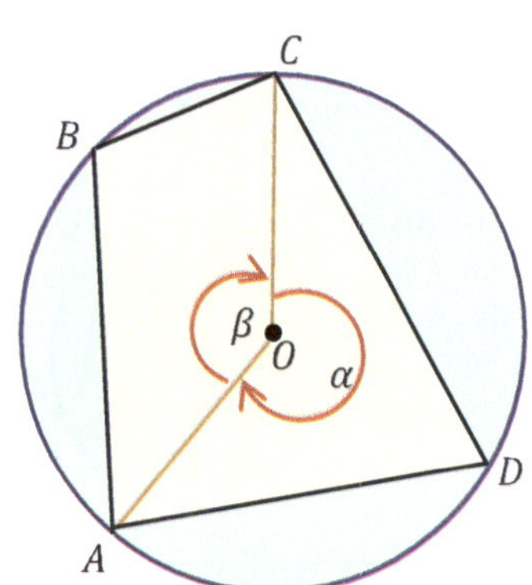

FURTHER ALGEBRA AND FACTORISATION

The 'Australian Curriculum Mathematics' (ACM) references for this sub-strand of 'Number and Algebra' (NA) are given below. This chapter may contain additional extension work which the author feels will be beneficial to the student.

- *Factorise algebraic expressions by identifying the HCF (ACMNA 191).*
- *Simplify algebraic expressions involving the 4 operations (ACMNA 192).*
- *Expanding binomial products (ACMNA 213).*
- *Simplify algebraic fractions involving the 4 operations (ACMNA 232).*
- *Special binomial products (ACMNA 233).*
- *Factorising trinomials and special binomial products (ACMNA 269).*
- *Factorising algebraic expressions involving fractions (NSW).*

Page

The first half of the chapter revises important Year 9 work. Therefore, in the exercises, more emphasis will be given to factorisation and algebraic fractions.

Sir Isaac Newton (1642 – 1726)

Newton was an English mathematician, physicist, astronomer, philosopher and theologian who is recognized as one of the most influential scientists of all time. His book Philosphiae Naturalis Principia Mathematica, first published in 1687, laid the foundation for the laws of motion and gravitational attraction that formed the cornerstone of physics until it was eventually superseded, almost 300 years later, by the theory of relativity. He was also credited with Leibnitz for developing ' infinitesimal differential and integral calculus', one of the most powerful mathematical concepts ever formulated. He contributed prolifically to so many areas of science…light, sound, optics, fluids, astronomy, mathematics (binomial theorem) and many other areas. He also built the first practical reflecting telescope.

ADDING AND SUBTRACTING IN ALGEBRA

'LIKE TERMS' in algebra are letters, or groups of letters, which are identical except for the leading number in front.

For Example:

$5a$, $-2a$, $7a$ are all like terms
$-3ab$, $6ab$, $2ab$ are all like terms
$8x^2$, $3x^2$, $-x^2$, are all like terms

'UNLIKE TERMS' contain different letters or groups of letters

For Example:

$7a$, $3b$, $-4ab$ are all unlike terms
$5c$, $-6d$, $2ac$ are all unlike terms
$4x^2$, $-3x$, 2 are all unlike terms

> In algebra, one can only add or subtract like terms.

Example 1: Simplify $8x^2 - 3x + 2x^2 + 5x - 4$

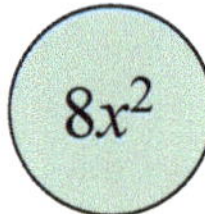

$- 3x$ $+ 2x^2$ $+ 5x$ -4 $= 10x^2 + 2x - 4$

$+ 8x^2$ and $+ 2x^2$ are like terms 

$- 3x$ and $+ 5x$ are like terms

Example 2: Simplify the following expressions by adding or subtracting like terms:

i) $2x + 3y + 5x + y = 7x + 4y$

ii) $6m - 3n + 2m + 5n = 8m + 2n$

iii) $6a - 3b - 2a - 6b = 4a - 9b$

iv) $8ab + 5b - 10ab - 2b = -2ab + 3b$

v) $5x^2 - 7x + x^2 + 4x - 8 = 6x^2 - 3x - 8$

vi) $7pq - 3m - 15qp + 8 - 7m = 8 - 8pq - 10m$

$7pq$ and $-15qp$ are like terms.

Note: The sign always belongs to the term directly after it.

REMOVING GROUPING SYMBOLS

'Grouping symbols' means exactly the same as 'BRACKETS'. You will often be asked to remove grouping symbols from a particular algebraic expression.
This is done by multiplying the term outside the brackets by each of the terms inside the brackets.
The word 'EXPAND' is often used instead of 'remove grouping symbols'.

THE GENERAL RULES

$$a(b+c) = a \times b + a \times c = ab + ac$$
$$a(b-c) = a \times b - a \times c = ab - ac$$

Examples: Expand the following:

i) $6(a-3) = 6 \times a - 6 \times 3$

$= 6a - 18$

ii) $4m(m+5) = 4m \times m + 4m \times 5$

$= 4m^2 + 20m$

iii) $-5(p+2) = -5 \times p + (-5) \times 2$

$= -5p - 10$

iv) $-3(2a-5) = -3 \times 2a + (-3) \times (-5)$

$= -6a + 15$

v) $-(2+3y) = -1(2+3y)$

$= -1 \times 2 + (-1) \times (3y)$

$= -2 - 3y$

Place 1 outside the bracket if there are no numbers.

The last three examples are important to understand because they occur regularly in algebra.

A minus sign outside of a bracket reverses the signs inside of the bracket.

MORE DIFFICULT EXAMPLES

In the examples on this page, it is first necessary to expand the brackets in the usual way. Then the resulting expressions are simplified by adding or subtracting the like terms.

Examples: Remove grouping symbols from the following, and simplify the resulting expressions:

i) $5(x + 3) + 4(x - 2)$
$= 5x + 15 + 4x - 8$ — Remove brackets
$= 9x + 7$ — Add like terms

ii) $4(2a - 3) + 3(a + 2)$
$= 8a - 12 + 3a + 6$ — Remove brackets
$= 11a - 6$ — Add like terms

iii) $x^2(2x - 5) + 3x(4 - 5x)$
$= 2x^3 - 5x^2 + 12x - 15x^2$ — Remove brackets
$= 2x^3 - 20x^2 + 12x$ — Add like terms

iv) $5(2m + 4) - 3(m + 5)$
$= 10m + 20 - 3m - 15$ — Remove brackets
$= 7m + 5$ — Note: $-3 \times +5 = -15$

v) $6(p - 5) - 4(p - 3)$
$= 6p - 30 - 4p + 12$ — Remove brackets
$= 2p - 18$ — Note: $-4 \times -3 = +12$

vi) $3x(2x - 4) - (3 - 2x)$
$= 3x(2x - 4) - 1(3 - 2x)$ — Place a 1 outside $-(3 - 2x)$
$= 6x^2 - 12x - 3 + 2x$ — Remove brackets
$= 6x^2 - 10x - 3$ — $-1 \times -2x = +2x$

Note: Once again, watch the signs carefully when there is a minus sign outside of the bracket, as shown in the last two examples. When the brackets are removed, the minus sign outside the bracket has reversed the sign inside the bracket.

EXPANDING BINOMIAL PRODUCTS

A binomial product is an expression which consists of two sets of brackets multiplied together.

Examples: $(x + 7)(x - 2)$
$(a + 3)(b - 5)$
$(2m + 7)(m + 4)$
$(y + 7)^2$

In order to expand (or remove the brackets) from a binomial product the following steps are required:

i) Multiply the first term in the first bracket by the whole of the second bracket.
ii) Multiply the second term in the first bracket by the whole of the second bracket.
iii) Expand each bracket separately.
iv) Simplify like terms where possible.

Example 1: Expand $(a + 3)(b + 7)$
$= a(b + 7) + 3(b + 7)$ — Multiply 'a' by $(b + 7)$ / Multiply '+ 3' by $(b + 7)$
$= ab + 7a + 3b + 21$ — Expand each bracket.

Example 2: Expand $(x + 5)(x - 2)$
$= x(x - 2) + 5(x - 2)$ — Multiply 'x' by $(x - 2)$ / Multiply '+ 5' by $(x - 2)$
$= x^2 - 2x + 5x - 10$ — Expand each bracket.
$= x^2 + 3x - 10$ — Simplify like terms, i.e. $-2x + 5x - + 3x$

Example 3: Expand $(3m - 2)(2m - 7)$
$= 3m(2m - 7) - 2(2m - 7)$ — Multiply '$3m$' by $(2m - 7)$ / Multiply '−2' by $(2m - 7)$
$= 6m^2 - 21m - 4m + 14$ — Expand each bracket
$- 6m^2 - 25m + 14$ — Simplify like terms, i.e. $-21m - 4m - -25m$

FURTHER EXAMPLES

The binomial expansion explained on the previous page could be summarised as follows:

$$\begin{aligned} &(a+b)(c+d) \\ &= a(c+d) + b(c+d) \\ &= ac + ad + bc + bd \end{aligned}$$

i.e. Both terms in the first bracket are multiplied by both terms in the second bracket

Make sure you fully understand each of the steps shown in the examples given below, because they represent each of the main types which occur so regularly in tests.

Example 1: Expand $(3x+7)(2x-5)$

$(3x+7)(2x-5)$	
$= 3x(2x-5) + 7(2x-5)$	Multiply '$3x$' by $(2x-5)$ Multiply '$+7$' by $(2x-5)$
$= 6x^2 - 15x + 14x - 35$	Expand each bracket.
$= 6x^2 - x - 35$	Simplify like terms.

Example 2: Expand $(m-6)(m+6)$

$$\begin{aligned} &= m(m+6) - 6(m+6) \\ &= m^2 + 6m - 6m - 36 \\ &= m^2 - 36 \end{aligned}$$

Example 3: Expand $(a+3)^2$

$$\begin{aligned} &= (a+3)(a+3) \\ &= a(a+3) + 3(a+3) \\ &= a^2 + 3a + 3a + 9 \\ &= a^2 + 6a + 9 \end{aligned}$$

Example 4: Expand $(2m-5)^2$

$$\begin{aligned} &= (2m-5)(2m-5) \\ &= 2m(2m-5) - 5(2m-5) \\ &= 4m^2 - 10m - 10m + 25 \\ &= 4m^2 - 20m + 25 \end{aligned}$$

SPECIAL BINOMIAL PRODUCTS

Binomial products which have the pattern $(a + b)^2$ or $(a - b)^2$ are called 'perfect squares'. Any product which has the pattern $(a + b)(a - b)$ is called 'the difference of two squares'. If we multiply out these special products using the steps explained in the previous section, we obtain the following results:

$$\begin{aligned}&(a+b)^2\\&=(a+b)(a+b)\\&=a(a+b)+b(a+b)\\&=a^2+ab+ab+b^2\\&=a^2+2ab+b^2\end{aligned}$$

$$\begin{aligned}&(a-b)^2\\&=(a-b)(a-b)\\&=a(a-b)-b(a-b)\\&=a^2-ab-ba+b^2\\&=a^2-2ab+b^2\end{aligned}$$

$$\begin{aligned}&(a+b)(a-b)\\&=a(a-b)+b(a-b)\\&=a^2-ab+ba-b^2\\&=a^2-b^2\end{aligned}$$

In summary:

$$(a+b)^2 = a^2 + 2ab + b^2$$
$$(a-b)^2 = a^2 - 2ab + \mathrm{b}^2$$
$$(a+b)(a-b) = a^2 - b^2$$

Rather than use the longer method shown on the previous pages, we can use the above patterns or rules to work out special expansions much more quickly.

Example 1: Expand $(x + 7)^2$

Using:	$(a+b)^2 = a^2$	+	$2ab$	+	b^2
	square 1st term	+	twice the product of the terms	+	square 2nd term

$\therefore$ $(x + 7)^2 = x^2 + 2 \times 7x + 7^2$
$= x^2 + 14x + 49$

Example 2: Expand $(m + 4)(m - 4)$

Using:	$(a+b)(a-b) = a^2$	–	b^2
	1st term squared	–	2nd term squared

$\therefore$ $(m + 4)(m - 4) = m^2 - 4^2$
$= m^2 - 16$

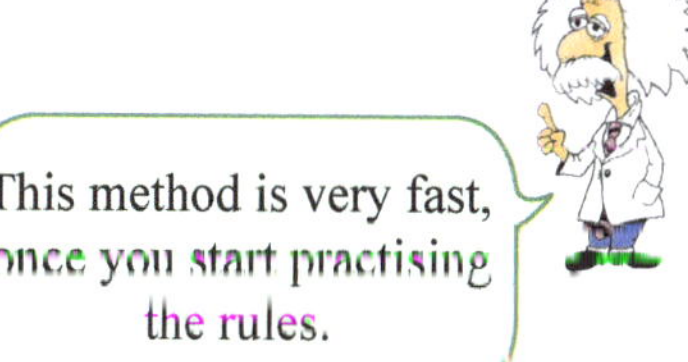

FURTHER EXAMPLES

Expand the following expressions and simplify wherever possible:

i) $(p-5)^2$

ii) $2(3m+5)^2$

iii) $(4m-5)(4m+5)$

iv) $(1-3x)^2+(x+7)(x-2)$

v) $(5x+2)^2-(3+x)(3-x)$

Solutions:

i) $(p-5)^2 = p^2 - 2\times 5p + 5^2$
$= p^2 - 10p + 25$

Using $(a-b)^2 = a^2 - 2ab + b^2$

ii) $2(3m+5)^2 = 2[(3m)^2 + 2\times 15m + 5^2]$
$= 2[9m^2 + 30m + 25]$
$= 18m^2 + 60m + 50$

Using $(a+b)^2 = a^2 + 2ab + b^2$
Multiply the expansion by 2

iii) $(4m-5)(4m+5) = (4m)^2 - 5^2$
$= 16m^2 - 25$

Using $(a+b)(a-b) = a^2 - b^2$

iv) $(1-3x)^2 + (x+7)(x-2)$
$= [1^2 - 2\times 3x + (3x)^2] + [x(x-2) + 7(x-2)]$
$= (1 - 6x + 9x^2) + (x^2 - 2x + 7x - 14)$
$= 10x^2 - x - 13$

$(x+7)(x-2)$ must be multiplied out the long way as it isn't a special product.

v) $(5x+2)^2 - (3+x)(3-x)$

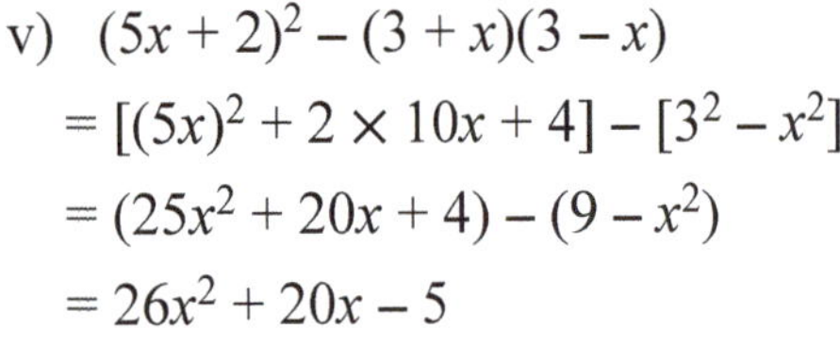

$= [(5x)^2 + 2\times 10x + 4] - [3^2 - x^2]$
$= (25x^2 + 20x + 4) - (9 - x^2)$
$= 26x^2 + 20x - 5$

Watch the effect of the subtract sign carefully, because this is a popular test question.

Note: Once the brackets have been expanded in the last two examples, then 'like terms' are added or subtracted in the usual way.

FACTORISING

'Factorising' involves the **reverse** process to 'removing brackets' which was explained on the previous pages. These two related ideas are important to understand, because they will be frequently tested in all major exams right through to the completion of Year 12.

> Take the highest common factor (HCF) out of the brackets. The HCF could be a number, or letter, or a combination of numbers and letters.

Note: You should always check your factorising by MENTALLY multiplying out the brackets. If correct, the answer should come back to the original expression.

Examples: Factorise the following expressions:

(i) $6a + 12 = 6(a + 2)$ HCF = 6

Mental check: $6(a + 2) = 6 \times a + 6 \times 2 = 6a + 12$

(ii) $15m + 18 = 3(5m + 6)$ HCF = 3

Mental check: $3(5m + 6) = 3 \times 5m + 3 \times 6 = 15m + 18$

(iii) $27x + 45 = 9(3x + 5)$ HCF = 9

Mental check: $9(3x + 5) = 9 \times 3x + 9 \times 5 = 27x + 45$

(iv) $am + 2a = a(m + 2)$ HCF = a

Mental check: $a(m + 2) = a \times m + a \times 2 = am + 2a$

(v) $5x^2 + 10x + 15 = 5(x^2 + 2x + 3)$ HCF = 5

Mental check: $5(x^2 + 2x + 3) = 5 \times x^2 + 5 \times 2x + 5 \times 3$

$= 5x^2 + 10x + 15$

Note that factorisation can also involve more than 2 terms as shown in the last example.

The examples on the previous page were relatively easy because the HCF in each case was a single number or letter. The first four examples on this page are slightly harder, because the HCF will include both a number and letter. The last three examples are also more difficult, because they involve a negative HCF.

Examples: Factorise the following expressions:

(i) $6x^2 + 8x$
$= 2x(3x + 4)$ HCF $= 2x$

(ii) $8ab - 12ac$
$= 4a(2b - 3c)$ HCF $= 4a$

(iii) $10x^2 + 15xy - 20x$
$= 5x(2x + 3y - 4)$ HCF $= 5x$

(iv) $15m^3 - 12m^2$
$= 3m^2(5m - 4)$ HCF $= 3m^2$

(v) $-6m + 12$
$= -6(m - 2)$ HCF $= -6$
When multiplied out, $-6 \times -2 = +12$

(vi) $-10p - 5$
$= -5(2p + 1)$ HCF $= -5$
When multiplied out, $-5 \times +1 = -5$

(vii) $-8a^2 + 12a$
$= -4a(2a - 3)$ HCF $= -4a$
When multiplied out, $-4a \times -3 = +12a$

As you can see, some of these are quite tricky. Watch the signs carefully, and check your answers by mentally multiplying out the brackets.

FACTORISING SPECIAL BINOMIALS

Algebraic expressions which contain only two terms are called 'binomials'. Factorisation involves the reverse procedure of removing brackets — an idea which has been explained on the previous pages.

If we expand $a(b + c)$, we obtain $ab + ac$.
If we factorise $ab + ac$, we obtain $a(b + c)$.
If we expand $(a + b)(a - b)$, we obtain $a^2 - b^2$.
If we factorise $a^2 - b^2$, we get $(a + b)(a - b)$.

With binomials, always look for the 'difference of two squares' pattern.
This pattern can always be written in this format:

$$(\ )^2 - (\ )^2$$

Then we factorise according to the formula:

$$a^2 - b^2 = (a + b)(a - b)$$

Examples:

(i) $m^2 - 9 = m^2 - 3^2 = (m + 3)(m - 3)$
(ii) $p^2 - 25 = p^2 - 5^2 = (p + 5)(p - 5)$
(iii) $k^2 - 100 = k^2 - 10^2 = (k + 10)(k - 10)$
(iv) $9x^2 - 16 = (3x)^2 - 4^2 = (3x + 4)(3x - 4)$
(v) $25p^2 - 36 = (5p)^2 - 6^2 = (5p + 6)(5p - 6)$

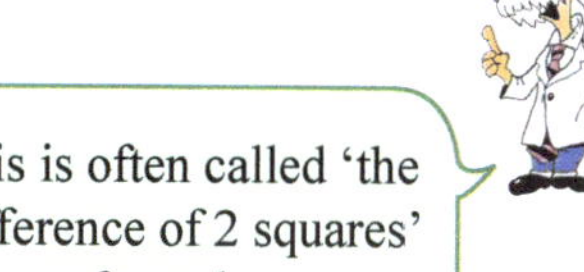

Keep in mind, that the very first step in ANY factorisation process is to always:

Look for the highest common factor (HCF) and write it outside the bracket.

Example 1: $2m^2 - 32$

$= 2(m^2 - 16)$ HCF = 2
$= 2(m^2 - 4^2)$ $a^2 - b^2$ pattern
$= 2(m + 4)(m - 4)$

Example 2: $4x^2 - 36$

$= 4(x^2 - 9)$ HCF = 4
$= 4(x^2 - 3^2)$ $a^2 - b^2$ pattern
$= 4(x + 3)(x - 3)$

FACTORISING EASIER TRINOMIALS

These are algebraic expressions which contain three terms expressed in the format $ax^2 + bx + c$. In this section, we shall only consider the factorisation of easier trinomials where the leading coefficient $a = 1$.

> To factorise $x^2 + bx + c$:
>
> Find 2 integers that multiply to give 'c' and also add up to give 'b'.

i.e. Find 2 integers that multiply to give the constant at the end AND add up to give the number in front of the x term.

Example 1: Factorise $x^2 - 2x - 8$

What two integers multiply to give -8, and also add up to give -2?

$$-4 \times +2 = -8$$
$$-4 + 2 = -2$$

The numbers are -4 and $+2$.

$\therefore\ x^2 - 2x - 8 = (x - 4)(x + 2)$

Example 2: Factorise $x^2 - 7x + 12$

What two integers multiply to give $+12$, and also add up to give -7?

$$-4 \times -3 = +12$$
$$-4 + (-3) = -7$$

The numbers are -4 and -3.

$\therefore\ x^2 - 7x + 12 = (x - 4)(x - 3)$

Example 3: Factorise $2a^2 - 2a - 12$

Remember from the last section, the very first step is to always look for a common factor!

$\therefore\ 2a^2 + 2a - 12 = 2(a^2 + a - 6)$

What two integers multiply to give -6, and also add up to give $+1$?

$$+3 \times -2 = -6$$
$$+3 + (-2) = +1$$

The numbers are $+3$ and -2.

$\therefore\ 2a^2 + 2a - 12 = 2(a + 3)(a - 2)$

Note: Once the common factor is taken out of the brackets, then we can factorise the trinomial using the steps shown at the top of the page.

FACTORISING FOUR TERM EXPRESSIONS

These might initially sound complicated, but in reality they are easier to factorise than trinomials. They must be understood, not only on their own account, but also in order to factorise harder trinomials in the next section. The steps, in order, are outlined below:

1. Split into 2 groups each containing two terms.
2. Factorise each group separately.
3. Both brackets should be the same.
4. Factorise again.

Examples: (i) Factorise $x^2 + 3x + bx + 3b$

$\boxed{x^2 + 3x} + \boxed{bx + 3b}$	Split into two groups.
$= x(x + 3) + b(x + 3)$	Factorise each group separately.
$= (x + b)(x + 3)$	Factorise again.

(ii) Factorise $x^2 - 2x - 5x + 10$

$\boxed{x^2 - 2x} - \boxed{5x + 10}$	Split into two groups.
$= x(x - 2) - 5(x - 2)$	Factorise each group separately.
$= (x - 5)(x - 2)$	Factorise again.
Note carefully above that:	$-5x + 10 = -5(x - 2)$

(iii) Factorise $a^3 + a^2 - a - 1$

$\boxed{a^3 + a^2} - \boxed{a - 1}$	Split into two groups.
$= a^2(a + 1) - 1(a + 1)$	Factorise each group separately.
$= (a^2 - 1)(a + 1)$	Factorise again.
$= (a + 1)(a - 1)(a + 1)$	$a^2 - 1$ can be further factorised
$= (a + 1)^2(a - 1)$	because it has the pattern $a^2 - b^2$.

FACTORISING HARDER TRINOMIALS

Many students have difficulty factorising trinomials such as '$ax^2 + bx + c$' where the leading coefficient 'a' is not equal to 1. The 'cross swords' method or a 'trial and error' approach can be used, but there is a much easier way which is best illustrated by the examples below.

Example 1: Factorise $3x^2 + 16x - 12$

Multiply 3 by -12 to give -36

Think of two numbers which:

Multiply to give -36

Add up to give $+16$

The numbers are $+18$ and -2.

Split up the middle term into $+18x$ and $-2x$.

The problem has now become a simple four term expression which can be factorised using theory from the previous section.

$\boxed{3x^2 + 18x} - \boxed{2x - 12}$	Split into two groups.
$= 3x(x + 6) - 2(x + 6)$	Factorise each group separately.
$= (3x - 2)(x + 6)$	Factorise again.

Example 2: Factorise $8k^2 + 18k - 5$

Multiply 8 by -5 to give -40

Find two numbers which:

Multiply to give -40

Add up to give $+18$

The numbers are $+20$ and -2.

Split up the middle term into $+20k$ and $-2k$.

$\boxed{8k^2 + 20k} - \boxed{2k - 5}$	Split into two group.
$= 4k(2k + 5) - 1(2k + 5)$	Factorise each group separately.
$= (4k - 1)(2k + 5)$	Factorise again.

THE CROSS SWORDS METHOD OF FACTORISING

This is another popular method which many students and teachers use to factorise trinomials. It doesn't matter which method you choose to use, as long as you are confident in factorising any trinomial. This is particularly important if you wish to study a higher level of Maths in Year 11, as trinomials occur frequently in many of the topics.

Example 1: Factorise $3x^2 + 14x + 8$

The four different possibilities which all give $3x^2$ and $+8$ are listed below. However, only one of them will give the correct middle term of $+14x$. The middle term is obtained by multiplying down each of the lines (or swords), and adding the like terms.

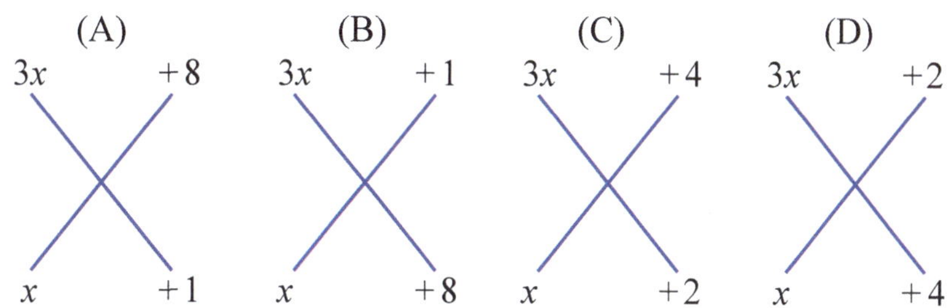

Middle terms: $3x + 8x = 11x$ $\quad 24x + x = 25x$ $\quad 6x + 4x = 10x$ $\quad 12x + 2x = 14x$

It can be seen that only (D) gives the correct middle term.
Therefore $3x^2 + 14x + 8 = (3x + 2)(x + 4)$

It is fairly obvious why it is called 'the cross swords' method.

Example 2: Factorise $2m^2 - 5m - 12$
Some of the possibilities which all give $2m^2$ and -12 are listed below.

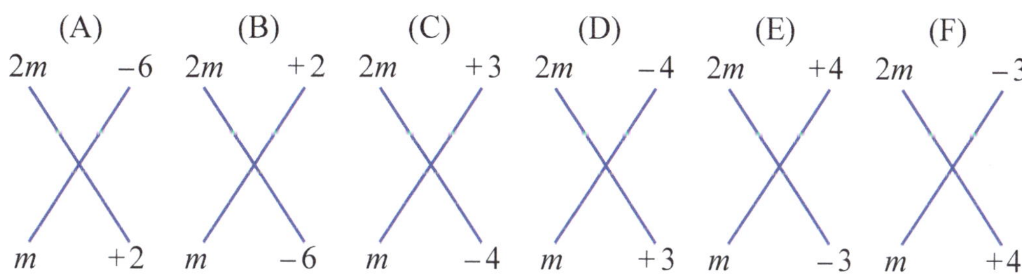

Middle terms: $4m - 6m = -2m$ $\quad -12m + 2m = -10m$ $\quad -8m + 3m = -5m$ $\quad 6m - 4m = 2m$ $\quad -6m + 4m = -2m$ $\quad 8m - 3m = +5m$

It can be seen that only (C) gives the correct middle term.
Therefore $2m^2 - 5m - 12 = (2m + 3)(m - 4)$.

ONE OF THE MAIN REASONS FOR FACTORISING

Factorisation is a method of changing an algebraic expression into a MULTIPLICATION. Once a multiplication is obtained, then fractions can be simplified in the usual way by cancelling down.

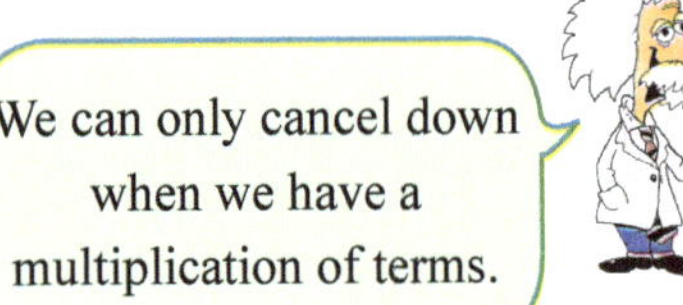

Factorising allows us to simplify algebraic fractions BY CANCELLING DOWN.

Examples: Simplify the following algebraic fractions:

(i) $\frac{5x + 10}{20}$ — Cannot cancel at this stage.

$= \frac{5(x + 2)}{20}$ — Factorise by taking out a common factor.

$= \frac{x + 2}{4}$ — The 5 now divides into 20 as shown.

(ii) $\frac{x^2 - 9}{x + 3}$ — Cannot cancel at this stage.

$= \frac{(x + 3)(x - 3)}{x + 3}$ — Factorise using $a^2 - b^2 = (a + b)(a - b)$.

$= x - 3$ — The $(x + 3)$ can now cancel in the numerator and denominator.

(iii) $\frac{x^2 - 2x - 8}{x - 4}$ — Cannot cancel at this stage.

$= \frac{(x - 4)(x + 2)}{x - 4}$ — Factorise the trinomial.

$= x + 2$ — The $(x - 4)$ can now cancel in the numerator and denominator.

(iv) $\frac{6x^2 + 30x + 36}{2x^2 + 8x + 6}$ — Cannot cancel at this stage.

$= \frac{6(x + 2)(x + 3)}{2(x + 1)(x + 3)}$ — Factorise the expressions.

$= \frac{3(x + 2)}{(x + 1)}$ — The brackets (x+3) cancel in the numerator and denominator and $\frac{6}{2} = 3$.

MULTIPLYING AND DIVIDING ALGEBRAIC FRACTIONS

Once the algebraic expressions have been factorised, multiplication and division is carried out in exactly the same way as ordinary numerical fractions.

(i) Factorise numerators and denominators where possible, and simplify by cancelling.

(ii) For division, remember to change the divide sign to a multiply sign, and invert the second fraction.

Example 1: Simplify $\dfrac{a^2-1}{a^2-6a+5} \times \dfrac{a^2-10a+25}{a^2-25}$

$= \dfrac{(a+1)(a-1)}{(a-5)(a-1)} \times \dfrac{(a-5)(a-5)}{(a+5)(a-5)}$ Factorise where possible.

$= \dfrac{a+1}{a+5}$ $(a-1)(a-5)(a-5)$ cancels in both numerator and denominator.

Example 2: Simplify $\dfrac{4x^2-36}{6x+12} \div \dfrac{x+3}{2}$

$= \dfrac{4x^2-36}{6x+12} \times \dfrac{2}{x+3}$ Multiply and invert second fraction.

$= \dfrac{4(x+3)(x-3)}{6(x+2)} \times \dfrac{2}{x+3}$ Factorise where possible.

$= \dfrac{4(x-3)}{3(x+2)}$ $(x+3)$ cancels and $\dfrac{8}{6} = \dfrac{4}{3}$

Example 3: Simplify

$\dfrac{5k^2+16k+3}{25k^2-1} \div \dfrac{2k^2+5k-3}{5k^2-k}$

$= \dfrac{5k^2+16k+3}{25k^2-1} \times \dfrac{5k^2-k}{2k^2+5k-3}$ Multiply and invert second fraction.

$= \dfrac{(5k+1)(k+3)}{(5k+1)(5k-1)} \times \dfrac{k(5k-1)}{(2k-1)(k+3)}$ Factorise where possible.

$= \dfrac{k}{2k-1}$ $(5k+1)(k+3)(5k-1)$ cancels out.

ADDING AND SUBTRACTING ALGEBRAIC FRACTIONS

Once again, these are done using the same steps and theory used in ordinary numerical fractions.

(i) If possible, factorise the denominators.
(ii) Find the lowest common denominator (LCD).
(iii) Express each fraction with this common denominator.
(iv) Simplify the numerators.

Example 1: Simplify $\frac{x}{x+1} - \frac{2x}{5x-1}$ — LCD = $(x+1)(5x-1)$.

$= \frac{x(5x-1) - 2x(x+1)}{(x+1)(5x-1)}$ — Rewrite each fraction with new LCD.

$= \frac{5x^2 - x - 2x^2 - 2x}{(x+1)(5x-1)}$ — Expand brackets in numerator.

$= \frac{3x^2 - 3x}{(x+1)(5x-1)}$ — Simplify numerator.

Example 2: Simplify $\frac{5}{a^2-1} + \frac{2}{a+1}$ — Factorise $a^2 - 1$.

$= \frac{5}{(a+1)(a-1)} + \frac{2}{a+1}$ — LCD = $(a+1)(a-1)$.

$= \frac{5 + 2(a-1)}{(a+1)(a-1)}$ — Rewrite each fraction with new LCD.

$= \frac{2a+3}{(a+1)(a-1)}$ — Simplify numerator.

Factorising allows us to obtain a lower and simpler LCD.

Example 3: Simplify

$\frac{5}{x^2 - x - 6} - \frac{3}{x^2 - 2x - 3}$ — Factorise both denominators.

$= \frac{5}{(x-3)(x+2)} - \frac{3}{(x+1)(x-3)}$ — LCD = $(x-3)(x+2)(x+1)$

$= \frac{5(x+1) - 3(x+2)}{(x-3)(x+2)(x+1)}$ — Rewrite each fraction with new LCD.

$= \frac{5x + 5 - 3x - 6}{(x-3)(x+2)(x+1)}$ — Expand brackets in numerator.

$= \frac{2x-1}{(x-3)(x+2)(x+1)}$ — Simplify numerator.

CHAPTER SUMMARY

REMOVING SINGLE BRACKETS: $a(b \pm c) = ab \pm ac$

Watch for a minus sign outside a bracket because it has the effect of reversing signs after multiplication.

Examples: i) $-3x\,(2x + 5) = -6x^2 - 15x$

ii) $-(3x - 7) = -3x + 7$

DOUBLE SETS OF BRACKETS

$$(a + b)(c + d) = a(c + d) + b(c + d) = ac + ad + bc + bd$$

Students can also use the faster FOIL method.

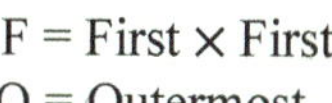

Method: $(a + b)(c + d) = ac + ad + bc + bd$

Example: $(2x + 3)(x - 5) = 2x^2 - 10x + 3x - 15$

THREE SPECIAL PRODUCTS

There are three products occurring frequently in Maths which can be done more quickly if you remember the following patterns:

$$(a + b)^2 = a^2 + 2ab + b^2$$

$$(a - b)^2 = a^2 - 2ab + b^2$$

$$(a + b)(a - b) = a^2 - b^2$$

TRIPLE SETS OF BRACKETS

Step 1: Multiply two of the brackets out using one of the special products shown above.

Step 2: Multiple this answer out by the third set of brackets using the longer FOIL method.

Step 3: Finally simplify by collecting like terms.

Example: $(a - 5)^3 = (a - 5)(a - 5)(a - 5)$

$= (a - 5)(a^2 - 10a + 25)$

$= a^3 - 10a^2 + 25a - 5a^2 + 50a - 125$

$= a^3 - 15a^2 + 75a - 125$

ALGEBRAIC FRACTIONS

All the rules relating to arithmetic fractions also hold true for algebraic fractions.

$\dfrac{a}{b} \pm \dfrac{c}{d} = \dfrac{ad \pm bc}{bd}$ With adding or subtracting, find a common denominator.

$\dfrac{a}{b} \times \dfrac{c}{d} = \dfrac{ac}{bd}$ Multiply out numerators and denominators, and cancel down where possible.

$\dfrac{a}{b} \div \dfrac{c}{d} = \dfrac{a}{b} \times \dfrac{d}{c} = \dfrac{ad}{bc}$ First multiply and invert second fraction. Then multiply out as shown.

FACTORISING

This process is the reverse process to 'removing the brackets'. Find the highest common factor (HCF) and write it outside the brackets. You can always check your factorising by MENTALLY multiplying out the brackets. If correct, the answer should come back to the original expression. For binomials also look for the pattern $a^2 - b^2 = (a + b)(a - b)$.

Examples: Factorise the following:

(i) $15m^3 - 12m^2 = 3m^2(5m - 4)$ HCF $= 3m^2$

Mental check: $3m^2(5m - 4) = 15m^3 - 12m^2$

(ii) $a^2 - 25 = (a + 5)(a - 5)$ Find the pattern $a^2 - b^2$

DIFFERENCE OF TWO SQUARES

From earlier work in algebra, you will recall that $(a + b)(a - b) = a^2 - b^2$. Look out for this pattern as we now use it in reverse.

i.e. $a^2 - b^2 = (a + b)(a - b)$

Examples: $m^2 - 49 = m^2 - 7^2 = (m + 7)(m - 7)$

$9x^2 - 1 = (3x)^2 - 1^2 = (3x + 1)(3x - 1)$

Note: It is impossible to factorise the sum of 2 squares. i.e. $a^2 + b^2$

FOUR-TERM EXPRESSIONS

Group in pairs and factorise each group separately.
Then factorise again.

Examples: $\boxed{x^2 + 3x} + \boxed{bx + 3b} = x(x + 3) + b(x + 3)$

$= (x + b)(x + 3)$

$\boxed{3x^2 - 21x} - \boxed{4x + 28} = 3x(x - 7) - 4(x - 7)$

$= (3x - 4)(x - 7)$

EASIER TRINOMIALS

In $ax^2 + bx + c$ (where $a = 1$), find two numbers α and β that multiply to give c, and add up to give b.

Examples:

$x^2 - 2x - 8$
$= (x - 4)(x + 2)$ $\left.\begin{array}{l}\alpha\beta = -8 \\ \alpha + \beta = -2\end{array}\right\} \Rightarrow$ $\alpha = -4$, $\beta = +2$

$m^2 - 10m + 24$
$= (m - 6)(m - 4)$ $\left.\begin{array}{l}\alpha\beta = +24 \\ \alpha + \beta = -10\end{array}\right\} \Rightarrow$ $\alpha = -6$, $\beta = -4$

$p^2 + p - 72$
$= (p + 9)(p - 8)$ $\left.\begin{array}{l}\alpha\beta = -72 \\ \alpha + \beta = +1\end{array}\right\} \Rightarrow$ $\alpha = +9$, $\beta = -8$

HARDER TRINOMIALS

In $ax^2 + bx + c$ (where $a \neq 1$), find two numbers a and B that multiply to give $a \times c$, and add up to give b. Then split the original trinomial into a simpler four term expression.

Example:

$$3p^2 + 4p - 7$$
$$= 3p^2 - 3p + 7p - 7$$
$$= 3p(p - 1) + 7(p - 1)$$
$$= (3p + 7)(p - 1)$$

$\alpha\beta = -21$, $\alpha + \beta = +4$ ⇨ $\alpha = -3$, $\beta = +7$

$\therefore$ Split $+4p$ into $-3p + 7p$

COMBINATION OF RULES

It is important to fully factorise expressions, and sometimes this may involve several of the rules already explained, as in the examples below.

Examples:

Factorise $2x^3 - 6x - 20x$ — Always look for the HCF first.

$= 2x(x^2 - 3x - 10)$ — Factorise the easier trinomial.

$= 2x(x - 5)(x + 2)$ — $(x^2 - 3x - 10)$in the usual way.

Factorise $8m^4 - 8n^4$ — Find HCF = 8.

$= 8(m^4 - n^4)$ — Use $a^2 - b^2 = (a + b)(a - b)$.

$= 8(m^2 + n^2)(m^2 - n^2)$ — Use $a^2 - b^2 = (a + b)(a - b)$ again.

$= 8(m^2 + n^2)(m - n)(m + n)$

TWO REASONS FOR FACTORISING

Factorisation is an important technique of changing an algebraic expression into a multiplication. There are two immediate reasons for obtaining a product in algebra.

1. **Cancelling down or simplifying algebraic fractions:**

$$\frac{x^2 - 4}{x^2 + 2x} = \frac{(x + 2)(x - 2)}{x(x + 2)} = \frac{x - 2}{x}$$

Must first factorise both expressions.

2. **Solving quadratic equations:**

$$x^2 - 2x - 63 = 0$$
$$\therefore \quad (x + 7)(x - 9) = 0$$
$$\therefore \quad x = -7 \text{ or } x = 9$$

Once again, must first factorise both expressions.

Note: The topic 'quadratic equations' will be explained in a later chapter.

LEVEL 1 — FURTHER ALGEBRA AND FACTORISATION

EASIER QUESTIONS

Note: Only turn back to page number shown if you have difficulty.

	Page
Q1. Simplify the following:	108
a) $5a + 2b - 3a + b$ b) $x + 2y - 5x - y$ c) $m - n + 4n - 3$	
d) $3x^2 + 5x - 2x + 6$ e) $4ab - a + 2b + ab$ f) $3m^2 - 3 + 5m - m^2$	
g) $3a^2 - 7a - a^2 + 2a$ h) $5x^2 - 7x - 6x^2 - 3x$ i) $4m^2 - 8 - 3m + 11 - m$	
Q2. Expand and simplify the following:	109 - 112
a) $5(a + 3) - 2(a + 1)$ b) $4(m - 7) - 3(m - 2)$ c) $8(x - 5) - (4 - 3x)$	
d) $(m + 7)(m - 2)$ e) $(x - 5)(x - 7)$ f) $(x + 3)(x - 8)$	
g) $(4a + 3)(a - 7)$ h) $(4k - 7)(3k - 2)$ i) $(3 - 5y)(4 + 2y)$	
Q3. Use the quicker methods and rules to expand the following special binomial products:	113, 114
a) $(m + 7)^2$ b) $(n - 8)^2$ c) $(x - 10)(x + 10)$	
d) $(2x - 3)^2$ e) $(5p + 4)^2$ f) $(3x - 7)(3x + 7)$	
g) $3(2x + 5)^2$ h) $4m(3m + 2)(3m - 2)$ i) $(5 - 3x)^2$	
Q4. Factorise the following binomials using the rule $a^2 - b^2 = (a + b)(a - b)$:	117
a) $m^2 - 81$ b) $49 - k^2$ c) $9n^2 - 25$	
d) $x^2 - 1$ e) $16 - 25x^2$ f) $2m^2 - 18$	
Q5. Factorise the following trinomials:	118
a) $x^2 + 9x + 20$ b) $x^2 + 7x + 12$ c) $x^2 - 3x + 2$	
d) $x^2 - 3x - 10$ e) $b^2 + 5b - 24$ f) $a^2 - 8a + 15$	
Q6. By first factorising the numerator, simplify the following algebraic expressions:	122
a) $\dfrac{6a + 27}{3}$ b) $\dfrac{10m + 30}{5}$ c) $\dfrac{3x^2 + x}{x}$	
d) $\dfrac{4m^2 - m}{6m}$ e) $\dfrac{x^2 - 4}{x + 2}$ f) $\dfrac{x^2 + 5x + 6}{x + 3}$	
g) $\dfrac{x^2 + x - 6}{x - 2}$ h) $\dfrac{x^2 - 2x - 35}{x - 7}$ i) $\dfrac{x^2 - 8x + 15}{x - 5}$	
Q7. Simplify the following:	124
a) $\dfrac{m - 5}{4} + \dfrac{4 + m}{3}$ b) $\dfrac{x + 4}{2} - \dfrac{8 + x}{5}$ c) $\dfrac{n + 1}{3} - \dfrac{n - 3}{5}$	
d) $\dfrac{2a + 1}{4} + \dfrac{3a - 5}{6}$ e) $\dfrac{2 - 5x}{3} - \dfrac{5 + 4x}{8}$ f) $\dfrac{6x + 5}{7} - \dfrac{4x + 8}{6}$	

LEVEL 2 — FURTHER ALGEBRA AND FACTORISATION

AVERAGE QUESTIONS

Note: Only turn back to page number shown if you have difficulty.

Page

Q1. Factorise these binomials: 115 - 117

a) $16m + 8$ b) $5x^2 - 15x$ c) $x^2 - 4$

d) $12a - 8a^2$ e) $9m^2 - 16$ f) $8b^2 - 18$

Q2. Factorise these trinomials: 118

a) $x^2 + 5x + 6$ b) $m^2 - m - 12$ c) $x^2 + 4x - 5$

d) $y^2 - 6y + 8$ e) $p^2 - 7p - 8$ f) $3a^2 + 15a - 18$

Q3. Factorise the following four term expressions: 119

a) $ab + ac + b^2 + bc$ b) $xy - xz + y^2 - yz$ c) $pr - p - qr + q$

d) $x^2 + xy - 3x - 3y$ e) $6ab + 2ac + 3b^2 + bc$ f) $5xy - 25x - y^2 + 5y$

Q4. Factorise these more difficult trinomials: 120, 121

a) $2m^2 - 9m - 5$ b) $3x^2 - 10x + 3$ c) $5a^2 + 22a + 8$

d) $6x^2 + 5x - 6$ e) $12p^2 - 19p + 4$ f) $16b^2 + 16b - 5$

Q5. Simplify the following expressions: 122

a) $\dfrac{8m-4}{2m-1}$ b) $\dfrac{m^2-4}{m+2}$ c) $\dfrac{9x^2+12x}{3x}$

d) $\dfrac{x^2+3x-4}{x-1}$ e) $\dfrac{6x-3}{10x^2-5x}$ f) $\dfrac{x^2-16}{x^2-8x+16}$

Q6. The area (A) of a trapezium is given by the formula $A = \frac{1}{2}(a + b)h$, where a and b are the lengths of the 2 parallel sides and h is the height of the trapezium in terms of x. Express the area of the trapezium on the right in terms of x:

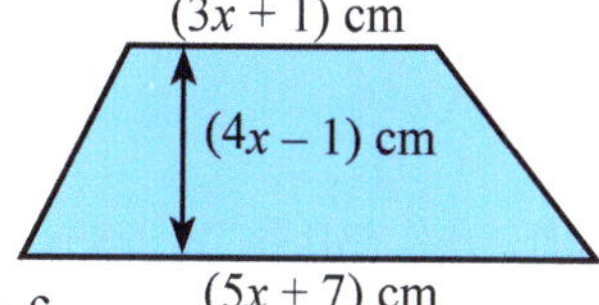

Q7. Find the simplest expression for the area of each figure below in terms of x:

a)

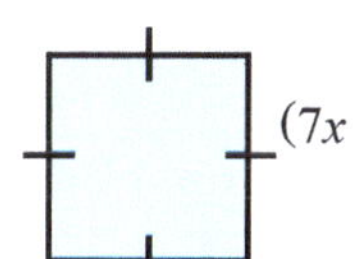

b)

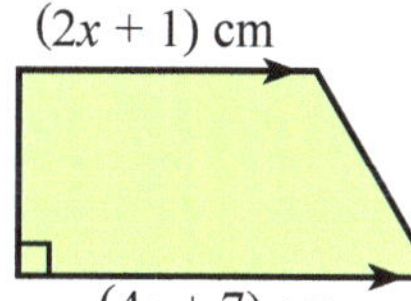

c)

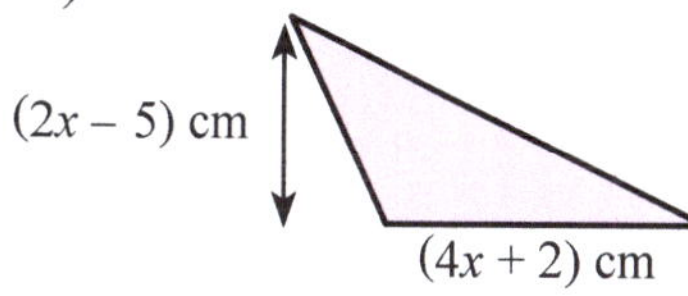

Q8. A car travels at x km/h for 12 km, then increases speed by 10 km/h and travels for a further 8 km. For how long did the car travel? (answer in terms of x)

Q9. A boy sits for 12 maths exams during the year and gains an average of y%. He sits for one more exam and receives a score of 65%. What is his new average?

Q10. At a concert, seated tickets cost $\$A$ and standing tickets cost $\$B$. The seated tickets are four times the price of standing tickets. If 150 seated tickets and 250 standing tickets are sold, and the total receipts from the concert is $\$Y$, write an expression to find the price of a standing ticket in terms of Y.

LEVEL 3 — FURTHER ALGEBRA AND FACTORISATION

AVERAGE QUESTIONS

Note: Only turn back to page number shown if you have difficulty.

	Page
Q1. Factorise the following expressions: a) $6x + 24$ b) $5 - 15x$ c) $18x + 30y$ d) $4x + 6x^2$ e) $9p - 15pq$ f) $14xy - 8xy^2$	115, 116
Q2. Factorise these: a) $x^2 - 16$ b) $4x^2 - y^2$ c) $49 - 25m^2$ d) $9a^2 - 4b^2$ e) $m^2n^2 - 36$ f) $2x^2 - 2$	117
Q3. Factorise following: a) $x^2 - 8x + 15$ b) $m^2 - m - 72$ c) $4y^2 + 4y - 80$ d) $(m + 3)^2 - 5^2$ e) $x^3 - x$ f) $p^4 - q^4$	118
Q4. Factorise the following four term expressions: a) $ab - ac + b^2 - bc$ b) $3xz + x + 6yz + 2y$ c) $12n + 6m - 2mn - m^2$ d) $x^2 + xy - 3x - 3y$ e) $6ab + bc + 3b^2 + 2ac$ f) $5xy - y^2 - 25x + 5y$	119
Q5. Factorise these trinomials: a) $2x^2 - 9x - 5$ b) $3x^2 - 10x + 3$ c) $5x^2 + 22x + 8$ d) $6x^2 + 5x - 6$ e) $12x^2 - 19x + 4$ f) $16x^2 + 16x - 5$	120, 121
Q6. Simplify the following expressions: a) $\dfrac{x^2 - 4x}{x^2 - 16}$ b) $\dfrac{6x - 3}{10x - 5}$ c) $\dfrac{xy + 2x - y^2 - 2y}{x^2 + 2x - xy - 2y}$ d) $\dfrac{x^2 + x - 6}{x^2 - 4x + 4}$ e) $\dfrac{3x^2 - 8x - 3}{3x^3 + x^2}$ f) $\dfrac{4x^2 - 4}{3x^2 - 9x - 12}$	122
Q7. Simplify: a) $\dfrac{x + 5}{x^2 - 4x - 5} \times \dfrac{x^2 - 5x}{x^2 + 10x + 25}$ b) $\dfrac{2x + 1}{2x^2 - 3x - 9} \div \dfrac{10x + 5}{2x^2 - 12x + 18}$ c) $\dfrac{a^2 - 36}{a^2 - 4a - 12} \times \dfrac{2a - 12}{3a + 18}$ d) $\dfrac{x^2 - x - 20}{x^2 - 25} \div \dfrac{x + 4}{x^2 + 5x}$	123
Q8. Simplify: a) $\dfrac{1}{2x - 1} - \dfrac{1}{2x + 1}$ b) $\dfrac{x}{x + 2} - \dfrac{8}{x^2 - 4}$ c) $\dfrac{6x}{x^2 + x - 6} + \dfrac{3x}{x - 2}$	124

Q1. Expand the brackets and simplify the following expressions:

a) $5(x+3)-3(x-1)$ b) $-3a(a-3)+2(a+1)$ c) $2m(1-m)-3m$
d) $b(b+1)-4(b-8)$ e) $-(m-1)-(3m-5)$ f) $x(y-6)+y(x+2)$
g) $8a(3a-1)-2(a+8)$ h) $xy(x-y)-x(xy-1)$ i) $-5(3-m)+(m-3)$

Q2. Expand and simplify:

a) $(2x+1)(x-5)$ b) $(4a+3)^2$ c) $(3b-2)(3b+2)$
d) $(5n-4)(9-n)$ e) $(3x+7)(3-4x)$ f) $(9p-1)^2$
g) $(6a-5)(6a+5)$ h) $(x+2y)(3x-5y)$ i) $(4m-3n)(4m+3n)$
j) $(7-5x)^2$ k) $(3m+5)^2$ l) $5x(2x-7)^2$

Q3. Factorise these expressions:

a) $24m^2+18m$ b) a^3-a^2b-a c) $81x^2-49y^2$
d) $8n^2+12n-6$ e) $-48x-16x^2$ f) p^2q-pq^2+pq
g) $6ab^2-15ab+12b$ h) $32x^2-2$ i) $-2m^2+50$

Q4. Factorise these trinomials:

a) $2x^2+10x-72$ b) $5a^2-30a+40$ c) $-x^2+x+12$
d) $6b^2+13b+6$ e) $15+11m-12m^2$ f) $a^2-2ab+b^2$
g) $8x^2-52x+80$ h) $2m^2+5mn+2n^2$ i) $20m^2-m-1$

Q5. Factorise these four term expressions:

a) $x^2-yz-xy+xz$ b) $8a+b-2-4ab$ c) n^3-n^2+n-1
d) $m-mn-m^2+n$ e) $ab+3bc-5a-15c$ f) $4xy-3yz+2xz-6y^2$

Q6. Simplify the following expressions:

a) $\dfrac{xy+2x-y^2-2y}{x^2+2x-xy-2y}$ b) $\dfrac{x^2+x-6}{x^2-4x+4}$ c) $\dfrac{3x^2-8x-3}{3x^3+x^2}$

d) $\dfrac{4x^2-4}{3x^2-9x-12}$ e) $\dfrac{18x^2-3x-3}{9x^2+6x+1}$ f) $\dfrac{2x^3-8x}{2x^3-8x^2+8x}$

g) $\dfrac{2a^2-5ab+2b^2}{2ab-b^2+4a-2b}$ h) $\dfrac{a+ab-a^2-b}{a-a^3}$ i) $\dfrac{x^3-x}{x-x^3}$

Q7. Simplify:

a) $\dfrac{3}{x^2-1}-\dfrac{1}{x+1}$ b) $x+\dfrac{5x}{x+6}$ c) $\dfrac{5}{27-3x}\div\dfrac{9}{18-2x}$

d) $\dfrac{2}{x^2-x}+\dfrac{3}{x-x^2}$ e) $\dfrac{3x-6}{4x^3}\times\dfrac{8}{4-2x}$ f) $\dfrac{x^3-x}{x^2-2x-8}\div\dfrac{x-1}{x^2-4x}$

LEVEL 5 — FURTHER ALGEBRA AND FACTORISATION

EXTENSION QUESTIONS

Q1. Find the value of a for the following:

a) $5x + ax + 3 - 5 = 4x - 2$

b) $5x - 4x + 1 + a = x + 10$

c) $ax + 3 - 5 + 6 = ax + a$

d) $ax + b - 5 + 6 = bx + 3$

Q2. Find the value of a for the following.

a) $-2(a - x) = 6 + 2x$

b) $4(a - x) = 15 - 4x$

c) $a(a + x) = 16 + ax$

d) $a(\frac{2}{3} + \frac{1}{5}x) = 3\frac{1}{3} + x$

e) $a(x - 3) + 2 = 2x - 4$

f) $3(2x + 4) = a(x + 2)$

Q3. Find the values of a and b for the following:

a) $a(x + 2) = 3x + b$

b) $4(x + a) = bx + 12$

c) $b(x + a) = -2x + 12$

d) $a(x + 3a) + b = 2x$

e) $3(2x - a) = 2bx + 15$

f) $-3(-2 - ax) = 2b - 6x$

Q4. For $a = \{0, 1, 2\}$ write the expression for y in terms of x.

a) $y = x^2 - ax - a$

b) $y = (x + a)^3 - (x + a)^2$

c) $y = \frac{x^2}{a} + \frac{x^2}{a} + \frac{a}{2}$

Q5. Find the values of a and b for the following:

a) $(x + 1)(x + a) = x^2 + bx + 2$

b) $a(x + 1)(x - b) = 3x^2 - 3x - 6$

c) $(x + 3)(x - a) = x^2 + bx - 6$

d) $a(x + a)(x - b) = 3x^2 - 3x - 36$

e) $a(2x + 1)(x - b) = 6x^2 - 3x - 3$

f) $a(2x + 1)(x + b) = x^2 + 5\frac{1}{2}x + 2\frac{1}{2}$

Q6. Expand the following brackets writing all expressions in simplest form.

a) $a(x + 1)(x - 1)$

b) $a(x + b)(x - b)$

c) $a(x + 2b)(x - 2b)$

d) $a(x - b)^2$

e) $(x + b + 1)(x + b - 1)$

f) $\frac{1}{4}(4x + 2a)(4x - 2a)$

Q7. a) Show that $(a + b)^2$ can be expressed as $a^2 + 2ab + b^2$.

b) (i) By expanding both sides show that $(x + 3)^2 = (x + 2)^2 + 2(x + 2) + 1$

(ii) Use the result $(a + b)^2 = a^2 + 2ab + b^2$ and expressing $(x + 3)^2 = [(x + 2) + 1)]^2$ show that $(x + 3)^2 = (x + 2)^2 + 2(x + 2) + 1$

c) (i) By expanding both sides show that $(x - 1)^2 = (x - 2)^2 + 2(x - 2) + 1$

(ii) Use the result $(a + b)^2 = a^2 + 2ab + b^2$ and expressing $(x - 1)^2 = [(x - 2) + 1]^2$ show that $(x - 1)^2 = (x - 2)^2 + 2(x - 2) + 1$

Q8. By factorising and cancelling show that $\dfrac{x^2 + (a + 1)x + a}{x^2 + (a + 2)x + 2a} = \dfrac{(x + 1)}{(x + 2)}$.

Q9. For $a = 2$, $b = 6$, $c = 3$ and $d = 5$, show that $\dfrac{x^2 + dx + b}{x^2 + cx + a} = \dfrac{x + c}{x + c - a}$.

Q10. Factorise the quadratic $x^2 \pm x - a$ for $a = \{2, 6, 12, 20\}$.

Q11. Factorise the following:

a) $x^2 + (a + b)x + ab$

b) $x^2 + (a - b)x - ab$

c) $x^2 + (b - a)x - ab$

d) $x^2 + 3ax + 2a^2$

e) $-x^2 - 3ax - 2a^2$

f) $x^2 + ax - 2a^2$

g) $-x^2 - ax + 2a^2$

h) $x^2 - ax - 2a^2$

i) $x^2 - 3ax + 2a^2$

SURDS AND INDICES

The 'Australian Curriculum Mathematics' (ACM) references for this sub-strand of 'Measurement and Geometry' (MG) are given below. This chapter may contain additional extension work which the author feels will be beneficial to the student.

- *Negative indices and summary of index laws (ACMNA 209).*
- *Scientific notation for very large and small numbers (ACMNA 210).*
- *Multiplying, dividing and the zero index (ACMNA 212).*
- *Pythagoras theorem to find the length hypotenuse or shorter side in the right angled triangle (ACMMG 222).*
- *Pythagorean problems and triads (ACMMG 222).*
- *Simplifying and investigating the 4 operations (+, −, ×, ÷) on surds (ACMNA 264).*
- *Binomial products involving surds (ACMNA 264).*
- *Rationalising the denominator (ACMNA 264).*
- *Fractional indices (ACMNA 264).*

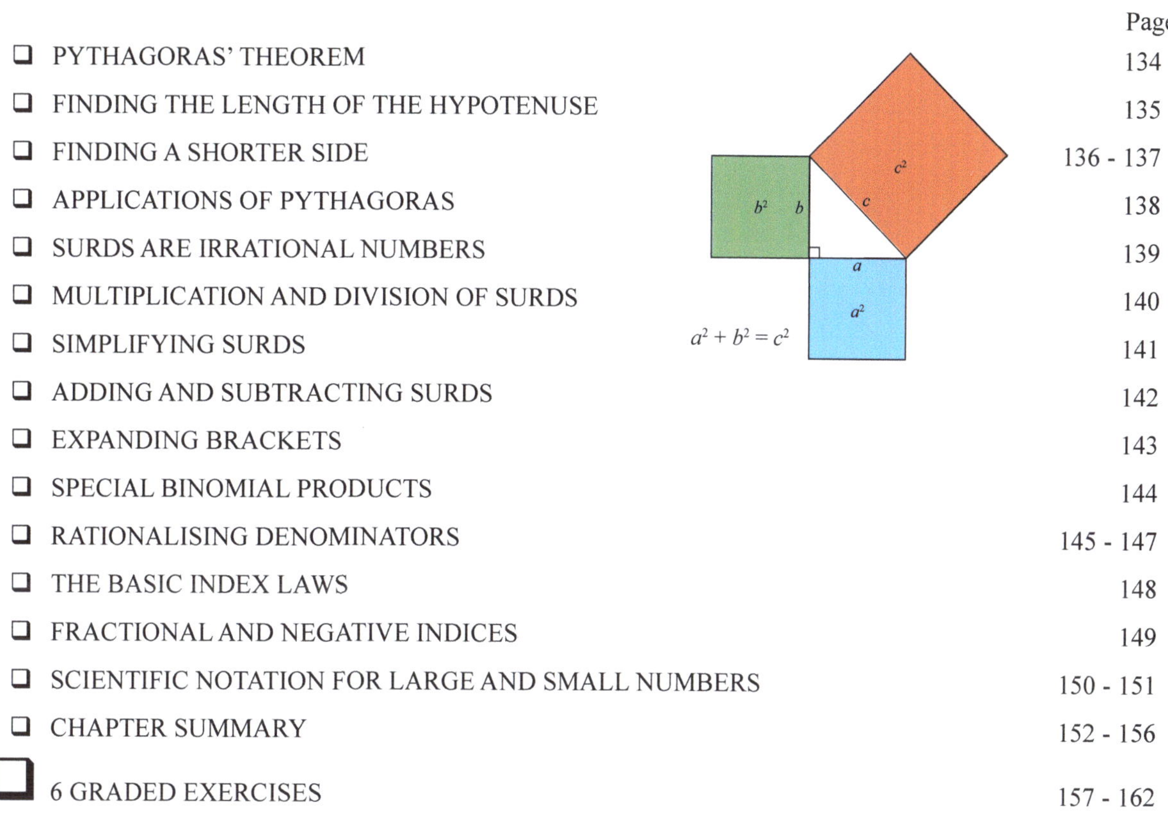

Pythagoras of Samos (570 - 495 BC)

He was an ancient Greek philosopher, mathematician and founder of the religious movement called Pythagoreanism. He made great contributions to philosophy and religion and is revered as a great mathematician, mystic and scientist. He is most famous for his Pythagorean theorem which is stated on the following page. It is said that he was the first man to call himself a philosopher, or lover of wisdom, and many of his ideas influenced Plato, and through him, all Western philosophy.

PYTHAGORAS' THEOREM

Pythagoras was a famous Greek mathematician who discovered a very important rule relating to the sides of a right angled triangle. This law or theorem is now named after him. This is what he discovered:

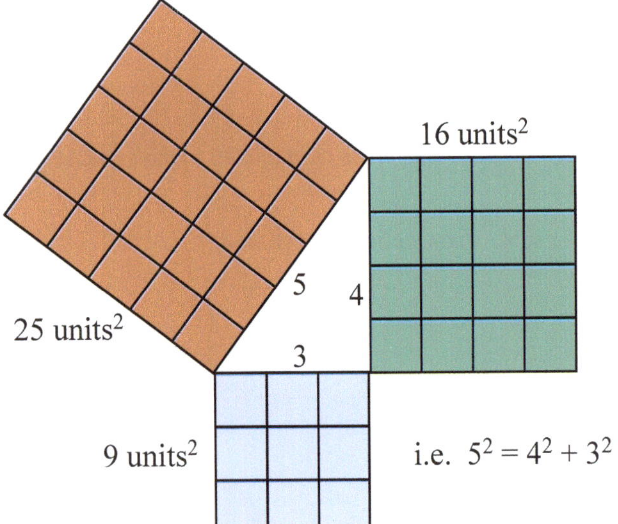

i.e. $5^2 = 4^2 + 3^2$

You will notice in the right angled triangle (at left) that the area of the large square (25 units^2) is equal to the sum of the areas of the two smaller squares.
(i.e. 9 units^2 + 16 units^2)
Pythagoras proved that this rule works for every right angled triangle.

PYTHAGORAS' THEOREM

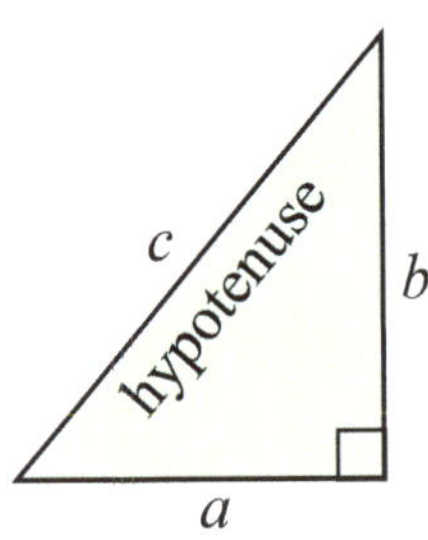

In any right angled triangle, the square of the longest side is equal to the sum of the squares of the 2 smaller sides. From the diagram:

$$c^2 = a^2 + b^2$$

Note: The longest side is called the HYPOTENUSE. The hypotenuse is always the side which faces directly opposite to the right angle.

Example: Use Pythagoras' Theorem to write an equation which relates the sides of the triangle shown.

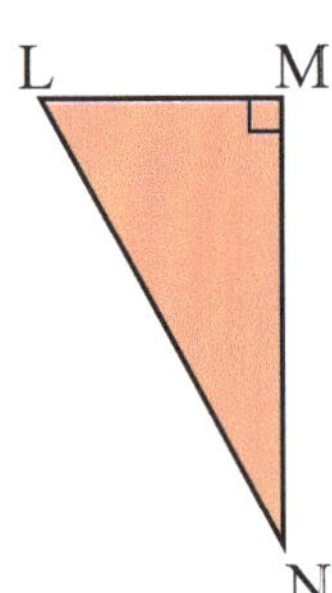

Solution:

Hypotenuse squared = sum of 2 smaller sides squared.

Therefore $LN^2 = LM^2 + MN^2$

FINDING THE LENGTH OF THE HYPOTENUSE

If we are given the lengths of the two shorter sides, it now becomes a simple calculation to work out the length of the hypotenuse.

(i) Form an equation using Pythagoras' Theorem.
(ii) Substitute in the lengths of the 2 shorter sides.
(iii) Complete the calculation by taking the square root of both sides.

Example 1: Find the length of BC in the diagram.

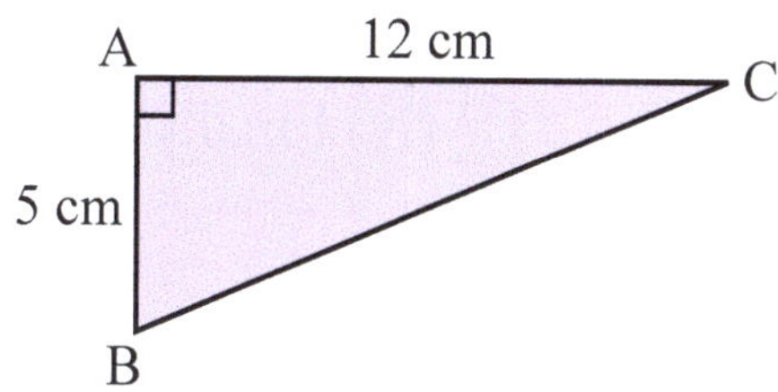

Solution: From Pythagoras' Theorem, the hypotenuse squared is equal to the sum of the squares of the 2 other sides.

$BC^2 = AB^2 + AC^2$	Form an equation
$\therefore\ BC^2 = 5^2 + 12^2$	Substitute the lengths of the shorter sides
$\therefore\ BC^2 = 25 + 144$	
$\therefore\ BC^2 = 169$	
$\therefore\ BC = \sqrt{169} = 13$ cm	Take the square root of both sides.

It is very important in tests to set your work out logically, step by step, as shown above.

Example 2: Find the value of x correct to 2 decimal places.

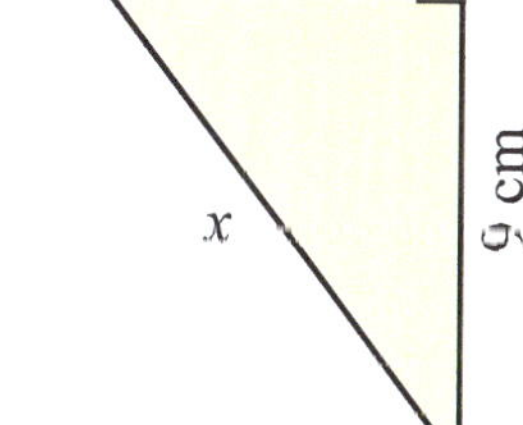

Solution:

$x^2 = 7^2 + 9^2$	Pythagoras' Theorem
$\therefore\ x^2 = 49 + 81$	
$\therefore\ x^2 = 130$	
$\therefore\ x = \sqrt{130}$	
$= 11.401745$	From calculator
$= 11.40$ to 2 d.p.	

Note: Often calculators will have to be used to square numbers, and also to find the square roots of numbers. The question should always state the accuracy required.

FINDING ONE OF THE SHORTER SIDES

So far, you have learned how to use Pythagoras' Theorem to find the hypotenuse when you have been given the lengths of the two shorter sides. It is a fairly simple procedure, because one can apply the theorem or formula directly.

However, you will also be asked to find the length of one of the shorter sides, and this requires an extra step which is important to understand.

STEPS:

(i) Use Pythagoras' Theorem to form an equation.
(ii) Substitute in the 2 given lengths.
(iii) A subtraction will now have to be done.
(iv) Finally take the square root of both sides.

The two examples given should clearly show you how these problems are worked out.

Example 1: Find the length of x correct to 2 significant figures (s.f.).

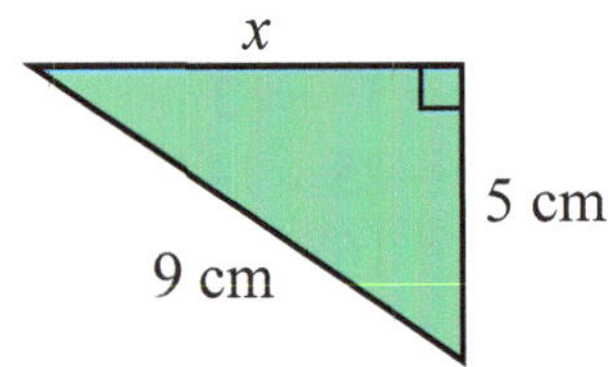

Solution:

$9^2 = 5^2 + x^2$	Pythagoras' Theorem
$\therefore \quad 81^2 = 25 + x^2$	Substitute in the 2 given lengths
$\therefore \quad 81 - 25 = x^2$	Subtract 25 from both sides
$\therefore \quad x^2 = 56$	so that x^2 is by itself
$\therefore \quad x = \sqrt{56}$	Finally take the square
$= 7.483\ 3148$	root of both sides.
$= 7.5$ to 2 s.f.	

If the question asked for the exact value of x, then the answer would be left in surd form i.e. $x = \sqrt{56}$

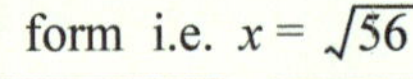

Example 2: Find the length of QR correct to 3 significant figures.

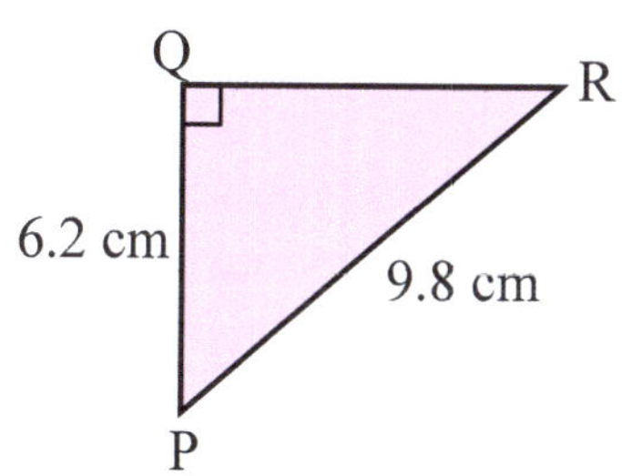

Solution:

	$PR^2 = QP^2 + QR^2$	Form an equation
$\therefore$	$9.8^2 = 6.2^2 + QR^2$	Substitute the 2 given lengths
$\therefore$	$96.04 = 38.44 + QR^2$	Use a calculator for the squares
$\therefore$	$96.04 - 38.44 = QR^2$	A subtraction must be done
$\therefore$	$QR^2 = 57.6$	
	$QR = \sqrt{57.6}$	Take the square root of both sides.
	$= 7.589\ 4664$	
	$= 7.59$ to 3 s.f.	

Most problems will require the use of a calculator, and the final answer will have to be rounded off to a given number of decimal places or significant figures. However there are a few famous Pythagorean Triads which only involve whole numbers. These should be remembered because they often occur in test questions.

There have been several formulas developed to discover Pythagorean Triads with sides of lengths a, b and c. If one number in the triad is given as a, then the other 2 numbers are $c = \frac{1}{2}(a^2 + 1)$ and $b = \frac{1}{2}(a^2 - 1)$.

The most well known Pythagorean triad is the (3, 4, 5) triangle, but all multiples of this will also form triads (6, 8, 10), (9, 12,15) etc.

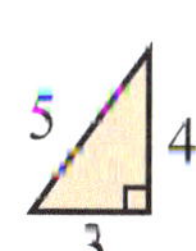

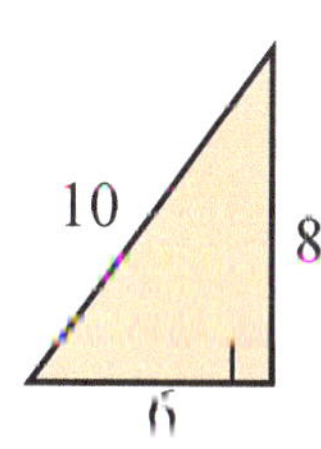

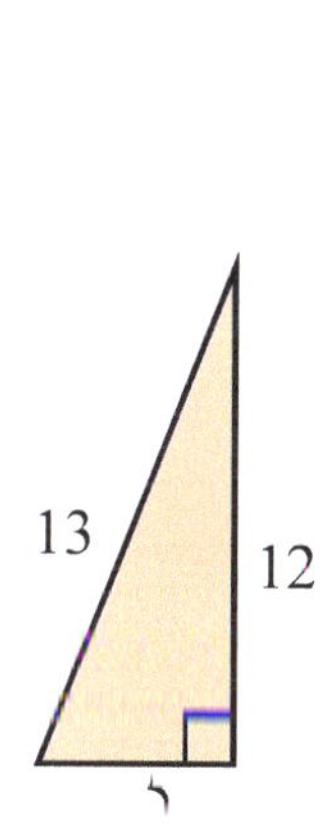

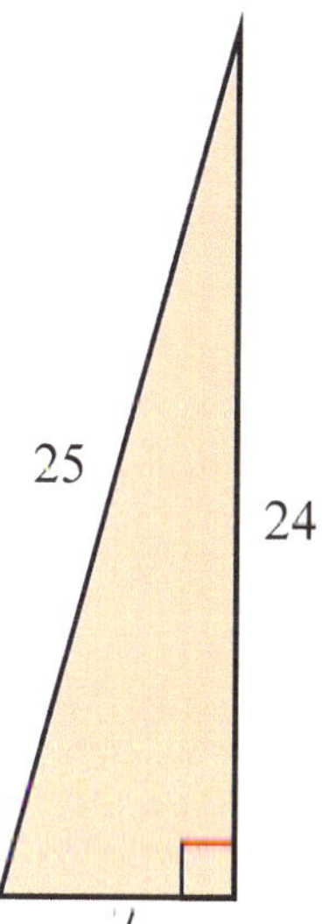

i.e. $5^2 = 3^2 + 4^2$ $\quad 10^2 = 6^2 + 8^2$ $\quad 13^2 = 5^2 + 12^2$ $\quad 25^2 = 7^2 + 24^2$

APPLICATIONS OF PYTHAGORAS' THEOREM

In some questions, you will be given the diagram and asked to find the length of one of the sides, as shown in all the previous examples. However, in other questions you will only be given a description of some practical application which occurs in everyday living.

> (i) Firstly convert the worded description into a diagram involving a right angled triangle.
> (ii) Then use Pythagoras' Theorem to calculate the length required.

Example: The foot of a ladder is 2 m from the base of a vertical wall. If the ladder reaches 6 m up the wall, find its length to the nearest cm.

Solution: Firstly convert the worded description into a diagram involving a right angled triangle.

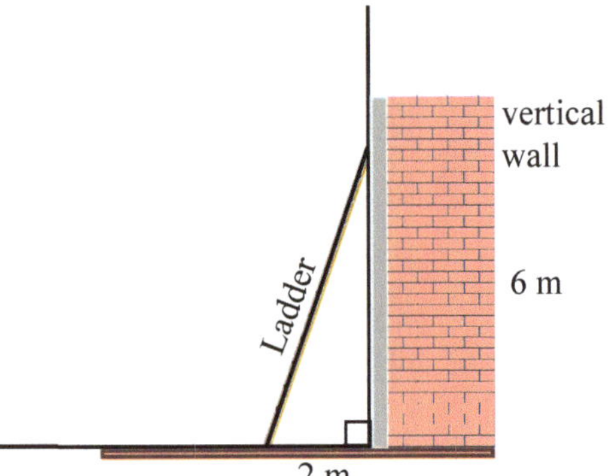

Let the unknown length of the ladder be x

$\therefore \quad x^2 = 6^2 + 2^2$

$\therefore \quad x^2 = 36 + 4$

$\therefore \quad x^2 = 40$

$\therefore \quad x^2 = \sqrt{40}$

$\quad\quad = 6.324\ 5553$

$\quad\quad = 6.32$ m

i.e. The length of the ladder is 6 m 32 cm.

Pythagoras' Theorem will also be required in other major topics such as trigonometry and coordinate geometry.

TESTING FOR RIGHT-ANGLED TRIANGLES

The converse of Pythagoras' Theorem is also true. If the 3 sides of a triangle are such that $c^2 = a^2 + b^2$, then the triangle is right-angled.

Example: Test if ΔPQR is right-angled

$2.5^2 = 6.25$

$1.5^2 + 2^2 = 6.25$

$\therefore \quad PR^2 = PQ^2 + RQ^2$

$\therefore \quad \Delta PQR$ is right-angled.

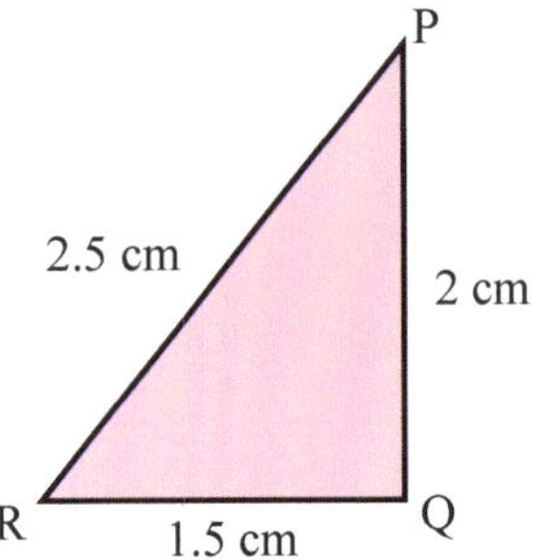

SURDS ARE IRRATIONAL NUMBERS

A **RATIONAL** number is any number that can be expressed as a fraction $\frac{p}{q}$ or as a decimal.

Examples: $\frac{3}{4}$, 5, $7\frac{1}{2}$, $-3\frac{1}{6}$, 4.3, 0.0056, $0.\dot{3}$. $1.\dot{4}\dot{7}$ etc.

Surds are **IRRATIONAL** numbers which contain square roots or cube roots, etc.

Examples: $\sqrt{2}$, $\sqrt{3}$, $4\sqrt{5}$, $-\sqrt{7}$, $\sqrt{101}$, $\sqrt[3]{10}$, $\sqrt[3]{20}$ etc.

An exact fraction or decimal equivalent cannot be obtained, and therefore many questions involving square roots require final answers to be left in their exact surd form.

Just as the set of **RATIONALS** is infinite, so too is the set of **IRRATIONAL** numbers. The set of rational numbers combined with the set of irrational numbers make up our complete number system called the set of **REAL** numbers.

SET OF REAL NUMBERS

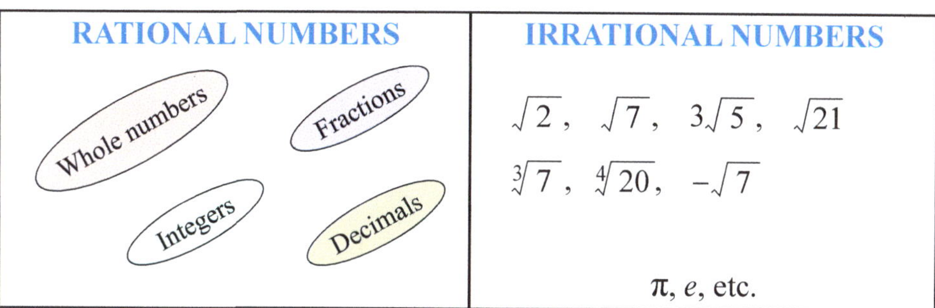

Although a surd doesn't have an exact value, nevertheless its position can be accurately plotted on a number line.

To plot $\sqrt{2}$ on a number line, draw a right angled triangle OAB so that OB = AB = 1 unit.

From Pythagoras' Theorem

$$OA = \sqrt{2}$$

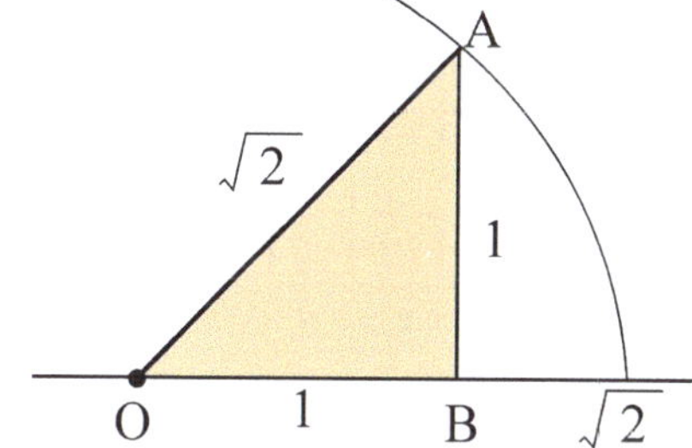

Now take a compass, with centre O and radius OA = $\sqrt{2}$ and draw an arc so that it cuts the number line as shown.

As you can see it cuts the number line at exactly $\sqrt{2}$.

Note: Other triangles can be constructed in a similar manner to plot $\sqrt{3}$, $\sqrt{5}$, $\sqrt{6}$ etc. on the number line.

MULTIPLICATION AND DIVISION OF SURDS

As with any new number system, it is important to understand and learn the rules for adding, subtracting, multiplying, dividing and simplifying surds.

The three rules for multiplication and division are very easy:

1. $\sqrt{a} \times \sqrt{b} = \sqrt{a \times b}$
2. $(\sqrt{a})^2 = a$
3. $\dfrac{\sqrt{a}}{\sqrt{b}} = \sqrt{\dfrac{a}{b}}$

Once again, it is important to understand each example shown below.

Examples: Simplify the following:

(i) $\sqrt{5} \times \sqrt{2} = \sqrt{5 \times 2} = \sqrt{10}$ (Rule 1)

(ii) $\sqrt{7} \times \sqrt{3} = \sqrt{7 \times 3} = \sqrt{21}$ (Rule 1)

(iii) $3\sqrt{5} \times 2\sqrt{7} = 6\sqrt{5 \times 7} = 6\sqrt{35}$ (Rule 1)

The whole numbers 3 and 2 are multiplied in the usual way to obtain the whole number 6.

(iv) $(\sqrt{5})^2 = 5$ (Rule 2)

(v) $(3\sqrt{2})^2 = 9 \times 2 = 18$ (Rule 2)

because $3^2 = 9$ and $(\sqrt{2})^2 = 2$

(vi) $\dfrac{\sqrt{10}}{\sqrt{5}} = \sqrt{\dfrac{10}{5}} = \sqrt{2}$ (Rule 3)

(vii) $\dfrac{\sqrt{20}}{\sqrt{5}} = \sqrt{\dfrac{20}{5}} = \sqrt{4} = 2$ (Rule 3)

(viii) $\dfrac{8\sqrt{15}}{2\sqrt{5}} = 4\sqrt{\dfrac{15}{5}} = 4\sqrt{3}$ (Rule 3)

Note: In all these sections relating to surds, the square roots of perfect squares should always be written as whole numbers.

i.e. $\sqrt{4} = 2$, $\sqrt{9} = 3$, $\sqrt{16} = 4$ etc.

SIMPLIFYING SURDS

A surd is expressed in its simplest form when the number under the square root sign is as small as possible. You should always look for a pair of factors, one of which is a **PERFECT SQUARE**.

i.e. 4, 9, 16, 25, 36, 49, 64, 81, 100, etc.

Then use the reverse of the multiplication rule shown on the opposite page:

$$\sqrt{a \times b} = \sqrt{a} \times \sqrt{b}$$

Example 1: Simplify $\sqrt{8}$

Look for a pair of factors of 8, which includes a perfect square.

$\therefore \quad \sqrt{8} = \sqrt{4} \times \sqrt{2}$

$= 2\sqrt{2}$

Example 2: Simplify $\sqrt{75}$

Look for pair of factors of 75, which includes a perfect square.

$\therefore \quad \sqrt{75} = \sqrt{25} \times \sqrt{3}$

$= 5\sqrt{3}$

Example 3: Simplify $2\sqrt{45}$

Look for a pair of factors of 45, which includes a perfect square.

$\therefore \quad 2\sqrt{45} = 2 \times \sqrt{9} \times \sqrt{5}$

$= 2 \times 3 \times \sqrt{5}$

$= 6\sqrt{5}$

Example 4: Write $3\sqrt{7}$ as an entire surd.

This follows the opposite steps to the examples above.

The whole number 3 written as a square root is equal to $\sqrt{9}$.

$3\sqrt{7} = \sqrt{9} \times \sqrt{7}$

$= \sqrt{63}$

ADDING AND SUBTRACTING SURDS

In algebra, one can only add or subtract 'like terms'.

$$5a + 3a - 4b = 8a - 4b$$

The same applies to surds:

$$5\sqrt{2} + 3\sqrt{2} - 4\sqrt{3} = 8\sqrt{2} - 4\sqrt{3}$$

> One can only ADD or SUBTRACT LIKE SURDS.

Therefore $\sqrt{8} + \sqrt{2}$ does NOT equal $\sqrt{10}$.

However $\sqrt{8}$ can be simplified to $2\sqrt{2}$ using the theory from the previous page.

$$\therefore \quad \sqrt{8} + \sqrt{2} = 2\sqrt{2} + \sqrt{2} = 3\sqrt{2}$$

Examples: Simplify the following expressions:

(i) $5\sqrt{2} + 7\sqrt{2} - 4\sqrt{3}$

$= 12\sqrt{2} - 4\sqrt{3}$

(ii) $7\sqrt{2} - \sqrt{32}$

$= 7\sqrt{2} - 4\sqrt{2}$

$= 3\sqrt{2}$

(iii) $5\sqrt{2} - 4\sqrt{3} + \sqrt{18}$

$= 5\sqrt{2} - 4\sqrt{3} + 3\sqrt{2}$

$= 8\sqrt{2} - 4\sqrt{3}$

(iv) $6\sqrt{a} - 2\sqrt{b} + \sqrt{81a} + \sqrt{36b}$

$= 6\sqrt{a} - 2\sqrt{b} + 9\sqrt{a} + 6\sqrt{b}$

$= 15\sqrt{a} + 4\sqrt{b}$

EXPANDING BRACKETS

When expanding brackets (or removing brackets), the same rules which are used in algebra apply in exactly the same way to surds.

IN ALGEBRA:

$$5(2a + 3) = 5 \times 2a + 5 \times 3$$
$$= 10a + 15$$

IN SURDS:

$$5(\sqrt{3} + 2) = 5 \times \sqrt{3} + 5 \times 2$$
$$= 5\sqrt{3} + 10$$

$$a(b + c) = a \times b + a \times c = ab + ac$$

$$a(b - c) = a \times b - a \times c = ab - ac$$

$$(a + b)(c + d) = a(c + d) + b(c + d) = ac + ad + bc + bd$$

Examples: Remove brackets from the following:

(i) $7(\sqrt{2} - 3) = 7 \times \sqrt{2} - 7 \times 3 = 7\sqrt{2} - 21$

(ii) $\sqrt{3}(\sqrt{2} + 7) = \sqrt{3} \times \sqrt{2} + \sqrt{3} \times 7 = \sqrt{6} + 7\sqrt{3}$

(iii) $\sqrt{5}(\sqrt{5} - 2) = (\sqrt{5})^2 - \sqrt{5} \times 2 = 5 - 2\sqrt{5}$

(iv) $2\sqrt{3}(4\sqrt{7} + 4) = 2\sqrt{3} \times 4\sqrt{7} + 2\sqrt{3} \times 4 = 8\sqrt{21} + 8\sqrt{3}$

(v) $(5 + \sqrt{3})(7 - \sqrt{2}) = 5(7 - \sqrt{2}) + \sqrt{3}(7 - \sqrt{2})$
$= 35 - 5\sqrt{2} + 7\sqrt{3} - \sqrt{6}$

(vi) $(\sqrt{6} - \sqrt{2})(\sqrt{5} - \sqrt{2}) = \sqrt{6}(\sqrt{5} - \sqrt{2}) - \sqrt{2}(\sqrt{5} - \sqrt{2})$
$= \sqrt{30} - \sqrt{12} - \sqrt{10} + \sqrt{4}$
$= \sqrt{30} - 2\sqrt{3} - \sqrt{10} + 2$

SPECIAL BINOMIAL PRODUCTS

There is a faster method of multiplying out surds which contain perfect squares $(a \pm b)^2$ or the pattern $(a + b)(a - b)$. The same rules used in algebra apply equally to the removal of brackets containing surds.

$$(a+b)^2 = a^2 + 2ab + b^2$$
$$(a-b)^2 = a^2 - 2ab + b^2$$
$$(a+b)(a-b) = a^2 - b^2$$

If you forget the above formulae, or don't feel confident using them, simply multiply out the brackets using the longer method shown in the previous section. However, it is particularly important to be able to apply the 'difference of two squares rule':
$(a + b)(a - b) = a^2 - b^2$.

Example 1: Use the quicker rules shown above to remove brackets and simplify where possible:

(i) $(3+\sqrt{2})^2 = 3^2 + 2\times 3\times\sqrt{2} + (\sqrt{2})^2$
$= 9 + 6\sqrt{2} + 2 = 11 + 6\sqrt{2}$

(ii) $(\sqrt{5}-\sqrt{2})^2 = (\sqrt{5})^2 - 2\times\sqrt{5}\times\sqrt{2} + (\sqrt{2})^2$
$= 5 - 2\sqrt{10} + 2 = 7 - 2\sqrt{10}$

(iii) $(6+\sqrt{7})(6-\sqrt{7}) = 6^2 - (\sqrt{7})^2 = 36 - 7 = 29$

(iv) $(3\sqrt{5}+2\sqrt{7})(3\sqrt{5}-2\sqrt{7}) = (3\sqrt{5})^2 - (2\sqrt{7})^2 = 45 - 28 = 17$

Important Note:

It is very important to note that the product of a surd and its conjugate surd always gives a rational (whole number) answer.

$(6+\sqrt{7})$ and $(6-\sqrt{7})$ are called conjugate surds.
$(3\sqrt{5}+2\sqrt{7})$ and $(3\sqrt{5}-2\sqrt{7})$ are called conjugate surds.
With conjugate surds, the brackets are identical except that the signs in the two brackets are opposite.

RATIONALISING DENOMINATORS

If a fraction contains a surd (i.e. an irrational number) in the denominator, then it must be rewritten so that the denominator only contains a rational number (i.e. whole number). We call this process **'RATIONALISING THE DENOMINATOR'**.

METHOD:

> To rationalise the denominator of $\frac{A}{B\sqrt{x}}$ then
>
> multiply by $\frac{\sqrt{x}}{\sqrt{x}}$ which equals 1.
>
> i.e. Multiply both the numerator and denominator by the surd part in the denominator.

Example 1: Rationalise the denominator in the following and simplify where possible:

(i) $\frac{1}{\sqrt{3}} = \frac{1}{\sqrt{3}} \times \frac{\sqrt{3}}{\sqrt{3}} = \frac{\sqrt{3}}{3}$

(ii) $\frac{6}{\sqrt{2}} = \frac{6}{\sqrt{2}} \times \frac{\sqrt{2}}{\sqrt{2}} = \frac{6\sqrt{2}}{2} = 3\sqrt{2}$

(iii) $\frac{\sqrt{3}}{\sqrt{5}} = \frac{\sqrt{3}}{\sqrt{5}} \times \frac{\sqrt{5}}{\sqrt{5}} = \frac{\sqrt{15}}{5}$

Whole number fractions are simplified by cancelling down in the usual way.

(iv) 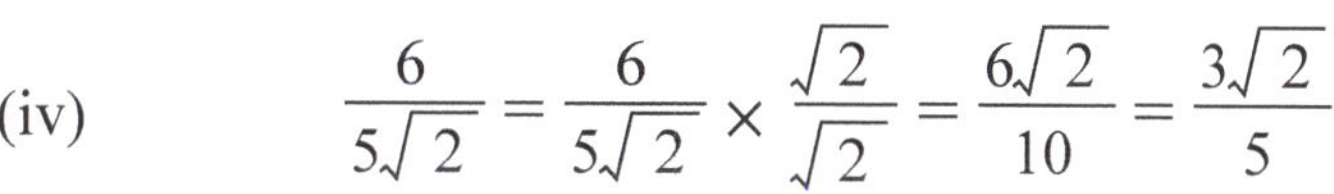

$\frac{6}{5\sqrt{2}} = \frac{6}{5\sqrt{2}} \times \frac{\sqrt{2}}{\sqrt{2}} = \frac{6\sqrt{2}}{10} = \frac{3\sqrt{2}}{5}$

(v) 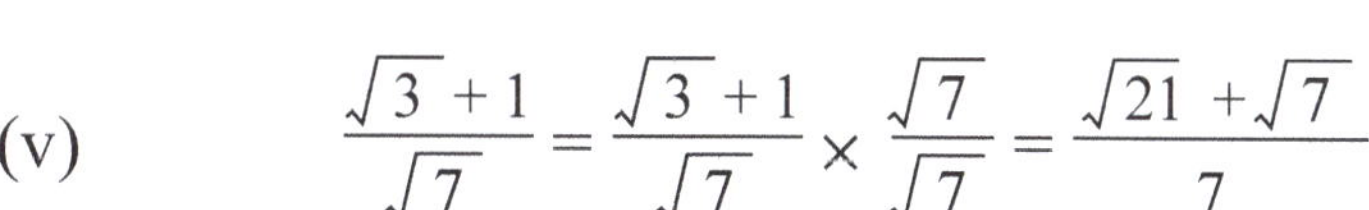

$\frac{\sqrt{3}+1}{\sqrt{7}} = \frac{\sqrt{3}+1}{\sqrt{7}} \times \frac{\sqrt{7}}{\sqrt{7}} = \frac{\sqrt{21}+\sqrt{7}}{7}$

(vi) $\frac{2\sqrt{3}}{\sqrt{6}} = \frac{2\sqrt{3}}{\sqrt{6}} \times \frac{\sqrt{6}}{\sqrt{6}} = \frac{2\sqrt{18}}{6} = \frac{6\sqrt{2}}{6} = \sqrt{2}$

The numerator $2\sqrt{18} = 2 \times \sqrt{9} \times \sqrt{2} = 2 \times 3 \times \sqrt{2} = 6\sqrt{2}$.

Note: Whole number fractions are simplified by cancelling down in the usual way.

RATIONALISING BINOMIAL DENOMINATORS

You have seen in the previous section how to rationalise a surd containing one term in the denominator. This page explains how to rationalise denominators which contain two terms. From the bottom of page 144 it was shown that if conjugate surds are multiplied, then the answer will always be a rational number.

REMEMBER!
A conjugate surd is obtained by reversing the sign in the brackets.

The conjugate surd of $3 - \sqrt{2}$ is $3 + \sqrt{2}$.

The conjugate surd of $\sqrt{5} + \sqrt{3}$ is $\sqrt{5} - \sqrt{3}$.

(i) Multiply both the numerator and denominator by the conjugate of the denominator.

(ii) Use the special product $(a + b)(a - b) = a^2 - b^2$ to multiply out the denominator.

(iii) Simplify where possible.

Examples: Rationalise the denominators in the expressions below:

(i) $\frac{5}{3-\sqrt{2}} = \frac{5}{(3-\sqrt{2})} \times \frac{(3+\sqrt{2})}{(3+\sqrt{2})}$ Multiply by the conjugate surd.

$= \frac{5(3+\sqrt{2})}{7}$ $(3-\sqrt{2})(3+\sqrt{2}) = 3^2 - (\sqrt{2})^2 = 7$

$= \frac{15+5\sqrt{2}}{7}$ Multiply out brackets.

(ii) $\frac{\sqrt{5}-\sqrt{3}}{\sqrt{5}+\sqrt{3}} = \frac{(\sqrt{5}-\sqrt{3})}{(\sqrt{5}+\sqrt{3})} \times \frac{(\sqrt{5}-\sqrt{3})}{(\sqrt{5}-\sqrt{3})}$ Multiply by the conjugate surd.

$= \frac{(\sqrt{5}-\sqrt{3})^2}{2}$ $(\sqrt{5}+\sqrt{3})(\sqrt{5}-\sqrt{3}) = (\sqrt{5})^2 - (\sqrt{3})^2 = 2$

$= \frac{8-2\sqrt{15}}{2}$ $(\sqrt{5}-\sqrt{3})^2 = (\sqrt{5})^2 - 2\times\sqrt{5}\times\sqrt{3} + (\sqrt{3})^2$

$= 5 - 2\sqrt{15} + 3$

$= 8 - 2\sqrt{15}$

$= \frac{2(4-\sqrt{15})}{2}$

$= 4 - \sqrt{15}$ Factorise and cancel.

Note: Students usually find this a more difficult section, because it involves a thorough understanding of all the different ideas covered in this chapter.

SOME HARDER APPLICATIONS

Example 1: Simplify (i) $5x\sqrt{x} - \sqrt{x^3}$ (ii) $\sqrt[3]{54}$

(i) $5x\sqrt{x} - \sqrt{x^3}$

$= 5x\sqrt{x} - x\sqrt{x}$

$= 4x\sqrt{x}$

$\sqrt{x^3} = \sqrt{x^2}\sqrt{x}$

$= x\sqrt{x}$

(ii) $\sqrt[3]{54}$

$= \sqrt[3]{27} \times \sqrt[3]{2}$

$= 3\sqrt[3]{2}$

With cubic questions, we must find a perfect cube root factor.

Example 2: Rationalise the denominator in $\dfrac{5\sqrt{2}}{3\sqrt{6} - 4\sqrt{2}}$

$\dfrac{5\sqrt{2}}{(3\sqrt{6} - 4\sqrt{2})} \times \dfrac{(3\sqrt{6} + 4\sqrt{2})}{(3\sqrt{6} + 4\sqrt{2})}$

$= \dfrac{15\sqrt{12} + 20\sqrt{4}}{22}$

$= \dfrac{30\sqrt{3} + 40}{22}$

$= \dfrac{15\sqrt{3} + 20}{11}$

Multiply by conjugate surd.

$(3\sqrt{6} - 4\sqrt{2})(3\sqrt{6} + 4\sqrt{2})$

$= (3\sqrt{6})^2 - (4\sqrt{2})^2 = 54 - 32 = 22$

$\sqrt{12} = \sqrt{4} \times \sqrt{3} = 2\sqrt{3}$

Dividing numerator and denominator by 2.

Example 3: AC is the arc of a circle with centre O. If OB = 6 cm and AB = 2 cm, find the exact length of BC.

$OA^2 = 6^2 + 2^2$

$\therefore \quad OA^2 = 40$

$\therefore \quad OA = \sqrt{40} = 2\sqrt{10}$

$\therefore \quad OC = 2\sqrt{10}$

$\therefore \quad BC = OC - OB$

$= (2\sqrt{10} - 6)$ cm

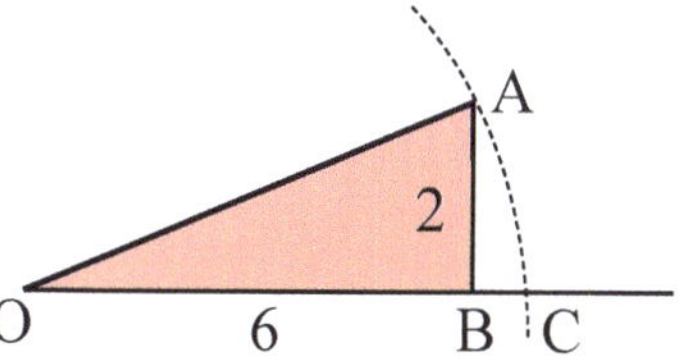

Pythagoras' Theorem

Example 4: Show that $\dfrac{1}{3 - \sqrt{2}} + \dfrac{1}{3 + \sqrt{2}}$ is a rational number.

$= \dfrac{1(3 + \sqrt{2}) + 1(3 - \sqrt{2})}{(3 - \sqrt{2})(3 + \sqrt{2})}$

$= \dfrac{3 + \sqrt{2} + 3 - \sqrt{2}}{7}$

$= \dfrac{6}{7}$

Find a common denominator.

$(3 - \sqrt{2})(3 + \sqrt{2}) = 3^2 - (\sqrt{2})^2 = 7$

This fraction is a rational number.

THE BASIC INDEX LAWS

Most students will already know all of these laws, as they should have been taught in earlier years. Rather than memorising them, it is very much more important to know how to apply them to typical questions.

1. $a^m \times a^n = a^{m+n}$
2. $a^m \div a^n = a^{m-n}$
3. $(a^m)^n = a^{mn}$
4. $(ab)^m = a^m b^m$
5. $a^0 = 1$

Examples:

(i)	$a^6 \times a^2$	$= a^{6+2}$	$=$	a^8	(law 1)
(ii)	$(a^2)^3$	$= a^{2\times 3}$	$=$	a^6	(law 3)
(iii)	$3m^0$	$= 3 \times 1$	$=$	3	(law 5)
(iv)	$(3m)^0$	$= 3^0 \times m^0$	$=$	1	(law 4, then law 5)
(v)	$a^{10} \div a^2$	$= a^{10-2}$	$=$	a^8	(law 2)
(vi)	$(3a^4)^2$	$= 3^2(a^4)^2$	$=$	$9a^8$	(law 4, then law 3)
(vii)	$\dfrac{12x^8}{4x^2}$	$= 3x^{8-2}$	$=$	$3x^6$	(law 2)
(viii)	$5m^2 \times 4m^3$	$= 20m^{2+3}$	$=$	$20m^5$	(law 1)
(ix)	$3^{m+2} \div 3^m = 3^{(m+2)-m}$	$= 3^2$	$=$	9	(law 2)
(x)	$\dfrac{(a+b)^2}{a+b}$	$= (a+b)^{2-1}$	$=$	$a+b$	(law 2)
(xi)	$\dfrac{6m^3}{8m^{-2}}$	$= \dfrac{3m^{3-(-2)}}{4}$	$=$	$\dfrac{3m^5}{4}$	(law 2)

(xii) $4^{x+1} \times 2^{3x-5} = 2^{2(x+1)} \times 2^{3x-5} = 2^{5x-3}$ (Must first get the same base of 2)

FRACTIONAL AND NEGATIVE INDICES

The 5 laws in this section are harder, but they are important to understand, because most test and exams will contain questions which require their application.

6.	$a^{\frac{1}{2}}$	=	$\sqrt{a}$
7.	$a^{\frac{1}{3}}$	=	$\sqrt[3]{a}$
8.	$a^{\frac{p}{q}}$	=	$\sqrt[q]{a^p}$
9.	a^{-n}	=	$\frac{1}{a^n}$
10.	$(\frac{a}{b})^{-1}$	=	$\frac{b}{a}$

$8^{\frac{1}{2}}$

The student is expected to know how to change from 'index notation' to 'surd notation' and vice versa.

Examples:

(i) $81^{\frac{1}{2}} = \sqrt{81} = 9$ (law 6)

(ii) $8^{\frac{1}{3}} = \sqrt[3]{8} = 2$ (law 7)

(iii) $64^{\frac{2}{3}} = \sqrt[3]{64^2} = 16$ (law 8)

(iv) $3^{-2} = \frac{1}{3^2} = \frac{1}{9}$ (law 9)

(v) $10^{-3} = \frac{1}{10^3} = \frac{1}{1\ 000}$ (law 9)

(vi) $16^{-\frac{3}{4}} = \frac{1}{16^{\frac{3}{4}}} = \frac{1}{\sqrt[4]{16^3}} = \frac{1}{8}$ (law 9 then law 8)

(vii) $\left(\frac{3}{4}\right)^{-1} = \frac{4}{3}$ law 10

(viii) $25^{2x+1} \div 5^{4x-1} = 5^{2(2x+1)-(4x-1)} = 5^3 = 125$ (Must first get the same base of 5)

Note: With negative fractions, the best method is to first change the negative fraction to a positive fraction by using law 9. Then apply law 8 to the positive fractional index in the denominator.

SCIENTIFIC NOTATION FOR LARGE NUMBERS

When a number is written as a number between 1 and 10, multiplied by a power of 10 – then it is said to be written in **SCIENTIFIC** or **STANDARD NOTATION**.

This is a very useful and quicker way of writing down very large or very small numbers.

Example 1: Write 7 240 000 in scientific notation.

7 240 000 = 7.24×10^6

7 240 000.

Example 2: Write 6.91×10^5 as an ordinary numeral.

6.91×10^5 = 691 000

6.91000

This is the reverse process to the above example. The decimal point must be moved 5 places over to the right as shown.

CALCULATORS

There are several different ways of implementing scientific notation on your calculator. Please note that the steps shown below may vary depending on the make and model of your particular calculator.

To enter 7.3×10^4 on the calculator:

Method i) 7.3 [$\times 10^x$] 4 [7.3×10^4] [=] [73 000]

Method ii) 7.3 [SHIFT] [10^x] 4 [7.3×10^4] [=] [73 000]

Note: The [$\times 10^x$] white botton on the bottom row of the calculator is different to the [10^x] yellow button near the top row of the calculator.

Calculators will also give the answer to a problem in scientific notation if the answer is very large (i.e. $> 10^{10}$).

If you wish all the answers to be displayed in scientific notation, then you can switch the calculator onto scientific mode using the following steps:

[SHIFT] [MODE] 7 [Sci 0 ~ 9] 4

This is asking for the number of significant figures required.

All answers will now be given in scientific notation correct to 4 significant figures.

SCIENTIFIC NOTATION FOR SMALL NUMBERS

These follow the same ideas as for larger numbers, except that the number 10 will be raised to a **NEGATIVE POWER**.

Example 1: Write 0.000 000 0957 in scientific notation.

0.000 000 0957 = 9.57×10^{-8}

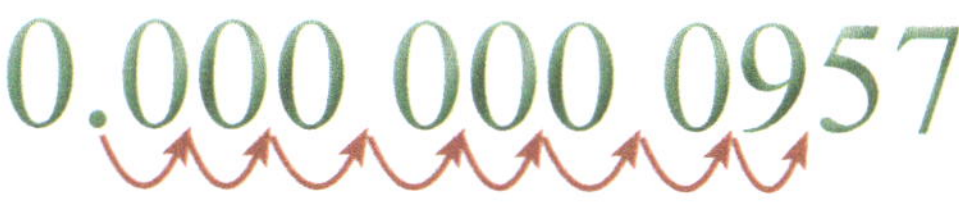

The decimal point must be moved 8 places to the right so that it comes between the 9 and 5 as shown.

Example 2: Write 6.28×10^{-4} as an ordinary number.

6.28×10^{-4} = 0.000 628

This is the reverse process to the example above.
The decimal point must be moved 4 places to the left as shown.

> (i) The number must always be written as a number between 1 and 10.
>
> (ii) You must be able to convert from 'ordinary numerals' to 'scientific notation' and from 'scientific notation' to 'ordinary numerals'.

CALCULATORS:

To enter 3.87×10^{-5} on the calculator:

Method i) 3.87 [$\times 10^x$] [(–)] 5 [=] [0.0000 387]

Method ii) 3.87 [SHIFT] [10^x] [(–)] 5 [=] [0.0000 387]

Calculators will also automatically give the answer to a problem in scientific notation if the answer is very small.

Example: Evaluate $3 \times 10^{-6} \times 7.82 \times 10^{-7}$

3 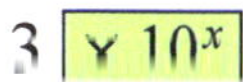 [$\times 10^x$]  [(–)] 6 [$\times$] 7.82  [$\times 10^x$]  [(–)]  7 [=] [2.346×10^{-12}]

CHAPTER SUMMARY

PYTHAGORAS' THEOREM

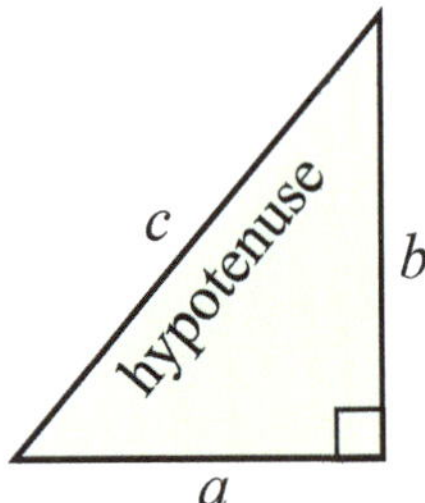

In any right angled triangle, the square of the longest side is equal to the sum of the squares of the 2 smaller sides. From the diagram:

$$c^2 = a^2 + b^2$$

The longest side which always faces directly opposite the right angle is called HYPOTENUSE.

FINDING THE LENGTH OF THE HYPOTENUSE

If we are given the lengths of the two shorter sides, it now becomes a simple calculation to work out the length of the hypotenuse.

Step 1: Form an equation using Pythagoras' Theorem.
Step 2: Substitute in the lengths of the two shorter sides.
Step 3: Complete the calculation by taking the square root of each side.

Note: If we are calculating the length of one of the shorter sides, then an extra subtraction step is required in the solution of the equation. Unless the exact answer is required, most problems will require the use of a calculator and the final answer will have to be rounded off to a given number of decimal places or significant figures.

TESTING FOR RIGHT-ANGLED TRIANGLES

The converse or reverse of Pythagoras' Theorem also holds true. If the 3 sides of a triangle are such that $c^2 = a^2 + b^2$, then the triangle is right-angled.

SOME FAMOUS PYTHAGOREAN TRIADS

Most problems will require the use of a calculator, and the final answer will have to be rounded off to a given number of decimal places. However, there are some famous Pythagorean Triads which involve only whole numbers. These are useful to memorise because they often occur in test questions.

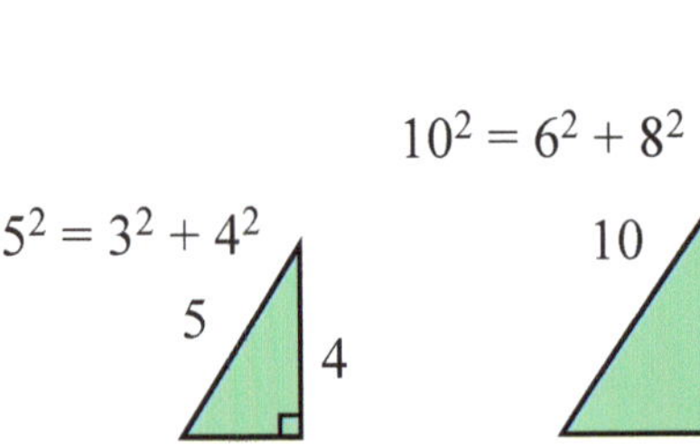

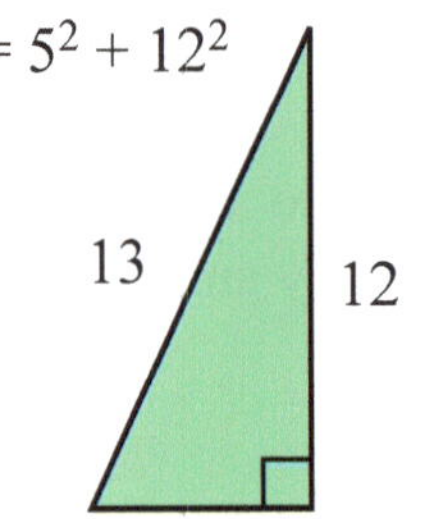

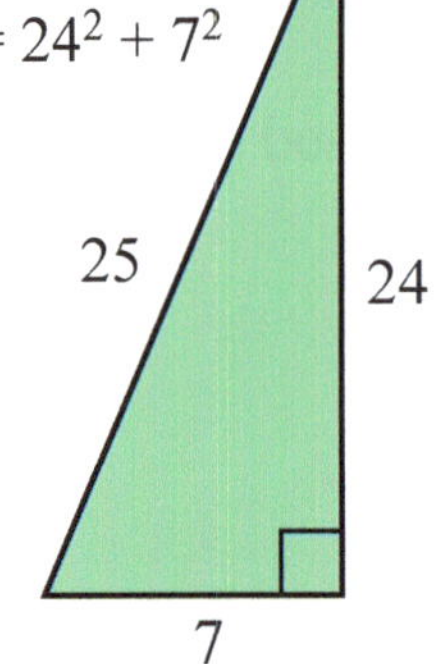

Special triads can be found using a formula. If one number in the triad is 'a', then the other 2 numbers are $c = \frac{1}{2}(a^2 + 1)$ and $b = \frac{1}{2}(a^2 - 1)$.

WHAT ARE SURDS?

A **RATIONAL** number is any number that can be expressed as a fraction $\frac{p}{q}$ or as a decimal.

Examples: $\frac{3}{4}, 5, 7\frac{1}{2}, -3\frac{1}{6}, 4.3, 0.0056, 0.3, \sqrt{4}, 0.\dot{6}, 5.\dot{7}\dot{9}$ etc.

Surds are **IRRATIONAL** numbers which contain square roots or cube roots.

Examples: $\sqrt{2}, \sqrt{3}, 4\sqrt{5}, -\sqrt{7}, \sqrt{101}, \sqrt[3]{10}, \sqrt[3]{20}$ etc.

An exact fraction or decimal equivalent cannot be obtained, and therefore many questions involving square roots require final answers to be left in their exact surd form (an example is the use of Pythagoras' theorem).

Just as the set of **RATIONALS** is infinite, so too is the set of **IRRATIONAL** numbers. The set of rational numbers combined with the set of irrational numbers make up our complete number system called the set of **REAL NUMBERS**.

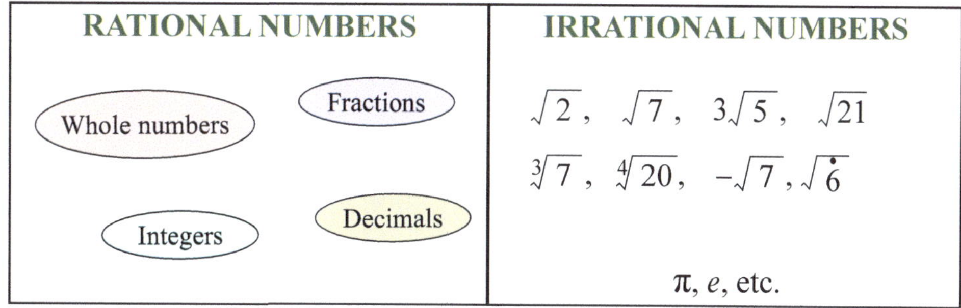

Note: Although an exact decimal equivalent of a surd cannot be obtained, their exact position can be plotted on a number line with the aid of Pythagoras' Theorem.

MULTIPLYING AND DIVIDING SURDS

The following three important properties need to be known:

1. $\sqrt{ab} = \sqrt{a} \times \sqrt{b}$ $\qquad \sqrt{8} = \sqrt{4 \times 2} = \sqrt{4} \times \sqrt{2} = 2\sqrt{2}$
2. $(\sqrt{a})^2 = a$ $\qquad (\sqrt{7})^2 = 7$
3. $\sqrt{\frac{a}{b}} = \frac{\sqrt{a}}{\sqrt{b}}$ $\qquad \sqrt{5\frac{4}{9}} = \sqrt{\frac{49}{9}} = \frac{\sqrt{49}}{\sqrt{9}} = \frac{7}{3}$

ADDING AND SUBTRACTING SURDS

Like any number system, it is essential to understand how to apply the four major operations (+, −, ×, ÷). Similar to algebra, one can only add or subtract **LIKE SURDS**.

Examples:

$3\sqrt{5} - \sqrt{10} + \sqrt{5}$
$= 4\sqrt{5} - \sqrt{10}$

$\sqrt{8} + 4\sqrt{3} - 5\sqrt{2}$
$= 2\sqrt{2} + 4\sqrt{3} - 5\sqrt{2}$
$= 4\sqrt{3} - 3\sqrt{2}$

SIMPLIFYING SURDS

A surd is expressed in its simplest form when the number under the square root sign is as small as possible. You should always look for a pair of factors, one of which is a **PERFECT SQUARE**.

i.e. 4, 9, 16, 25, 36, 49, 64, 81, 100, etc.

Then use the multiplication property i.e. $\sqrt{ab} = \sqrt{a} \times \sqrt{b}$

Example: $\sqrt{75} = \sqrt{25} \times \sqrt{3}$

$= 5\sqrt{3}$

EXPANDING BRACKETS AND SPECIAL BINOMIAL PRODUCTS

The removal of brackets, expansion of 'perfect squares', and 'difference of two squares' is also done using the same steps and rules as in algebra.

$(a \pm b)^2 = a \pm 2ab + b^2$

$(3 + \sqrt{5})^2 = 9 + 6\sqrt{5} + 5$

$= 14 + 6\sqrt{5}$

$(a + b)(a - b) = a^2 - b^2$

$(5 + \sqrt{2})(5 - \sqrt{2}) = 5^2 - (\sqrt{2})^2$

$= 25 - 2$

$= 23$

CONJUGATE SURDS

These are obtained by reversing the sign in the original surdic expression.

Therefore, the conjugate of $\sqrt{5} - \sqrt{3}$ is $\sqrt{5} + \sqrt{3}$.

$(\sqrt{5} - \sqrt{3})(\sqrt{5} + \sqrt{3}) = (\sqrt{5})^2 - (\sqrt{3})^2$

$= 5 - 3$

$= 2$

Important to note that a surd multiplied by its conjugate surd always gives a RATIONAL number.

RATIONALISING THE DENOMINATOR

This means to obtain a rational (integer) number in the denominator.

1. With 1 term in the denominator, multiply both top and bottom by the surd in the denominator.

$$\frac{5}{\sqrt{3}} = \frac{5}{\sqrt{3}} \times \frac{\sqrt{3}}{\sqrt{3}} = \frac{5\sqrt{3}}{3}$$

2. With 2 terms in the denominator, multiply both top and bottom by the conjugate surd in the denominator.

$$\frac{7}{6 - \sqrt{2}} = \frac{7}{6 - \sqrt{2}} \times \frac{(6 + \sqrt{2})}{(6 + \sqrt{2})}$$

$$= \frac{7(6 + \sqrt{2})}{34}$$

INDEX NOTATION

Many multiplication expressions in algebra can be shortened or made simpler by using a system called 'Index notation'. This is done by placing the **INDEX** (or exponent) on the top right hand corner of the letter, which is called the **BASE**.

For example: $a \times a = a^2$

The index or exponent is obtained by counting how many times the particular letter has been multiplied by itself.

Remember:

$a \times a = a^2$	Read as 'a squared'
$a \times a \times a = a^3$	Read as 'a cubed'
$a \times a \times a \times a = a^4$	Read as 'a to the power of 4'
$\underbrace{a \times a \times a \ldots \times a}_{n \text{ factors}} = a^n$	Read as 'a to the power of n'

THE TEN INDEX LAWS

A thorough understanding of how to apply each of the 10 index laws is required, because they will be used extensively in many later topics, such as Algebra and Logarithms.

1.	$a^m \times a^n$	=	a^{m+n}	Add powers when multiplying with the same base.
2.	$a^m \div a^n$	=	a^{m-n}	Subtract powers when dividing with the same base.
3.	$(a^m)^n$	=	a^{mn}	When a power in brackets is raised to a further power then one must multiply the indices.
4.	$(ab)^m$	=	$a^m b^m$	The power is applied to both the terms in the brackets.
5.	a^0	=	1	Any number raised to the power zero is equal to 1.
6.	a^{-n}	=	$\frac{1}{a^n}$	The base raised to a negative index is equivalent to the reciprocal of that base to the positive index.
7.	$\left(\frac{a}{b}\right)^{-1}$	=	$\frac{b}{a}$	A fraction raised to the power of –1 is the same as taking the reciprocal of the fraction.
8.	$a^{\frac{1}{2}}$	=	$\sqrt{a}$	The power of $\frac{1}{2}$ is the same as taking the square root of that base.
9.	$a^{\frac{1}{3}}$	=	$\sqrt[3]{a}$	The base raised to the power of $\frac{1}{3}$ means taking the cube root of that base.
10.	$a^{\frac{p}{q}}$	=	$\sqrt[q]{a^p}$ = $\left(\sqrt[q]{a}\right)^p$	The base raised to a fractional power is equivalent to the qth root of the base raised to the power of p, either way.

SCIENTIFIC NOTATION FOR LARGE NUMBERS

When a number is written as a number between 1 and 10, multiplied by a power of 10 — then it is said to be written in **SCIENTIFIC** or **STANDARD NOTATION**.

This is a very useful and quick way of writing down very large or very small numbers.

Students must be able to convert from 'ordinary numerals' to 'scientific notation' and vice versa.

Example: Write 7 240 000 in scientific notation.

$7\ 240\ 000 = 7.24 \times 10^6$

The decimal point must be moved 6 places to the left so it comes between the 7 and 2.

The number 10 must therefore be raised to the power of 6 as shown.

CALCULATORS:

Problems involving scientific notation can also be done on most calculators, although the process may vary depending on the make and model.

Method: i) Use $\times 10^x$ white button

ii) Use SHIFT 10^x yellow button

If you wish all the answers to be displayed in scientific notation, then you can switch the calculator onto scientific mode and obtain final answers to a given number of significant figures.

SCIENTIFIC NOTATION FOR SMALL NUMBERS

These follow the same ideas as for larger numbers, except that the number 10 will be raised to a NEGATIVE POWER.

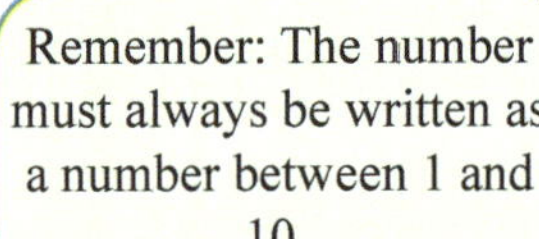

Example: Write 0.000 000 0957 in standard notation.

$0.000\ 000\ 0957 = 9.57 \times 10^{-8}$

The decimal point must be moved 8 places to the right so that it comes between the 9 and 5 .

PRACTICAL APPLICATIONS OF INDICES

Indices can be used to model excessive rates of 'growth' or 'decay' under certain favourable conditions.

Example: Rates at which mice breed; rates of growth of certain bacteria; compound interest.
Rates of decrease of radioactive substances; rates of cooling; depreciation.

 EASIER QUESTIONS

Note: Only turn back to page number shown if you have difficulty.

	Page

Q1. Find the length of the hypotenuse in each of the following: 135

a) 7.8 cm, 10.4 cm, h cm
b) 24.8 m, 18.6 m, h m
c) h cm, 20.8 cm, 15.6 cm
d) 8.4 cm, 11.2 cm, h cm

Q2. Find the value of the pronumeral: 136, 137

a) 11.25 cm, 6.75 cm, b cm
b) a cm, 17.5 cm, 10.5 cm
c) a m, 28.6 m, 26.4 m
d) b m, 38.4 m, 41.6 m

Q3. Find the value of the pronumeral correct to 1 decimal place: 135- 137

a) 15 cm, 45 cm, a m, 42 cm
b) 20 m, 12 m, 9 m, a m
c) 16 cm, x cm
d) 24 m, a m, b m, 26 m, 5 m

Q4. Are the following triangles right-angled? 135-138

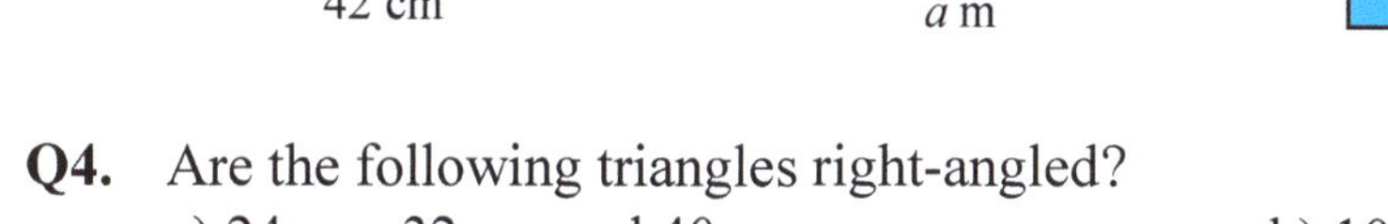

a) 24 cm, 32 cm and 40 cm
b) 10 m, 14 m and 17 m
c) 25.8 m, 34.4 m and 43 m
d) 15 cm, 25 cm and 30 cm

Q5. A boat leaves port and sails 7 nautical miles due west. It then changes course and sails 16.8 nautical miles due north. How far is the boat from port? 135-138

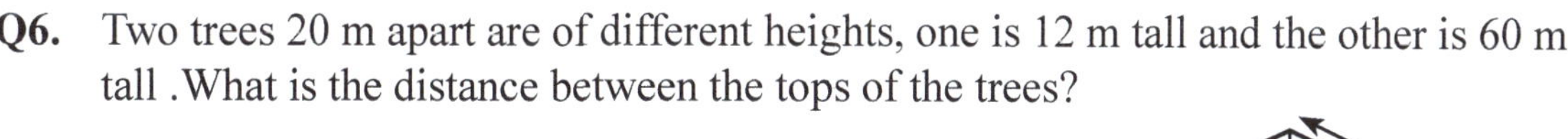

Q6. Two trees 20 m apart are of different heights, one is 12 m tall and the other is 60 m tall .What is the distance between the tops of the trees? 135-138

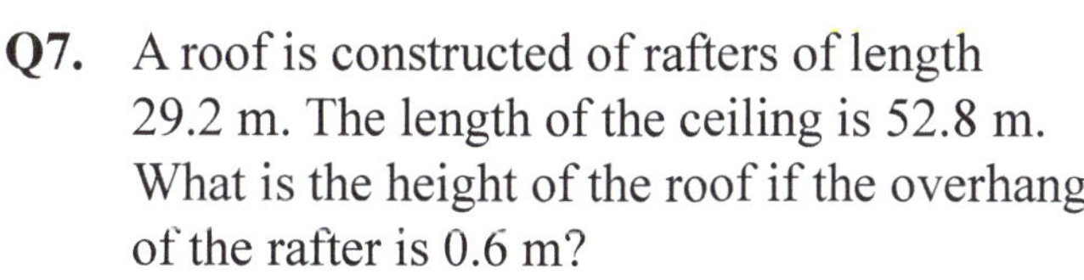

Q7. A roof is constructed of rafters of length 29.2 m. The length of the ceiling is 52.8 m. What is the height of the roof if the overhang of the rafter is 0.6 m? 135-138

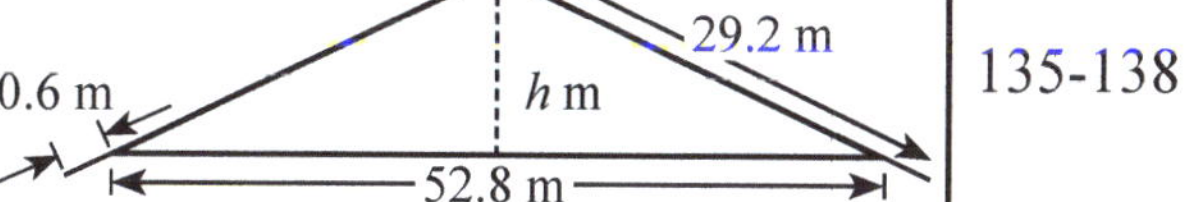

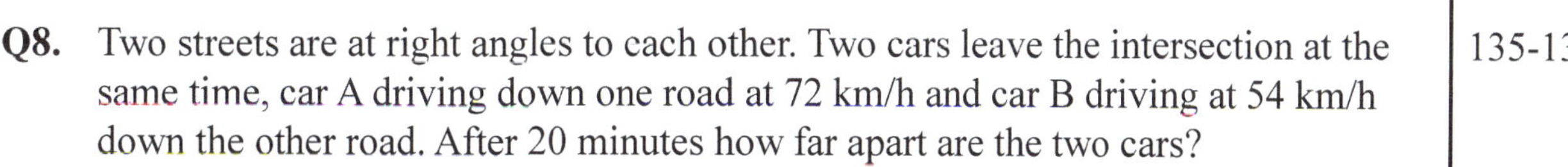

Q8. Two streets are at right angles to each other. Two cars leave the intersection at the same time, car A driving down one road at 72 km/h and car B driving at 54 km/h down the other road. After 20 minutes how far apart are the two cars? 135-138

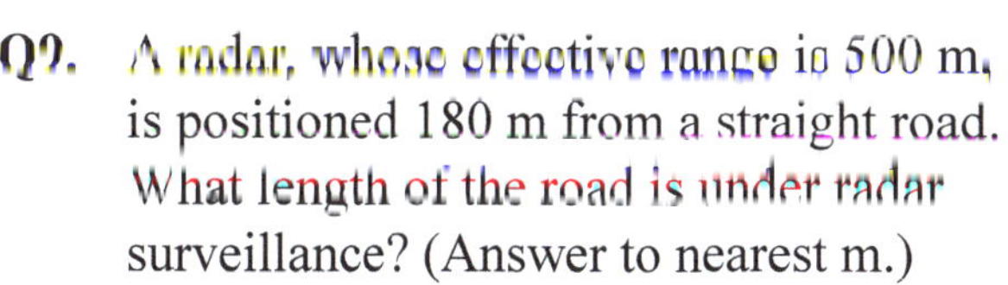

Q9. A radar, whose effective range is 500 m, is positioned 180 m from a straight road. What length of the road is under radar surveillance? (Answer to nearest m.) 135-138

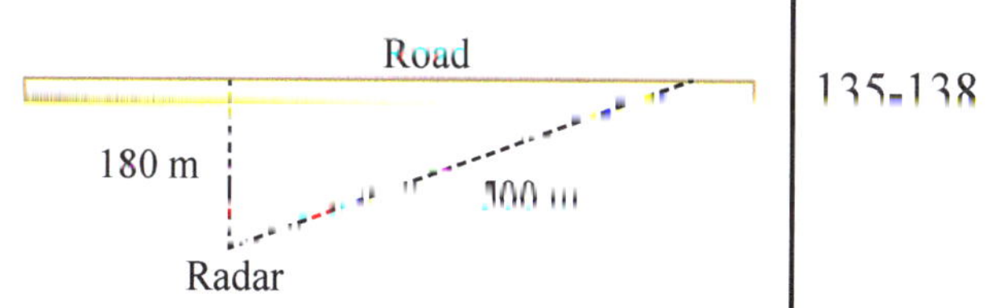

EASIER QUESTIONS

Note: Only turn back to page number shown if you have difficulty.

	Page
Q1. Simplify: a) $5a^3b^2 \times 2ab$ b) $3p^4q^2 \times 3p^2q^3$ c) $7mn^3 \times 4m^3n^2$ d) $16x^5y^3 \times \frac{1}{2}x^2y$ e) $2x^2 \times 3x^4 \times xy$ f) $5a^2b^2 \times 3ab^0 \times b^2$	148
Q2. Simplify: a) $5m^2n^3 \div 20m^0n^2$ b) $16x^2y^5 \div 12x^2y^2$ c) $3a^5b^2c \div a^2bc$ d) $18p^4q^5 \div 3p^2q^3$ e) $m^5n^4 \div 2m^2n$ f) $48x^7y^5 \div 12x^0y^4$	148
Q3. Simplify: a) $(x^4)^3 \times 2x^5$ b) $(ab)^2 \times a^3b^2$ c) $(3m^3)^2 \times m^0$ d) $(2y^2)^2 \times 2y^3$ e) $5(x^2y^3)^2 \times 3xy^2$ f) $(3a^3)^2 \times (4a^2)^2$	148
Q4. Simplify: a) $(4m^3)^2 \div 2m^4$ b) $(2a^4)^3 \div 4a^5$ c) $(6x^4)^2 \div 4x^0$ d) $(2y^5)^3 \div 4y^5$ e) $24a^4b^5 \div (2a^2b^2)^2$ f) $(5m^3n^2)^2 \div 5(mn)^3$	148
Q5. Simplify these expressions: a) $\frac{3x^5 \times 4x^2}{(2x^2)^2}$ b) $\frac{5m^4 \times 12m^5}{(5m^3)^2}$ c) $\frac{(2a^5)^3 \times (a^2)^3}{8a^4 \times 2a^2}$ d) $\frac{(ab)^3 \times 2a^3b^4}{(a^2b)^3}$ e) $\frac{(x^3y^2)^0 \times 10x^2y^4}{5x \times y^2}$ f) $\frac{(2m^3n^2)^2 \times (3mn)^2}{6mn^2 \times 4m^2n}$	148
Q6. Change any negative indices to positive indices: a) $5m^{-2}$ b) $(3a)^{-1}$ c) 9×10^{-2} d) $(4x)^{-3}$	149
Q7. Evaluate: a) $25^{\frac{1}{2}}$ b) $8^{\frac{1}{3}}$ c) $9^{-\frac{1}{2}}$ d) 1^{-2} e) $27^{-\frac{1}{3}}$ f) 4^{-3} g) $16^{-\frac{3}{2}}$ h) $81^{-\frac{3}{4}}$ i) $3m^0$ j) $(5p)^0$ k) 10^{-6} l) $64^{-\frac{2}{3}}$	149
Q8. Expand and simplify: a) $\frac{6.348 \times 10^5}{2.3 \times 10^{-3} \times 3 \times 10^4}$ b) $\frac{1.9 \times 10^{-3} \times 2.4 \times 10^4}{8 \times 10^4}$ c) $\frac{3.98 \times 10^4 \times 6.42 \times 10^{-5}}{1.592 \times 10^{-3} \times 1.07 \times 10^7}$ d) $\frac{9.81 \times 10^{-3} \times 5.74 \times 10^4}{2.87 \times 10^2 \times 1.635 \times 10^{-4}}$ Leave final answers in scientific notation.	150-151
Q9. Simplify: a) $8^x \times 2^{4x}$ b) $9^{\frac{x}{2}} \times 27^{\frac{x}{3}}$ c) $32^{x+2} \div 8^{2x-1}$	148, 149

AVERAGE QUESTIONS

Note: Only turn back to page number shown if you have difficulty.

	Page
Q1. Simplify the following: a) $\sqrt{6} \times \sqrt{3}$ b) $\sqrt{15} \div \sqrt{3}$ c) $2\sqrt{8} \times \sqrt{7}$ d) $2\sqrt{8} \div \sqrt{2}$ e) $4\sqrt{2} \times 3\sqrt{3}$ f) $6\sqrt{12} \div 2\sqrt{3}$	140, 141
Q2. Simplify these surds: a) $\sqrt{50}$ b) $\sqrt{18}$ c) $\sqrt{45}$ d) $\sqrt{72}$ e) $\sqrt{12}$ f) $\sqrt{20}$ g) $4\sqrt{24}$ h) $3\sqrt{27}$	141
Q3. Simplify these expressions: a) $5\sqrt{2} - 3\sqrt{2}$ b) $2\sqrt{5} + 3\sqrt{3} - \sqrt{5}$ c) $6\sqrt{10} - 4\sqrt{5} - 3\sqrt{10}$ d) $4\sqrt{3} + 2\sqrt{12}$ e) $9\sqrt{12} - 2\sqrt{75}$ f) $5\sqrt{80} + 5\sqrt{125}$	141, 142
Q4. Expand and simplify: a) $5(\sqrt{6} + 3)$ b) $\sqrt{2}(\sqrt{3} - 1)$ c) $\sqrt{10}(4 - 2\sqrt{5})$ d) $2\sqrt{2}(\sqrt{5} + 5)$ e) $\sqrt{7}(2\sqrt{7} + 3)$ f) $2\sqrt{6}(7 - 3\sqrt{6})$	143
Q5. Expand and simplify: a) $(2 + \sqrt{7})(3 - \sqrt{5})$ b) $(1 - \sqrt{2})(1 + \sqrt{2})$ c) $(2\sqrt{2} + 3)(\sqrt{3} - 1)$ d) $(\sqrt{6} - 4)^2$ e) $(2 - \sqrt{5})(4 + 2\sqrt{3})$ f) $(3\sqrt{2} + 2)(3\sqrt{2} - 2)$	143, 144
Q6. Express the following with rational denominators: a) $\frac{1}{\sqrt{2}}$ b) $\frac{5}{\sqrt{5}}$ c) $\frac{10}{2\sqrt{3}}$ d) $\frac{\sqrt{2}}{\sqrt{6}}$ e) $\frac{7}{2\sqrt{7}}$ f) $\frac{2\sqrt{3}}{3\sqrt{10}}$ g) $\frac{3 - \sqrt{2}}{\sqrt{2}}$ h) $\frac{\sqrt{5} + 4}{2\sqrt{3}}$	145
Q7. Rationalise the denominator in each expression: a) $\frac{1}{1 + \sqrt{2}}$ b) $\frac{5}{6 - \sqrt{3}}$ c) $\frac{\sqrt{5}}{\sqrt{3} + 3}$ d) $\frac{7}{3\sqrt{6} + 2}$ e) $\frac{\sqrt{3}}{5 - 3\sqrt{3}}$ f) $\frac{2}{\sqrt{5} - \sqrt{3}}$ g) $\frac{10}{2\sqrt{3} - 2\sqrt{2}}$ h) $\frac{\sqrt{5} + \sqrt{3}}{\sqrt{5} - \sqrt{3}}$	146

HARDER QUESTIONS

Q1. Simplify:

a) $3\sqrt{48}$ b) $2\sqrt{54}$ c) $4\sqrt{96}$ d) $2\sqrt{108}$

e) $3\sqrt{162}$ f) $\sqrt{\frac{3}{4}}$ g) $\sqrt{\frac{24}{25}}$ h) $\sqrt{x^3}$

Q2. Simplify:

a) $5\sqrt{20} - 2\sqrt{18} - \sqrt{45}$ b) $\sqrt{98} + \sqrt{32} - \sqrt{63}$ c) $\sqrt{108} - \sqrt{24} + \sqrt{27}$

d) $\sqrt{500} + 2\sqrt{45} - \sqrt{80}$ e) $2\sqrt{4x} + 5\sqrt{9x} - \sqrt{x}$ f) $x\sqrt{16y} - \sqrt{x^2y} + 3x\sqrt{9y}$

Q3. Expand and simplify:

a) $(2\sqrt{5} + 3\sqrt{3})^2$ b) $(2\sqrt{3} + \sqrt{6})(3\sqrt{2} - \sqrt{3})$

c) $(4\sqrt{2} - 2\sqrt{3})(4\sqrt{2} + 2\sqrt{3})$ d) $(3\sqrt{6} - 2\sqrt{7})(2\sqrt{6} - 3\sqrt{3})$

e) $x\sqrt{y}\,(6\sqrt{x} - 2\sqrt{y})$ f) $(x\sqrt{x} - y\sqrt{y})(x\sqrt{x} + y\sqrt{y})$

Q4. Express the following with rational denominators:

a) $\dfrac{\sqrt{3} - \sqrt{5}}{3\sqrt{5} + 2\sqrt{3}}$ b) $\dfrac{\sqrt{6} - \sqrt{2}}{2\sqrt{6} - 4\sqrt{2}}$ c) $\dfrac{2\sqrt{5} - 3\sqrt{3}}{2\sqrt{5} + 3\sqrt{3}}$

d) $\dfrac{3}{2\sqrt{x} - 3\sqrt{y}}$ e) $\dfrac{2 - \sqrt{5}}{3\sqrt{x} + 4\sqrt{y}}$ f) $\dfrac{\sqrt{x} + \sqrt{y}}{4\sqrt{x} - 2\sqrt{y}}$

Q5. Express the following as single fractions with rational denominators:

a) $\dfrac{1}{2\sqrt{5}} + \dfrac{1}{5\sqrt{2}}$ b) $\dfrac{\sqrt{3}}{\sqrt{2}} - \dfrac{\sqrt{2}}{\sqrt{3}}$ c) $\dfrac{\sqrt{5}}{2\sqrt{6}} + \dfrac{\sqrt{5}}{\sqrt{7}}$

d) $\dfrac{1}{\sqrt{3} + 2} - \dfrac{1}{\sqrt{3} - 2}$ e) $\dfrac{3}{\sqrt{6} - 2} + \dfrac{3}{\sqrt{6} + 3}$ f) $\dfrac{2}{2\sqrt{2} + 3} - \dfrac{1}{5\sqrt{2} + 3}$

Q6. If $x = \dfrac{1}{\sqrt{2} + \sqrt{5}}$ evaluate $x^2 + \dfrac{1}{x^2}$

Q7. Simplify: $\dfrac{x\sqrt{x} \times \sqrt{x^3}}{2\sqrt{x} \times \sqrt{2x}}$ (Express answer with a rational denominator)

Q8. Expand and simplify: $(3\sqrt{2} + \sqrt{3})^2 - (3\sqrt{2} - \sqrt{3})^2$

Q9. Show that $\dfrac{2\sqrt{3} - 2\sqrt{2}}{3\sqrt{3} - 3\sqrt{2}}$ is a rational number.

HARDER QUESTIONS

Q1. Write the following in their simplest form:

a) $(x^2y^{-2})^{-1}$ b) $\left(\frac{1}{x}\right)^{-2} \times y^{-\frac{1}{2}}$ c) $\left(x^2\right)^{-\frac{3}{2}} \div \left(y^{-\frac{1}{2}}\right)^3$

d) $(x^{-1} \div y^2)^{-2}$ e) $x^{-2} - y^{-2}$ f) $(x + y)^{-1} - (x - y)^{-1}$

Q2. Evaluate:

a) $27^{\frac{5}{3}} \div 9^{\frac{3}{2}}$ b) $8^{\frac{4}{3}} \times 4^{\frac{5}{2}}$ c) $125^{-\frac{2}{3}} \times 625^{\frac{3}{4}}$

d) $125^{-\frac{5}{3}} \div 25^{-\frac{3}{2}}$ e) $16^{-\frac{3}{4}} - 8^{-\frac{5}{3}}$ f) $81^{-\frac{3}{4}} + 27^{-\frac{4}{3}}$

Q3. Simplify:

a) $x^{-1} \times 2x^{\frac{1}{2}}$ b) $\left(27x^2\right)^{\frac{1}{3}} \div \frac{1}{3}\left(x^3\right)^{\frac{1}{2}}$ c) $\left(3x^{\frac{2}{3}}\right) \times \left(4x^4\right)^{\frac{1}{2}}$

d) $(9x)^{\frac{1}{2}} \times \left(8x^{-\frac{1}{2}}\right)^{\frac{1}{3}}$ e) $\left(8x^{\frac{3}{4}}\right)^{-2} \div \left(\frac{1}{2}x^{-1}\right)^2$ f) $\left(5x^{\frac{1}{4}}\right)^2 \div \left(125x^3\right)^{\frac{1}{3}}$

Q4. The number of carbon atoms in 12g of carbon is called Avogadro's number which is 6.022×10^{23}.

a) Find the number of carbon atoms in (i) 6 grams (ii) 3 grams (iii) 20 grams of carbon.

b) Find the weight of carbon for the following number of atoms:
(i) 6.022×10^{20} (ii) 6.022×10^{10} (iii) 6.022×10^5 (iv) 6.022×10^2

Q5. Light travels at 3×10^8 m/s.

a) Find the distance that light would travel in:
(i) one minute (ii) one hour (iii) one day

b) If the radius of the Earth is approximately 6 370 km, find the time that it would take for a beam of light to make one cycle around the Earth.

c) The distance called a 'light year' is the distance that light would travel in a year. The star system (Alpha Centauri) is 4.85 light years from the Earth. Express the distance for a light year in kilometres and hence, state the distance from the Earth to Alpha Centauri in kilometres.

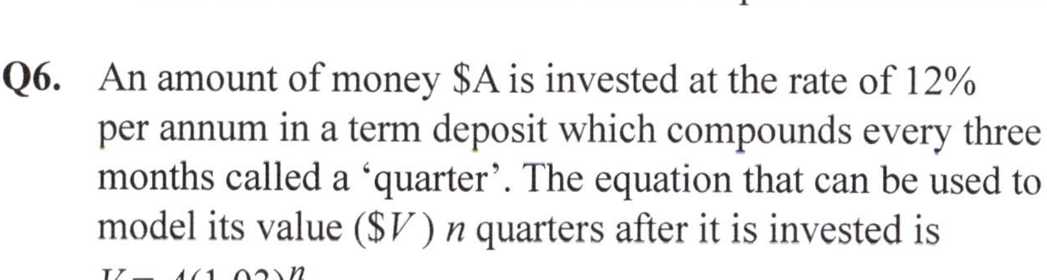

Q6. An amount of money $A is invested at the rate of 12% per annum in a term deposit which compounds every three months called a 'quarter'. The equation that can be used to model its value ($$V$) n quarters after it is invested is $V = A(1.03)^n$.

a) Justify why the value of 1.03 is used in the formula.

b) If $1 000 was invested, find the value of the investment at the end of the first year.

c) Find the length of time, to the nearest month that it will take for the investment to double in value.

Q1. A piece of cardboard in the shape of a square has an area of 252 cm^2.

a) Find the length of an edge of the square in simplest surd form.

b) Use Pythagoras' Theorem to find the exact length of the diagonal of the square in simplest surd form.

c) Six of these squares are glued together to form a perfect cube. Find the exact volume of the cube in simplest surd form.

d) Using your answer in part (a), find the total length of all the edges of the cube in simplest surd form.

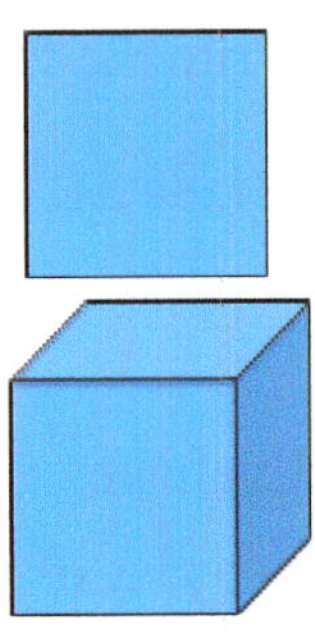

Q2. a) Use Pythagoras' Theorem to calculate the exact height (marked h) of this equilateral triangle, hence find its exact area.

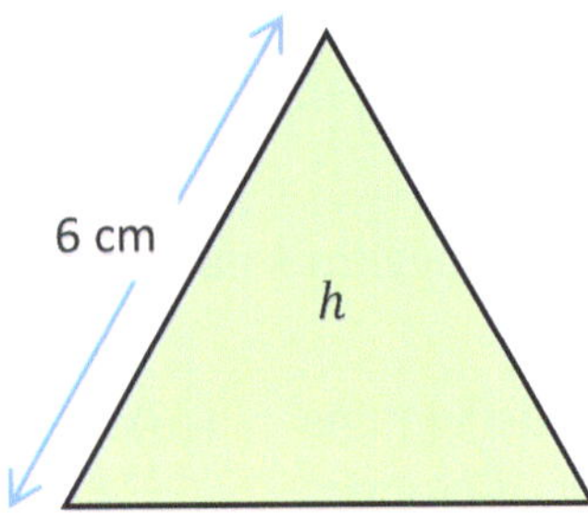

b) Calculate the exact perimeter and area of these rectangles.

(i)

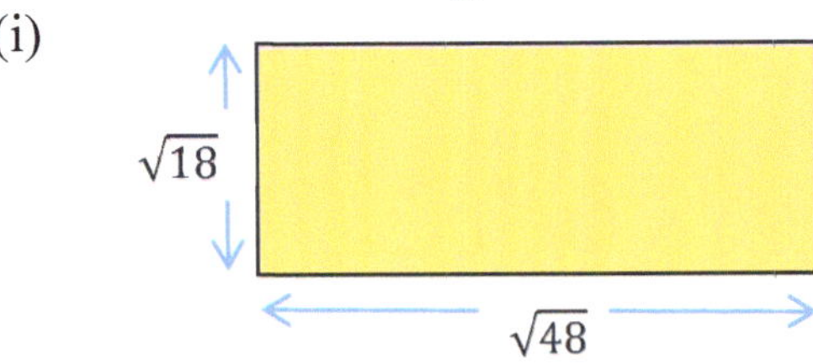

(ii)

$2\sqrt{72}$

$10\sqrt{128}$

Q3. This shape has some irrational side lengths.

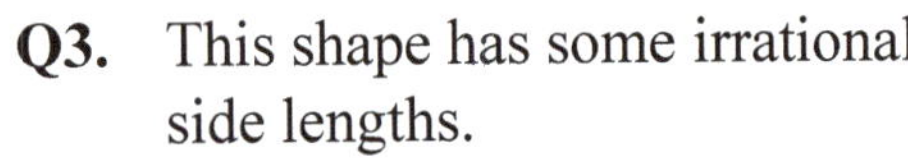

a) Find the exact side lengths shown as (i) x (ii) y.

b) Express the (i) perimeter and (ii) area in exact form.

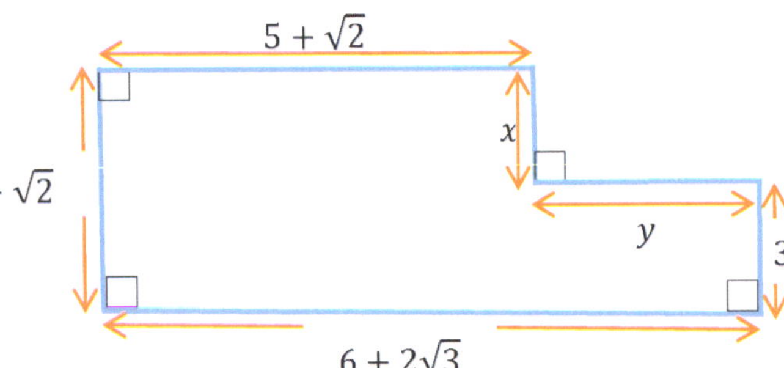

Q4. A gift box has the dimensions length: $10\sqrt{6}$ cm, width: $8\sqrt{3}$ cm and height: $(4\sqrt{2} + 5)$ cm.

a) If the box is to have its six sides covered by paper find the area of paper required to cover the box. Expressed as a simplified surd.

b) Express the volume of the box in simplest exact form.

Q5. A geometric sequence with first term, $2ab^{-2}$ and common ratio, $b\sqrt{a}$ is generated by multiplying each term by $b\sqrt{a}$ to give the next term as follows: First term = $2ab^{-2}$,

Second term = $2ab^{-2} \times b\sqrt{a} = 2ab^{-1}\sqrt{a} = \dfrac{2a\sqrt{a}}{b}$ or $\dfrac{2a^{\frac{3}{2}}}{b}$, …

a) Find the fourth and fifth terms in the sequence expressed in simplest surd form and index form where appropriate.

b) Find the product of the first five terms in the sequence expressed in index form.

EQUATIONS AND FORMULAE

The 'Australian Curriculum Mathematics' (ACM) references for this sub-strand of 'Number and Algebra' (NA) are given below. This chapter may contain additional extension work which the author feels will be beneficial to the student.

- *Applications of equations involving the solution of word problems (ACMNA 194).*
- *Equations and formulae (ACMNA 234).*
- *Equations with algebraic fractions (ACMNA 241).*
- *Changing the subject of the formula (NSW).*
- *Solve simultaneous equations using algebraic and graphic techniques (ACMNA 237).*
- *Quadratic equations (ACMNA 269).*

Leonard Euler (1707 – 1783)

Euler was a Swiss mathematician, physicist, astronomer, geographer, logician and engineer who made important and influential discoveries in many branches of mathematics, such as calculus, graph theory, topology and number theory. He is also famous for his work in mechanics, fluid dynamics , optics, astronomy and music theory He was one of the most eminent mathematicians of the 18th century and some people consider him to be the greatest of all time. He was also the most prolific with his collected works filling 92 volumes, more than any other mathematician. He spent much of his adult life in St Petersburg, which was at that time the capital of Russia.

SUMMARY OF LINEAR EQUATIONS FROM PREVIOUS YEARS

Linear equations (sometimes also called simple equations) have been thoroughly explained in Year 9 and also in earlier years.
Therefore I shall only give a summary of the main points in the next few pages.

INTRODUCTION AND THE BASIC RULES

An equation is a balanced number sentence where one of the numbers is represented by a pronumeral or variable (letter). To SOLVE an equation means to find the value of this unknown letter.

Whatever operation (+, –, ×, ÷) is done to one side of the equation, exactly the same operation must be done to the other side. The final aim is to obtain the unknown letter by itself by doing one or more inverse operations.

INVERSE OPERATION simply means the opposite operation, and this is best illustrated by the different types of equations shown in the following sections.

ONE STEP EQUATIONS

Solve $\frac{x}{3} = 7$

Solution: $\therefore \quad 3 \times \frac{x}{3} = 3 \times 7$ — Multiply both sides by 3.

$\therefore \quad x = 21$

TWO STEP EQUATIONS

The aim is to get the unknown letter (3*x*, 5*a* or 8*w* etc.) on one side of the equation and the pure numbers (16, 20 or – 10 etc.) on the other side.

Solve $6m - 10 = 20$ — Add 10 to both sides of the equation.

Solution: $\therefore \quad 6m = 30$ — Divide both sides by 6.

$\therefore \quad \frac{6m}{6} = \frac{30}{6}$ — Write down the steps.

$\therefore \quad m = 5$ — The final answer.

EQUATIONS INVOLVING BRACKETS

Firstly remove the grouping symbols in the usual way by multiplying out the brackets, and then continue as in the previous section.

Solve $3(x + 2) = 24$ — First remove grouping symbols.

Solution: $\therefore \quad 3x + 6 = 24$ — Subtract 6 from both sides.

$\therefore \quad 3x = 18$ — Divide both sides by 3.

$\therefore \quad \frac{3x}{3} = \frac{18}{3}$ — Write down the steps.

$\therefore \quad x = 6$ — The final answer.

EQUATIONS INVOLVING FRACTIONS

The first step is to multiply each side of the equation by the lowest common denominator in order to clear the fractions. Then continue as in the previous sections.

Solve $\dfrac{x-1}{3} = \dfrac{x+5}{4}$ — Multiply each side by 12 to clear fractions.

Solution:

$\therefore \quad 12\left(\dfrac{x-1}{3}\right) = 12\left(\dfrac{x+5}{4}\right)$ — The fractions in the denominators both cancel.

$\therefore \quad 4(x-1) = 3(x+5)$ — Expand out the brackets.

$\therefore \quad 4x - 4 = 3x + 15$ — Add 4 to both sides.

$\therefore \quad 4x = 3x + 19$

$\therefore \quad x = 19$

INEQUATIONS

Inequations always contain one of the 4 inequality signs shown below:

'greater than or equal to'	'greater than'	'less than or equal to'	'less than'

Inequations are solved in exactly the same way as ordinary equations except that **the inequality sign is reversed whenever we multiply or divide both sides by a negative number**, **or when we take the reciprocal of both sides.**

$-5x > 20$  $\dfrac{-5x}{-5} < \dfrac{20}{-5}$  $x < -4$

INEQUATIONS AND THE NUMBER LINE

Because there are usually an infinite number of solutions to an inequation, you will often be required to graph the final answer on a number line.

A solid dot ● plus a line and arrow is used for $\geq$ and $\leq$.

An open circle ○ plus a line and arrow is used for $>$ and $<$.

Example: Solve $3(2 - x) \geq 9$ and graph your answer on a number line.

Solution:

$\therefore \quad 6 - 3x \geq 9$

$\therefore \quad -3x \geq 3$

$\therefore \quad \dfrac{-3x}{-3} \leq \dfrac{3}{-3}$

$\therefore \quad x \leq -1$

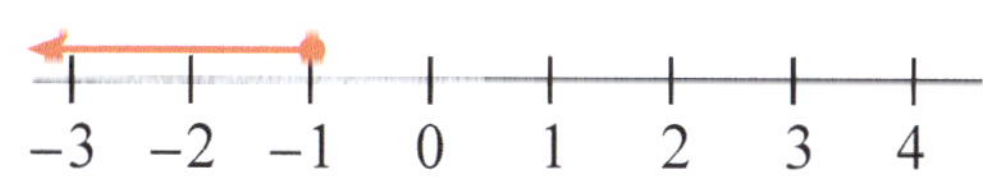

When we divide both sides by −3, then the inequality sign must be reversed.

APPLICATIONS OF EQUATIONS

Equations can be used to solve problems in many areas of Mathematics. Although some problems could be solved by using a 'trial and error' approach, it is however much easier and quicker if you make up an equation using basic known theorems. Variations of the questions shown below frequently occur in examinations right up to Year 12.

Example 1: Find the value of x and hence find the size of $\angle ABC$.

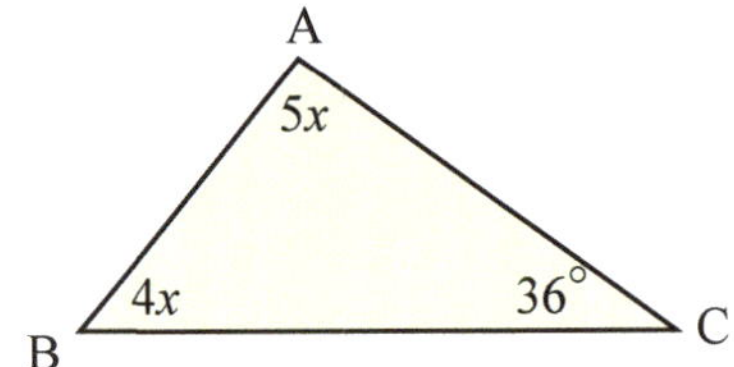

Solution: Firstly form an equation using the fact that the 3 angles of any triangle always add up to 180°.

Form an equation.

$\therefore \quad 5x + 4x + 36 = 180$

Add like x terms.

$\therefore \quad 9x + 36 = 180$

Subtract 36 from both sides.

$\therefore \quad 9x = 144$

Divide both sides by 9.

$\therefore \quad \frac{9x}{9} = \frac{144}{9}$

$\therefore \quad x = 16$

$\angle ABC = 4x$

$\therefore \quad = 4 \times 16$

$\therefore \quad = 64^\circ$

Example 2: If the sum of a number and 3 is multiplied by 4, the answer is the same as twice the number plus 26. What is the number?

Solution: Let the unknown number be x.

$\therefore$	$4(x + 3) = 2x + 26$	Form an equation.
$\therefore$	$4x + 12 = 2x + 26$	Expand the brackets.
$\therefore$	$4x = 2x + 14$	Subtract 12 from both sides.
$\therefore$	$2x = 14$	Subtract $2x$ from both sides.
$\therefore$	$x = 7$	Divide both sides by 2.

The unknown number is 7.

Example 3: a) Write an algebraic expression for the area of the rectangle shown.

b) If the area is 78 cm^2, write an equation and solve it to find x.

c) Hence find the perimeter of the rectangle.

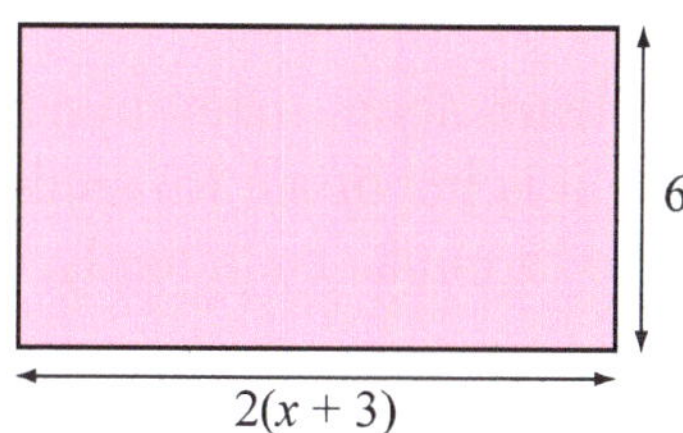

Solutions: a) Area of rectangle = length × breadth

$\therefore$ Area = $6 \times 2(x + 3)$

$\therefore$ Area = $12x + 36$

b) $12x + 36 = 78$

Subtract 36 from both sides.

$\therefore$ $12x = 42$

Divide both sides by 12.

$\therefore$ $x = 3.5$ cm

c) Perimeter = $2(x + 3) + 6 + 2(x + 3) + 6$

= $4x + 24$

Substitute $x = 3.5$ into the above expression

$\therefore$ Perimeter = 38 cm.

Example 4: The sum of 3 consecutive odd numbers is 81. Find the numbers.
(**Note:** consecutive means 'one after the other'.)

Solution: Let the smallest odd number be x.

$\therefore$ Next consecutive odd number is $(x + 2)$

$\therefore$ Next consecutive odd number is $(x + 4)$

Now form an equation and solve it to find x.

$x + (x + 2) + (x + 4) = 81$

$\therefore$ $3x + 6 = 81$ Add like terms.

$\therefore$ $3x = 75$ Subtract 6 from both sides.

$\therefore$ $x = 25$ Divide both sides by 3.

The 3 consecutive odd numbers are 25, 27 and 29.

Example 5: Solve the non-linear equation $5x^2 - 27 = 378$.

Solution:

$\therefore$ $5x^2 = 405$ Add 27 to both sides.

$\therefore$ $x^2 = 81$ Divide both sides by 5.

$\therefore$ $x = \sqrt{81}$ Take square root of both sides.

$= +9$ or -9

$= \pm 9$

SUBSTITUTING INTO FORMULAE

You will learn about many hundreds of different formulae during the next 4 years of high school. Therefore it is important for students, not only to memorise many formulae for exams, but also to be able to use formulae confidently.

Listed below are some formulae which you should have already seen:

$A = \ell b$	Area of rectangle = length × breadth
$V = \ell bh$	Volume of rectangular prism = length × breadth × height
$P = 2\ell + 2b$	Perimeter of rectangle = twice length + twice breadth
$a^2 = b^2 + c^2$	Pythagoras' Theorem
$D = S \times T$	Distance = speed × time

It doesn't matter whether or not you have seen a particular formula before, but a popular question involves substituting numbers in for the different letters or pronumerals.

STEPS:

(i) The question will give you a formula which you may, or may not have seen before.
(ii) The question will also give you the values of all the letters in the formula, except for one.
(iii) Simply substitute these given values in for the various letters, and hence find out the value of the one unknown letter.

Example 1: If $v = u + at$ find the value of v when $u = 7$, $a = 4$ and $t = 3$.
Simply substitute $u = 7$, $a = 4$ and $t = 3$ into the formula.

$\therefore \quad v = 7 + 4 \times 3$

$\therefore \quad v = 19$

Example 2: If $h = \sqrt{a^2 + b^2}$ find the value of h when $a = 12$ and $b = 5$.

$\therefore \quad h = \sqrt{12^2 + 5^2}$

$\therefore \quad h = \sqrt{144 + 25}$

$\therefore \quad h = \sqrt{169}$

$\therefore \quad h = 13$

Simply substitute $a = 12$ and $b = 5$ into the given formula.

CHANGING THE SUBJECT OF THE FORMULA

In the last section, the unknown letter whose value you had to find was always the subject of the formula.

> The subject of a formula is the pronumeral (or letter) which is by itself on the left hand side.

For example, in $v = u + at$ the subject of the formula is the letter v. So it was fairly easy to find the value of v when given the values of the other 3 pronumerals. However you will also be asked to find the value of pronumerals which aren't the subject of the formula.

> (i) Substitute in the values of the letters you are given in the usual way.
> (ii) Now find the value of the unknown pronumeral using exactly the same steps you have learned to solve equations.

Example 1: If $v = u + at$ find the value of t when $v = 18$, $u = 9$ and $a = 2$.
Substitute $v = 18$, $u = 9$, $a = 2$ into $v = u + at$.

$\therefore \quad 18 = 9 + 2t$ — This is an equation.

$\therefore \quad 9 = 2t$ — Subtract 9 from both sides.

$\therefore \quad t = \dfrac{9}{2} = 4\tfrac{1}{2}$ — Divide both sides by 2.

Example 2: Given the formula $v^2 = u^2 + 2as$ find the value of 'a' given that $v = 9$, $u = 7$ and $s = 3$.

$\therefore \quad 9^2 = 7^2 + 2 \times a \times 3$

$\therefore \quad 81 = 49 + 6a$ — This is an equation.

$\therefore \quad 32 = 6a$ — Subtract 49 from both sides.

$\therefore \quad a = \dfrac{32}{6} = 5\tfrac{1}{3}$ — Divide both sides by 6.

TYPICAL EXAM QUESTIONS

Example 1: a) If the volume of a cylinder is given by the formula $V = \pi r^2 h$, make 'r' the subject of the formula.

b) Hence find the radius (r) of a cylinder of volume (V) 200 cm^3 and height (h) 8 cm. (Answer to 1 d.p.)

Solution:

a) $V = \pi r^2 h$

$\therefore \dfrac{V}{\pi h} = r^2$ — Divide both sides by πh.

$\therefore r = \sqrt{\dfrac{V}{\pi h}}$ — Take the square root of both sides.

b) $r = \sqrt{\dfrac{200}{\pi \times 8}}$

$\therefore$ radius = 2.8 cm

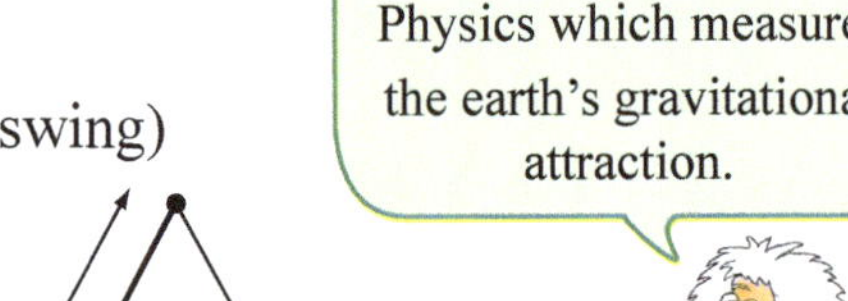

Example 2: The time (T) taken for one complete oscillation (swing) of a pendulum arm is given by the formula

$T = 2\pi\sqrt{\dfrac{\ell}{g}}$ where

ℓ = length of pendulum

g = 9.8 m/s^2

If a clockmaker wants the time taken (T) for one swing to be 1 second, find what length he must make the pendulum (to the nearest cm).

Solution:

$T = 2\pi\sqrt{\dfrac{\ell}{g}}$ — Make ℓ the subject of the formula.

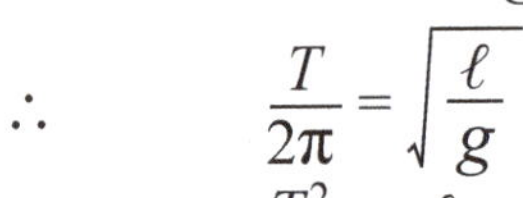

$\therefore \dfrac{T}{2\pi} = \sqrt{\dfrac{\ell}{g}}$ — Divide both sides by 2π.

$\therefore \dfrac{T^2}{4\pi^2} = \dfrac{\ell}{g}$ — Square both sides.

$\therefore \ell = \dfrac{T^2 g}{4\pi^2}$ — Multiply both sides by g.

$\therefore \ell = \dfrac{1^2 \times 9.8}{4\pi^2}$ — Substitute in data.

$\therefore \ell = 0.248$ m

$= 25$ cm

Example 3: If $y = \dfrac{x-2}{x-5}$ make x the subject of the equation.

$\therefore y(x-5) = x-2$ — Multiply both sides by $x - 5$.

$\therefore yx - 5y = x - 2$ — Remove brackets.

$\therefore yx - x = 5y - 2$ — Collect terms involving x on left.

$\therefore x(y-1) = 5y - 2$ — Factorise.

$\therefore x = \dfrac{5y-2}{y-1}$ — Divide both sides by $y - 1$.

CONSTRUCTING FORMULAE

You will often be required to construct a formula from the information given in the question. This process involves trying to find a relationship or formula which connects two pronumerals (or variables).

Example 1: From the table of values, find a relationship of y in terms of x.

x	2	3	4
y	5	8	11

Solutions:

$$5 = 3 \times 2 - 1$$
$$8 = 3 \times 3 - 1$$
$$11 = 3 \times 4 - 1$$
$$\therefore \quad y = 3 \times x - 1$$
$$y = 3x - 1$$

In this problem, we must find a pattern which connects y to x. Notice that x increases by 1 and y increases by 3. This explains the positive of 3 in the formula $y = 3x - 1$.

Example 2: A straight river together with 400 m of fencing for the other three sides is used to enclose a rectangular field. If the width of the field is x metres, find the area (A) of the field in terms of x.

RIVER

FIELD

x x

Solution: The length of the field = $400 - 2x$

$\therefore$ Area (A) = length $\times$ breadth

$\therefore$ $A = (400 - 2x)x$

$\therefore$ $A = 400x - 2x^2$

Example 3: Construct a formula which relates the angle sum of a polygon (S) to the number of sides (n) of the polygon.

$n = 3$  $S = 180° = 1 \times 180° = (3 - 2) \times 180°$

$n = 4$ $S = 360° = 2 \times 180° = (4 - 2) \times 180°$

$n = 5$ $S = 540° = 3 \times 180° = (5 - 2) \times 180°$

$n = 6$ $S = 720° = 4 \times 180° = (6 - 2) \times 180°$

Solution: Can you see that the angle sum (S) is equal to (the number of sides $-$ 2) $\times$ 180°,

i.e. $S = (n - 2) \times 180°$

The above formula enables us to find the angle sum of any polygon.

INTRODUCTION TO SIMULTANEOUS EQUATION

Two equations are said to be solved SIMULTANEOUSLY, when you find an ordered pair (x, y) which **satisfies both equations**. This can be done graphically by drawing graphs of the two equations and finding their point of intersection.

The dictionary definition of simultaneous means 'at the same time'.

Example: Solve the following pair of simultaneous equations graphically:

$2x + y = 5$ (i)

$y = x + 2$ (ii)

Solution: It is first necessary to draw up a table of values for each equation, and then plot these ordered pairs on the number plane.

$2x + y = 5$

x	0	1	2
y	5	3	1

$y = x + 2$

x	0	1	2
y	2	3	4

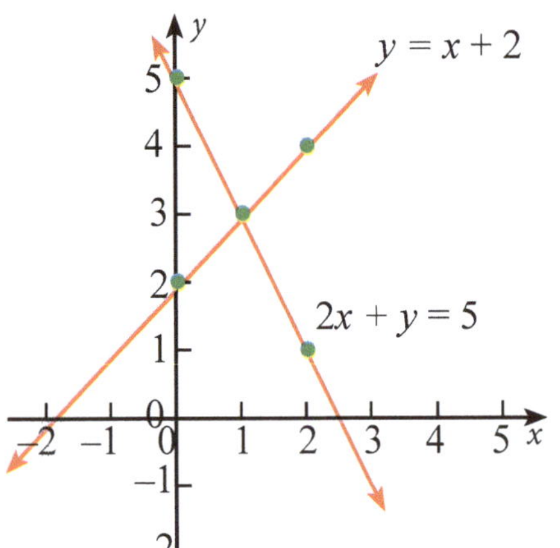

From the graph, the point of intersection is (1, 3).

Therefore the solution which satisfies both equations simultaneously are $x = 1$ and $y = 3$.

$x = 1$ and $y = 3$ are the only 2 numbers that satisfy both equations at the same time. This can easily be tested by substituting these solutions back into the original 2 equations.

Checking the answer:

$2 \times 1 + 3 = 5$ True

$3 = 1 + 2$ True

Solving 'Simultaneous equations' is very useful when you are given 2 distinct and separate pieces of information.

Examples: (i) The sum of 2 numbers is 46 and their difference is 12. Find the numbers.

(ii) 3 pencils and 2 rulers cost \$1.06 and 2 pencils and 3 rulers cost \$1.29. Find the cost of each.

SUBSTITUTION METHOD

This is one of two main methods involving algebra. It is not only faster than the graphical method described, but it also gives the exact answers — particularly when fractions are involved. However, it is first necessary to clearly understand what 'subject of the equation' means.

Consider the equation $2x + y = 6$

If we subtract $2x$ from both sides, we make y the subject of the equation,

i.e. $y = -2x + 6$

If we subtract y from both sides, and then divide both sides by 2, we make x the subject of the equation,

i.e. $x = \dfrac{-y + 6}{2}$

As you see, it was much easier to make the single letter 'y' the subject in the example above. The steps below show how to solve two equations simultaneously using the method of substitution:

(i) Make one letter (either x or y) the subject of one of the equations.
(ii) Substitute the expression for this letter into the second equation.
(iii) Solve the resulting simple equation.
(iv) Substitute this value back into either equation (1) or (2) to find the value of the other letter.

Example: Solve

$2x - y = 6$ (1)

$x + 3y = 10$ (2)

by the substitution method.

Solution: First, make 'y' the subject of equation (1),

i.e. $y = 2x - 6$

Now substitute $2x - 6$ for the letter 'y' in equation (2).

$$\therefore \quad x + 3(2x - 6) = 10$$
$$x + 6x - 18 = 10$$
$$7x = 28$$
$$x = 4$$

Substitute $x = 4$ into either equation (1) or (2) to find y.

Hence $y = 2$

Note: This could also have been solved by making x the subject of equation (2), i.e. $10 - 3y$. This expression is then substituted for the letter x in equation (1).

ELIMINATION METHOD

This is the other main method involving algebra. Both methods (substitution or elimination) will give the correct answers, but in some cases the choice of method used will be much easier and faster.

The 'coefficient' of a letter is the number in front of the letter. Therefore in the equation $2x + 3y = 8$, the coefficient of x is 2, and the coefficient of y is 3.

The steps below show how to solve two equations simultaneously using the method of elimination:

(i) By multiplication, make the coefficient of one of the letters in both equations the same.
(ii) Add or subtract the two equations to eliminate one of the letters.
(iii) Solve the resulting simple equation.
(iv) Substitute this value back into either equation (1) or (2) to find the value of the other letter.

Example: Solve $2x - y = 6$ (1) by the elimination method.
$x + 3y = 10$ (2)

Solution: Multiply equation (1) by 3 to obtain the same y coefficient.

$\therefore \quad 6x - 3y = 18 \quad (3)$

$x + 3y = 10 \quad (2)$

Now simply add equation (3) to equation (2).

$7x = 28$

$\therefore \quad x = 4$

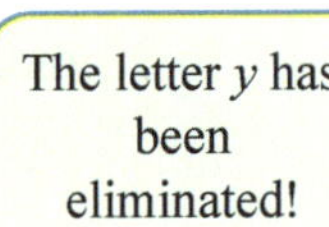

Substitute $x = 4$ into either equation (1) or (2) to find y.

$\therefore \quad y = 2$

Note: As you can see, this is exactly the same question, with the same answers, as on the previous page.

PROBLEM SOLVING APPLICATIONS

The solution of simultaneous equations has many applications in Maths, some of which are illustrated by the examples on these two pages.

Example 1: The sum of two numbers is 46 and their difference is 12. Find the numbers.

Solution: Let x and y be the two numbers.

$\therefore \quad x + y = 46 \qquad (1) \qquad$ Sum is 46

$\therefore \quad x - y = 12 \qquad (2) \qquad$ Difference is 12

Add equation (1) to equation (2) to eliminate y.

$\therefore \quad 2x = 58$

$\therefore \quad x = 29$

Substitute $x = 29$ back into equation (1) or (2) to find y.

$\therefore \quad y = 17$

Therefore the two numbers are 29 and 17.

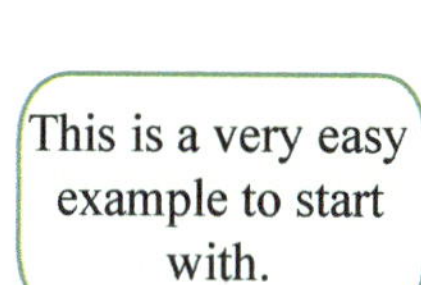

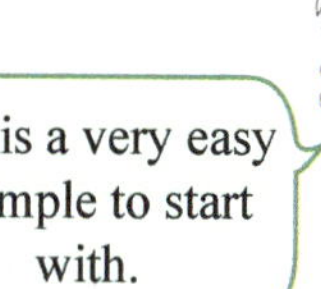

Example 2: Three pencils and two rulers cost \$2.28 and two pencils and three rulers cost \$2.82. Find the cost of each.

Solution: Let p = cost of pencil and r = cost of ruler.

$\therefore \quad 3p + 2r = 228 \qquad (1) \qquad$ Form two separate equations from

$\therefore \quad 2p + 3r = 282 \qquad (2) \qquad$ the information given.

Multiply equation (1) by 3 and equation (2) by 2 to obtain the same coefficient of 6 for the letter 'r'.

$(1) \times 3 \quad \therefore \quad 9p + 6r = 684 \qquad (3)$

$(2) \times 2 \quad \therefore \quad 4p + 6r = 564 \qquad (4)$

Subtract equation (4) from equation (3) to eliminate 'r'.

$\therefore \quad 5p = 120$

$\therefore \quad p = 24$

Substitute $p = 24$ back into equation (1) to find 'r'.

$\therefore \quad 72 + 2r = 228$

$\therefore \quad r = 78$

Therefore a pencil costs 24 c and a ruler costs 78 c.

Example 1: Form two equations from the isosceles triangle shown in the figure, and solve them simultaneously to find the values of x and y.

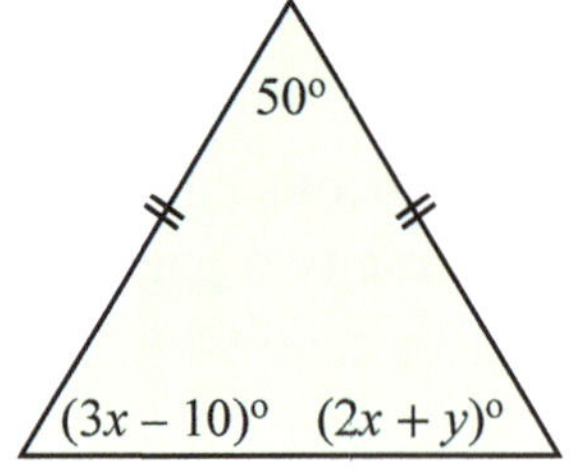

Solution:

$3x - 10 = 2x + y$	(1)	Base angles of isosceles $\triangle$ are equal
$(3x - 10) + (2x + y) + 50° = 180°$	(2)	Angle sum of $\triangle = 180°$
$\therefore \quad x - y = 10$	(1)	Rewriting and simplifying the original
$\therefore \quad 5x + y = 140$	(2)	equation (1) and (2)

Add equation (1) to equation (2) to eliminate the letter y.

$\therefore \quad 6x = 150$

$\therefore \quad x = 25$

Substitute $x = 25$ into $x - y = 10$ to find the value of y.

$\therefore \quad y = 15$

Example 2: Determine whether or not the 3 lines below are concurrent:

$2x + y = 7$ (1) $\qquad x + 2y = -1$ (2) $\qquad 3x - 4y = 27$ (3)

Solution: Three or more lines are said to be 'concurrent' if they pass through the same point.

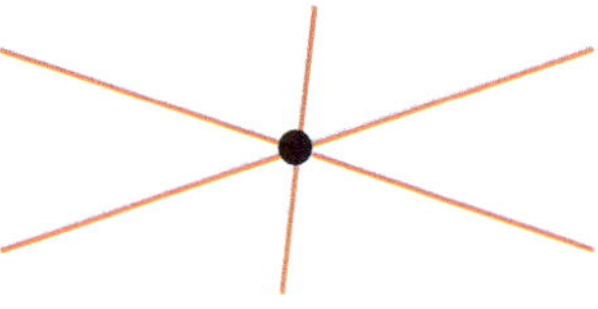

'Concurrent lines'

Solve equation (1) and (2) simultaneously to find the point of intersection of the first two lines.

$2x + y = 7$ (1)

$x + 2y = -1$ (2)

Make 'y' the subject of equation (1).

$\therefore \quad y = 7 - 2x$

Substitute this expression of y into equation (2).

$\therefore \quad x + 2(7 - 2x) = -1$

$\therefore \quad -3x + 14 = -1$

$\therefore \quad x = 5$

Substitute $x = 5$ into equation (1) to find y.

$\therefore \quad y = -3$

The point of intersection of the first two lines is $(5, -3)$.

Substitute $(5, -3)$ into equation (3) to see if it satisfies it.

$3 \times 5 - 4 \times (-3) = 27$

$\therefore$ Left hand side = 27 which equals right hand side = 27

$\therefore$ Point $(5, -3)$ satisfies this line.

$\therefore$ The 3 lines are concurrent.

Note: Note if the point of intersection doesn't satisfy the third line, then the 3 lines are not concurrent.

HARDER EXAMPLES

Example 1: Solve the following simultaneous equations, expressing the variables x and y in terms of c.

$2x + 3y = 4c$ — (1)

$x - y = 7c$ — (2)

Solution: Multiply (2) by 3 to obtain the same y coefficient.

$\therefore \quad 3x - 3y = 21c$ — (3)

$2x + 3y = 4c$ — (1)

Now add equation (3) to equation (1) to eliminate y.

$\therefore \quad 5x = 25c$

$x = 5c$

Substitute $x = 5c$ into either equation (1) or (2) to find y.

$\therefore \quad y = -2c$

Example 2: Solve the following simultaneous equations, expressing the answers in exact form.

$x + 2y = 3\sqrt{3}$ — (1)

$2x - y = \frac{7\sqrt{3}}{2}$ — (2)

Solution: Multiply (2) by 2 to obtain same y coefficient.

$\therefore \quad 4x - 2y = 7\sqrt{3}$ — (3)

$x + 2y = 3\sqrt{3}$ — (1)

Now add equation (1) to equation (3) to eliminate y.

$\therefore \quad 5x = 10\sqrt{3}$

$\therefore \quad x = 2\sqrt{3}$

Substitute $x = 2\sqrt{3}$ into either (1) or (2) to find y.

$\therefore \quad y = \frac{\sqrt{3}}{2}$

Example 3: Solve the following simultaneously to find the values of a, b and c.

$a + b + c = 6$ — (1)

$a - 2b - 3c = 5$ — (2)

$a + 2b - c = 13$ — (3)

Simultaneous equations may involve 3 unknown variables.

Solution: Multiply equation (1) by 2 to obtain same b coefficient.

$\therefore \quad 2a + 2b + 2c = 12$ — (4)

$a - 2b - 3c = 5$ — (2)

Now add equation (4) to equation (2) to eliminate b.

$3a - c = 17$ — (5)

Also add equation (2) and (3) to eliminate b.

$2a - 4c = 18$ — (6)

Now using (1) and (6) we have only 2 unknown variables 'a' and 'c'.

Multiply (5) by 4

$12a - 4c = 68$ — (7)

Subtract equation (6) from (7)

$\therefore \quad 10a = 50$

$a = 5$

Substitute $a = 5$ into (5) or (6)

$\therefore \quad c = -2$

and finally substitute a = 5 and c = –2 into (1).

$\therefore \quad b = 3$

SOLVING QUADRATIC EQUATIONS BY FACTORISATION

Quadratic equations are different to 'simple equations' and 'simultaneous equations' because they always have a 'squared term'.

They have the format or pattern:

$$ax^2 + bx + c = 0$$

where a, b, c are constants and $a \neq 0$.

If two numbers multiplied together give an answer of zero, then either one of the numbers must be equal to zero.

i.e. If $a \times b = 0$ then either $a = 0$ or $b = 0$.

Similarly if two brackets multiplied together give an answer of zero, then either the first bracket or the second bracket must equal zero.

i.e. (FIRST) × (SECOND) = 0 then either (FIRST) = 0 or (SECOND) = 0

We use the above simple logic to solve quadratic equations.

(i) Make sure the right hand side of the equation is equal to zero.
(ii) Factorise the quadratic expression.
(iii) Solve each bracket equal to zero.
(iv) There are usually two distinct solutions, although occasionally the two answers are the same.

Example: Solve $x^2 = 2x + 8$

Solution:

$x^2 - 2x - 8 = 0$	Make the right hand side of the equation equal to zero.
$\therefore \quad (x - 4)(x + 2) = 0$	Factorise the quadratic expression.
$\therefore \quad x - 4 = 0$ or $x + 2 = 0$	Let each bracket equal zero.
$\therefore \quad x = 4$ or $x = -2$	Solve each simple equation.

FURTHER EXAMPLES

This chapter assumes that you are proficient and confident in factorising all types of binomial and trinomial expressions which occur in algebra. If you are unsure of how to factorise, please thoroughly revise Chapter 1 before continuing with this section.

Examples: Solve the following quadratic equations.

(i) $x^2 - 16 = 0$

(ii) $3x^2 - 2x - 5 = 0$

(iii) $x^2 - 6x = 0$

(iv) $x^2 = 6x - 9$

v) $x = \frac{60}{x} - 4$

Solutions:

(i) $\therefore\ (x + 4)(x - 4) = 0$ — Factorising using $a^2 - b^2 = (a + b)(a - b)$.

$\therefore\ x + 4 = 0$ or $x - 4 = 0$ — Let each bracket equal to zero.

$\therefore\ x = -4$ or $x = 4$ — Solve each simple equation.

(ii) $\therefore\ (3x - 5)(x + 1) = 0$ — Factorise this harder trinomial.

$\therefore\ 3x - 5 = 0$ or $x + 1 = 0$ — Let each bracket equal to zero.

$\therefore\ x = \frac{5}{3}$ or $x = -1$ — Solve each simple equation.

(iii) $\therefore\ x(x - 6) = 0$ — Factorise this expression.

$\therefore\ x = 0$ or $x = -6$ — x is the same as $(x + 0)$.

$\therefore\ x = 0$ or $x = 6$ — Solve each simple equation.

(iv) $\therefore\ x^2 - 6x + 9 = 0$

$\therefore\ (x - 3)(x - 3) = 0$

$\therefore\ x - 3 = 0$ or $x - 3 = 0$

$\therefore\ x = 3$ is the only solution.

(v) $x^2 = 60 - 4x$

$\therefore\ x^2 + 4x - 60 = 0$

$\therefore\ (x - 6)(x + 10) = 0$

$\therefore\ x = 6$ or $x = -10$

Note: Firstly, multiply by x to clear fractions. Then rearrange the equation into the format $ax^2 + bx + c = 0$.

COMPLETING THE SQUARE

The easiest and fastest way of solving quadratic equations is by the method of factorisation explained on the previous pages. However, there are two alternative methods you can use if the quadratic cannot be factorised easily by inspection.

'Completing the square' involves making a perfect square on one side of the equation.

i.e. $x^2 + 2ax + a^2 = (x + a)^2$

or $x^2 - 2ax + a^2 = (x - a)^2$

Before outlining the steps involved in this method, it is important you understand the meaning of the word 'coefficient'.

The coefficient of a term is the number directly before the term.

For example:

In $5x^2 - 8x + 4$

The constant term is + 4 or 4.

The coefficient of x^2 is 5.

The coefficient of x is – 8.

Half the coefficient of x is – 4.

(Half the coefficient of x)^2 is $(-4)^2 = 16$.

STEPS FOR COMPLETING THE SQUARE METHOD:

(i) Ensure that the constant term is on the right hand side.

(ii) Ensure the coefficient of x is 1 by dividing through.

(iii) Add $\left(\frac{1}{2}\text{ the coefficient of } x\right)^2$ to both sides of the equation.

(iv) Complete the square using either $(x + a)^2$ or $(x - a)^2$. The constant term in the bracket is found by halving the coefficient of x.

(v) Take the square root of both sides of the equation.

(vi) Finally solve for x.

This method is not as difficult as it might first appear, and the examples over the page will help to illustrate and clarify the steps involved.

EXAMPLES

Example: Solve $x^2 + 6x - 10 = 0$ by completing the square.

Step (i)	$x^2 + 6x = 10$	Constant term must be on right.
Step (ii)	$x^2 + 6x = 10$	The coefficient of x^2 is already 1.
Step (iii)	$x^2 + 6x + \left(\frac{6}{2}\right)^2 = 10 + \left(\frac{6}{2}\right)^2$	Add $\left(\frac{1}{2}\text{ the coefficient of } x\right)^2$ to both sides.
	$\therefore\ x^2 + 6x + 9 = 19$	
Step (iv)	$(x + 3)^2 = 19$	Complete the square on the left side. by halving the coefficient of x.
Step (v)	$x + 3 = \pm\sqrt{19}$	Take square root of both sides.
Step (vi)	$\therefore\ x = -3 + \sqrt{19}$ or $x = -3 - \sqrt{19}$	Finally solve for x, taking into account the 2 solutions for $-\sqrt{19}$ or $+\sqrt{19}$.

ANSWERS ARE ALWAYS LEFT IN EXACT SURD FORM UNLESS THE QUESTION STATES OTHERWISE.

Example: Solve $2x^2 - 3x - 1 = 0$ by completing the square, and give the answer correct to 2 decimal places.

Step (i)	$2x^2 - 3x = 1$	Constant term must be on right.
Step (ii)	$x^2 - \frac{3}{2}x = \frac{1}{2}$	Divide through by 2 to make coefficient of x^2 equal 1.
Step (iii)	$x^2 - \frac{3}{2}x + \left(\frac{3}{4}\right)^2 = \frac{1}{2} + \left(\frac{3}{4}\right)^2$	Add $\left(\frac{1}{2}\text{ the coefficient of } x\right)^2$ to both sides.
	$\therefore\ x^2 - \frac{3}{2}x + \frac{9}{16} = 1\frac{1}{16}$	
Step (iv)	$\left(x - \frac{3}{4}\right)^2 = \frac{17}{16}$	Complete the square on the left side by halving the coefficient of x.
Step (v)	$x - \frac{3}{4} = \pm\sqrt{\frac{17}{4}}$	Take square root of both sides.
Step (vi)	$\therefore\ x = \frac{3}{4} + \sqrt{\frac{17}{4}}$ or $x = \frac{3}{4} - \sqrt{\frac{17}{4}}$	Solve for x, noting the 2 solutions.
	i.e. $x = 1.78$ or $x = -0.28$	Answers to 2 decimal places.

THE QUADRATIC FORMULA

A formula has been worked out which automatically takes into account all the steps outlined in the previous section on 'completing the square'. If you can remember the formula, then it is a very much simpler and faster method of solving quadratic equations. The first step is to compare the given quadratic equation to the general formula '$ax^2 + bx + c = 0$' in order to determine the values of the constants a, b and c. Then simply substitute these values into the quadratic formula below:

$$x = \frac{-b \pm \sqrt{b^2 - 4ac}}{2a}$$

Example 1: Solve $x^2 + 6x - 10 = 0$

Compare $x^2 + 6x - 10 = 0$ to $ax^2 + bx + c = 0$

$\therefore\ a = 1, b = 6, c = -10$

using $x = \dfrac{-b \pm \sqrt{b^2 - 4ac}}{2a}$ Formula.

$= \dfrac{-6 \pm \sqrt{6^2 - 4 \times 1 \times (-10)}}{2 \times 1}$ Substitute in values.

$= \dfrac{-6 \pm \sqrt{76}}{2}$ Simplify.

$= -3 \pm \sqrt{19}$ $\sqrt{76} = 2\sqrt{19}$

Example 2: Solve $2x^2 - 3x - 1 = 0$ to 2 decimal places.

Compare $2x^2 - 3x - 1 = 0$ to $ax^2 + bx + c = 0$

$\therefore\ a = 2, b = -3, c = -1$

using $x = \dfrac{-b \pm \sqrt{b^2 - 4ac}}{2a}$

$= \dfrac{3 \pm \sqrt{(-3)^2 - 4 \times (2) \times (-1)}}{2 \times 2}$

$= \dfrac{3 \pm \sqrt{17}}{4}$

$= 1.78$ or -0.28

Note: The examples on this page are identical to the examples on the previous page, which uses the method of 'completing the square'.

APPLICATIONS OF QUADRATIC EQUATIONS

This is an important chapter to thoroughly understand, because many major topics in Years 11 and 12 involve quadratic equations. In summary, there are 3 main methods of solution:

(i)	Factorising
(ii)	Completing the square
(iii)	Using the quadratic formula

If the quadratic can be factorised, then the first method is the best and fastest to use. If it can't be factorised, then it is easier to use the quadratic formula – unless the question specifically asks you to solve the equation by 'completing the square'. When using either of the last two methods, the final answers should always be left in their exact surd form, unless the question asks you to approximate the answers to a certain number of decimal places. The examples to follow all involve applications of quadratic equations.

Example 1: When a number is subtracted from its square, the result is 36. FInd the number(s).

Solution: Let the unknown number be x.

$\therefore\ x^2 - x = 56$	Form an equation from the question.
$\therefore\ x^2 - x - 56 = 0$	Make right hand side equal zero.
$\therefore\ (x - 8)(x + 7) = 0$	Factorise the trinomial.
$\therefore\ x - 8 = 0$ or $x + 7 = 0$	Let each bracket equal zero.
$\therefore\ x = 8$ or $x = -7$	Solve each simple equation.

The numbers are 8 and –7.

CHECK: $8^2 - 8 = 56$

$(-7)^2 - (-7) = 56$

Example 2: Calculate where the parabola y $= x^2 - x - 6$ cuts the x-axis.

Solution: The x-axis and the line $y = 0$ are exactly the same.

$\therefore$ Solve $0 = x^2 - x - 6$

$\therefore\ 0 = (x - 3)(x + 2)$

$\therefore\ x = 3$ or $x = -2$

$\therefore$ Parabola cuts the x-axis at $(3, 0)$ and $(-2, 0)$.

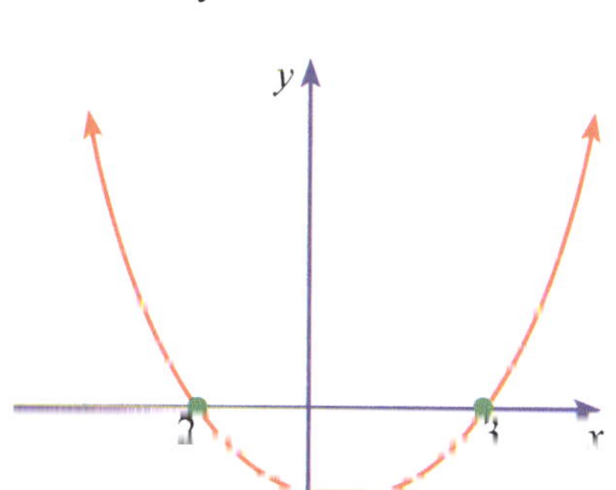

The values of $x = 3$ and $x = -2$ are often called the 'roots' of the quadratic expression.

Example 3: A rectangle is 3 m longer than its breadth. If the area is 54 cm^2, find its dimensions.

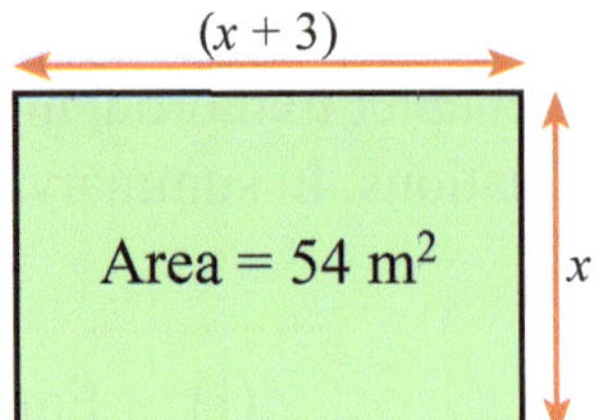

Solution:

Let the breadth = x.

$\therefore$ the length = $(x + 3)$ m

length $\times$ breadth = area

$\therefore\ x(x + 3) = 54$ — Form an equation from the question.

$\therefore\ x^2 + 3x - 54 = 0$ — Make right hand side equal zero.

$\therefore\ (x + 9)(x - 6) = 0$ — Factorise the trinomial.

$\therefore\ x + 9 = 0$ or $x - 6 = 0$ — Let each bracket equal zero.

$\therefore\ x = -9$ or $x = 6$ — Solve each simple equation.

$\therefore$ Breadth = x = 6 m
Length = $x + 3$ = 9 m

Must ignore the solution $x = -9$ because a negative dimension doesn't make sense.

Example 4: A ball is thrown upwards, and its height, h metres, after t seconds is given by the formula $h = 8t - t^2$.

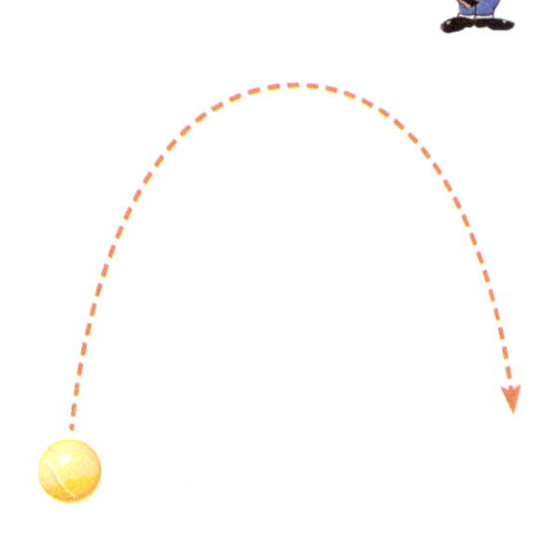

a) What is the height of the ball after 5 seconds?
b) At what times will the height be 12 m? Explain why there are 2 answers.
c) How long will it take for the ball to land on the ground?

Solution:

a) When $t = 5$, $h = 8 \times 5 - 5^2 = 15$ m high

b) Substitute $h = 12$ into $h = 8t - t^2$ and solve.

$\therefore\ 12 = 8t - t^2$

$\therefore\ t^2 - 8t + 12 = 0$

$\therefore\ (t - 2)(t - 6) = 0$

$\therefore\ t = 2$ or $t = 6$

There are 2 answers: one for the way up and one for the way back down.

The height will be 12 m after 2 seconds and 6 seconds.

c) Substitute $h = 0$ for ground level and solve.

$\therefore\ 0 = 8t - t^2$

$\therefore\ 0 = t(8 - t)$

$\therefore\ t = 0$ or $t = 8$

i.e. The ball will land back on the ground after 8 seconds.

CHAPTER SUMMARY

SUBSTITUTING INTO FORMULAE

Step 1: The question will give you a formula which you may, or may not, have seen before.

Step 2: Substitute in the values of the given letters.

Step 3: Now find the value of the unknown letter using exactly the same steps you have learned to solve equations.

Example: If $v = u + at$ find the value of t when $v = 18$, $u = 9$ and $a = 2$.
Substitute $v = 18$, $u = 9$ and $a = 2$ into $v = u + at$.

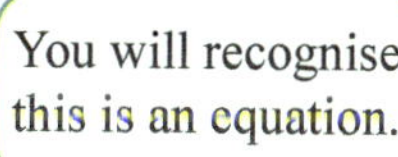

Solution: $\therefore \quad 18 = 9 + 2t$ Subtract 9 from both sides.

$\therefore \quad 9 = 2t$ Divide both sides by 2.

$\therefore \quad t = \frac{9}{2}$ or $4\frac{1}{2}$

CHANGING THE SUBJECT OF A FORMULA

The subject of a formula is the pronumeral (or letter) which is by itself on the left hand side.

For example, if $v = u + at$ then the subject of the formula is the letter v. You may be asked to make one of the other letters the subject of the formula. This is done by rearranging the formula using exactly the same theory used to solve equations.

Example: If $v = u + at$,   Make t the subject of the formula.

Solution: $\therefore \quad v - u = at$ Subtract u from both sides.

$\therefore \quad \frac{v - u}{a} = t$ Divide both sides by the letter a.

$\therefore \quad t = \frac{v - u}{a}$ The letter t is now the subject of the formula.

USING EQUATIONS TO SOLVE PROBLEMS

We can often make use of equations to solve problems more easily. There are 4 important steps which should always be followed:

Step 1: Introduce a pronumeral or variable (letter).

Step 2: Write down the equation.

Step 3: Solve the equation.

Step 4: Answer the original question.

It doesn't matter which letter or pronumeral you introduce!

SIMULTANEOUS EQUATIONS

GRAPHICAL METHOD OF SOLUTION

Two equations are said to be solved **SIMULTANEOUSLY**, when you find an ordered pair (x, y) which **satisfies both equations**. This could be done graphically by drawing graphs of the two equations and finding their point of intersection.

The disadvantage in using graphs is that they are time consuming to plot, and also they can be inaccurate if the answers involve fractions.

THE SUBSTITUTION METHOD

This is one of two main methods involving algebra. It is not only faster than the graphical method described, but it also gives the exact answer - particularly when fractions are involved.

Step 1: Make one letter (either x or y) the subject of one of the equations.

Step 2: Substitute the expression for this letter into the second equation.

Step 3: Solve the resulting simple equation.

Step 4: Substitute this value back into either equation to find the value of the other letter.

ELIMINATION METHOD

This is the other main method involving algebra. Both methods (substitution or elimination) will give the correct answers, but in some cases the choice of method used will be much easier and faster.

The **COEFFICIENT** of a letter is the number in front of the letter. Therefore in the equation $2x + 3y = 8$, the coefficient of x is 2, and the coefficient of y is 3.

Step 1: By multiplication make the coefficient of one of the letters in both equations the same.

Step 2: Add or subtract the two equations to eliminate one of the letters.

Step 3: Solve the resulting simple equation.

Step 4: Substitute this value back into either equation to find the value of the other letter.

PROBLEM SOLVING APPLICATIONS

The solution of simultaneous equations has many applications in Maths. Students should be able to apply their knowledge to solve typical problems involving 2 or more sets of different data or information. In later chapters, you will be required to solve a variety of simultaneous equations in order to find the points of intersection of lines, parabolas and other graphs. Sometimes simultaneous equations will involve the solution of 3 unknown variables.

QUADRATIC EQUATIONS

THE FORMAT OR PATTERN

Quadratic equations are different to 'simple equations' and 'simultaneous equations' because they always have a 'squared term'. The general formula is:

$ax^2 + bx + c = 0$ where a, b and c are constants and $a \neq 0$

SOLVING BY FACTORISATION

Step 1: Make sure the right hand side of the equation is equal to zero.

Step 2: Factorise the quadratic expression.

Step 3: Solve each bracket equal to zero.

Step 4: There are usually two distinct solutions, although occasionally the two answers are the same.

Example: Solve

$x^2 = 2x + 8$	
$\therefore\ x^2 - 2x - 8 = 0$	Make right hand side = 0.
$\therefore\ (x - 4)(x + 2) = 0$	Factorise the quadratic equation.
$\therefore\ x - 4 = 0$ or $x + 2 = 0$	Let each bracket equal zero.
$\therefore\ x = 4$ or $x = -2$	Solve both simple equations.

COMPLETING THE SQUARE

The easiest and fastest way of solving quadratic equations is by the method of factorisation explained above. However, there are two alternative methods you can use if the quadratic cannot be factorised easily by inspection.

'Completing the square' involves making a perfect square on one side of the equation:

i.e. $x^2 + 2ax + a^2 = (x + a)^2$ or $x^2 - 2ax + a^2 = (x - a)^2$

STEPS FOR COMPLETING THE SQUARE

Step 1: Ensure that the constant term is on the right hand side.

Step 2: Ensure the coefficient of x^2 is 1 by dividing through.

Step 3: Add (half the coefficient of x)^2 to both sides of the equation.

Step 4: Complete the square using either $(x + a)^2$ or $(x - a)^2$. The constant term in the bracket is found by halving the coefficient of x.

Step 5: Take the square root of both sides of the equation.

Step 6: Finally solve for x.

THE QUADRATIC FORMULA

A formula has been worked out which automatically takes into account all the steps outlined in the previous section on 'completing the square'. If you can remember the formula, then it is a very much simpler and faster method of solving quadratic equations.

The first step is to compare the given quadratic equation to the general formula $ax^2 + bx + c = 0$ in order to determine the values of the constants a, b and c. Then simply substitute these values into the quadratic formula below:

$$x = \frac{-b \pm \sqrt{b^2 - 4ac}}{2a}$$

Example: Solve $3x^2 + 6x - 10 = 0$
Compare $3x^2 + 6x - 10 = 0$ to $ax^2 + bx + c = 0$
Now substitute $a = 3$, $b = 6$, $c = -10$ into the formula above.

SOLVING PROBLEMS USING QUADRATIC EQUATIONS

Example:

The sides of a right angled triangle are $(x + 1)$ cm, $(x + 3)$ cm and $(x + 5)$ cm. Find the perimeter of the triangle.

Using Pythagoras' theorem:

A, B, C; $(x + 3)$, $(x + 5)$, $(x + 1)$

$AC^2 = AB^2 + BC^2$

$\therefore\ (x + 5)^2 = (x + 3)^2 + (x + 1)^2$

$\therefore\ x^2 + 10x + 25 = x^2 + 6x + 9 + x^2 + 2x + 1$

$\therefore\ x^2 - 2x - 15 = 0$

$\therefore\ (x - 5)(x + 3) = 0$

$\therefore\ x = 5$ or $x = -3$ Ignore -3 because we can't have negative dimensions.

$\therefore$ Sides are 6 cm, 8cm, 10 cm

$\therefore$ Perimeter = 6 + 8 + 10 = 24 cm

SOME OTHER APPLICATIONS

The quadratic or parabola $y = ax^2 + bx + c$ has many practical applications in Science, Engineering and the Military.

Some examples:

(i) If you throw a ball it will follow a parabolic trajectory.
(ii) Bullets, mortar shells or any other type of military cannon will follow a parabolic trajectory.
(iii) The cables of a well-designed suspension bridge follow a parabolic curve.
(iv) Mirrors for high grade astronomical telescopes are parabolic in shape.
(v) Radar antennas, the 'dishes' used to pick up satellite television signals, are all parabolic in shape.

LEVEL 1 — EQUATIONS AND FORMULAE — EASIER QUESTIONS

Note: Only turn back to page number shown if you have difficulty.

	Page
Q1. Solve these equations: a) $3x - 8 = 6x + 4$ b) $4p + 7 = 6p - 11$ c) $9 + 10x = 15x + 7$ d) $12 - 5a = 3a + 2$ e) $4b - 9 = -2b - 5$ f) $16 - 3a = 8 - 5a$ g) $2(5 - 2x) = -1\ (7x + 5)$ h) $5(3m - 8) = 4(3m - 4)$ i) $6(3 - 2y) = 2(3y + 4)$	164, 165
Q2. Solve for x: a) $\frac{x}{3} - \frac{x}{4} = 2$ b) $\frac{2x}{5} + \frac{x}{2} = 3$ c) $\frac{3x+1}{2} - 3 = -7$ d) $\frac{6}{2x} + \frac{5}{x} = 2$ e) $\frac{4}{3x-2} = -2$ f) $\frac{6}{6x+3} + 4 = 8$	164, 165
Q3. Solve these inequations and graph them on a number line: a) $\frac{x}{9} + \frac{x}{6} < 5$ b) $x - \frac{x}{3} > 4$ c) $2x + \frac{2x}{3} \leq 2$ d) $\frac{x-1}{2} < \frac{x+5}{5}$ e) $\frac{x+2}{3} > \frac{2x+3}{4}$ f) $\frac{3-2x}{3} \leq \frac{x+9}{2}$	165
Q4. a) A woman is twice as old as her daughter. Five years ago the sum of their ages was 62. What are their present ages? b) A bottle is half full. After adding 275 mL to the bottle it is three quarters full. How much does the bottle hold when full? c) A father is five times as old as his son. Three years ago he was eight times as old. Find the father's present age. d) The hire of a car is \$53 per day plus 84c per kilometre driven. If the total cost for hiring the car for 3 days was \$349.68, find out how many kilometres were driven. e) Zac saves all his 20c coins and 50c coins. When he counts them he has a total of 112 coins amounting to \$37.10. How many of each type of coins does he have?	166, 167
Q5. Given the equation $X = \frac{1}{x^2} - \frac{1}{y^2}$ find: a) X when $x = \frac{1}{5}, y = \frac{1}{4}$ b) x when $X = 11\frac{1}{4}, y = \frac{1}{3}$ c) y when $X = 1\frac{3}{4}, x = \frac{1}{2}$	168
Q6. In the formula $v^2 = u^2 + 2as$, make the following letters the subject of the formula: i) a ii) s iii) v iv) u	170

Note: Only turn back to page number shown if you have difficulty.

	Page
Q1. Solve these equations using the substitution method: a) $2x + y = 4$, $x + y = 10$ b) $x + 2y = 1$, $2x + y = 5$ c) $7x + 3y = 8$, $4x - y = -9$ d) $5x + 2y = -15$, $3x - y = 2$ e) $3x + 2y = 8$, $2x - 4y = -8$ f) $5x + 2y = -16$, $2x - 3y = 5$	173
Q2. Solve these equations using the elimination method: a) $x - 3y = 7$, $x - y = 3$ b) $3x - 2y = 3$, $x - 2y = -15$ c) $4x - 3y = -1$, $2x + 3y = 13$ d) $5x + y = -6$, $x + 2y = 24$ e) $5x + 2y = 25$, $4x - 3y = -3$ f) $3x - 4y = -11$, $7x + 6y = -18$	174
Q3. Form two separate equations using the given information and solve them simultaneously to calculate the solution: a) The difference between two numbers is 11 and twice the smaller number, minus four, equals the larger number. Find the two numbers. b) Six small marbles and four large marbles have a combined weight of 30 g, whilst seven small marbles and two large marbles have a weight of 23 g. Find the weight of a large marble. c) A rectangle is 5 cm longer than it is wide. If the length and breadth are increased by 2 cm each, the area increases by 50 cm^2. Find the dimensions of the original rectangle. d) A man is five times as old as his son. Four years ago he was nine times as old. Find their present ages. e) Rectangle ABCD is twice as long and four times as wide as rectangle MNOP. Find the dimensions of each rectangle, if their perimeters are 48 cm and 20 cm respectively. f) A square is transformed into a rectangle of perimeter 30 cm, by increasing one side by 3 cm and reducing the adjacent side by 2 cm. Find the length of the square's side. g) Find the point of intersection of the line $y = 3x + 2$ and the line $y = 7x - 6$. h) Find the point of intersection of the line $3x + 2y - 6 = 0$ and the line $2x - 3y - 17 = 0$.	175, 176
Q4. Show that the lines $4x - 3y + 23 = 0$, $3x + 2y - 4 = 0$ and $5x + 4y - 10 = 0$ are concurrent.	176
Q5. The line $y = mx + b$ passes through the points (3, 2) and (2, –1). Form two equations and solve them simultaneously to find the values of 'm' and 'b' and hence the equation of the line.	175, 176

Note: Only turn back to page number shown if you have difficulty. | Page

Q1. Solve the following quadratic equations by first factorising: 178, 179

a) $x^2 + 5x + 6 = 0$ b) $x^2 - 2x - 24 = 0$ c) $x^2 - 8x + 16 = 0$

d) $x^2 + x - 12 = 0$ e) $x^2 + x - 56 = 0$ f) $x^2 - 5x - 24 = 0$

Q2. Solve the following quadratic equations by first factorising: 178, 179

a) $6x^2 - 8x - 8 = 0$ b) $8x^2 - 10x - 3 = 0$

c) $20x^2 + 19x + 3 = 0$ d) $6x^2 - 10x - 24 = 0$

e) $12x^2 + 31x + 7 = 0$ f) $9x^2 - 70x - 16 = 0$

Q3. Solve the following quadratic equations by first factorising: 178, 179

a) $x^2 - 25 = 0$ b) $x^2 + 6x = 27$ c) $10x^2 = x + 2$ d) $x^2 - x = 0$

e) $5x - x^2 = 0$ f) $9x = x^2$ g) $x^2 = 1$ h) $9x^2 - 4 = 0$

Q4. Solve the following by completing the square: 180, 181

a) $x^2 + 4x - 3 = 0$ b) $x^2 - 6x + 1 = 0$ c) $x^2 + 2x - 5 = 0$

d) $x^2 - 3x - 2 = 0$ e) $x^2 - 7x + 2 = 0$ f) $x^2 + 9x - 8 = 0$

Q5. Solve the following by completing the square: 180, 181

a) $2x^2 - 5x + 2 = 0$ b) $4x^2 + 10x + 3 = 0$

c) $6x^2 - 5x - 2 = 0$ d) $5x^2 - 3x - 4 = 0$

e) $3x^2 - 3x - 1 = 0$ f) $3x^2 - 8x + 3 = 0$

Q6. Use the quadratic formula to solve these: 182

a) $3x^2 - 9x + 2 = 0$ b) $2x^2 + 2x - 1 = 0$

c) $4x^2 + 7x + 3 = 0$ d) $2x^2 + 11x - 5 = 0$

e) $3x^2 + 9x + 5 = 0$ f) $5x^2 - 3x - 4 = 0$

Q7. Solve: 182

a) $3x + 5 = -\frac{2}{x}$ b) $x = \frac{3}{5x + 8}$ c) $x^2 + 4x = -\frac{4}{3}$

d) $x = \frac{1}{2x} + \frac{1}{4}$ e) $x - 1 = -\frac{1}{4x}$ f) $4x = 7 - \frac{3}{2x}$

Note: Firstly, rewrite each equation in the format $ax^2 + bx + c = 0$

Q8. a) Two consecutive even numbers have a product of 288. What are the numbers? 183, 184

b) The sum of a negative number and its square equals 156. What is the number?

c) The base of a triangle is 6 cm longer than its height. If the area of the triangle is 140 cm^2, what is the length of the base?

d) The sum of three consecutive numbers is squared and the result is 441. What are the numbers?

 HARDER QUESTIONS

Q1. Write down an equation which satisfies each table of values below:

a)

x	0	1	2	3
y	–1	2	5	8

b)

x	0	1	2	3
y	5	4	3	2

c)

x	0	2	4	6
y	–1	0	1	2

Q2. Work out the equation for each graph below: (**Hint:** A table of values may be useful.)

a) 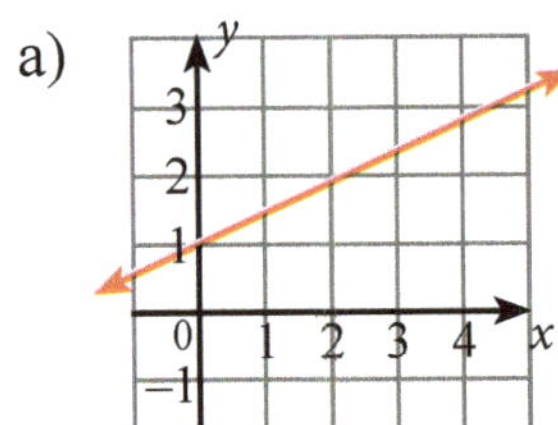

b) 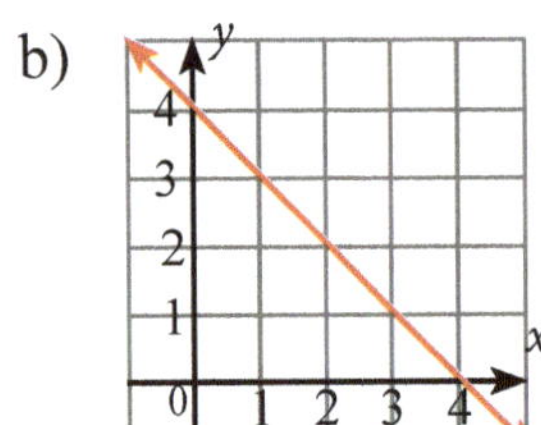

c) 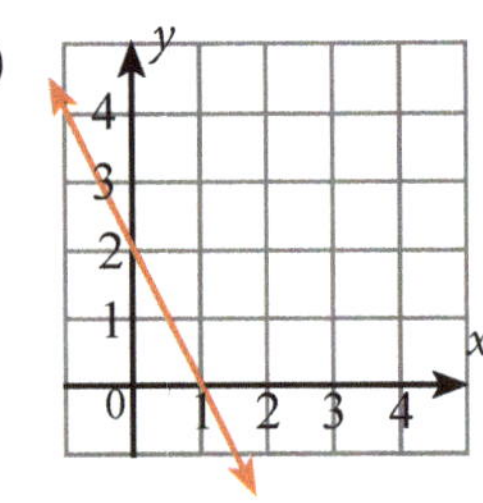

Q3. Solve these equations by first expanding the brackets on both sides:

a) $3(2x - 3) = 9(4x - 21)$
b) $3(3x + 4) = 3(2x - 5)$
c) $5(2y - 3) = 2(3y + 1)$
d) $2(x - 3) = 4(3x - 1)$
e) $10(2m + 1) = 6(3m + 4)$
f) $7(5 - 3b) = 8(2b - 5)$
g) $3(9x - 2) = 5(3x + 7)$
h) $12(8 - 5a) = 5(8a + 12)$

Q4. In the following equations make the letter shown in brackets the subject of the formula:

i) $V = \pi r^2 h$ [h]
ii) $V = \frac{4}{3}\pi r^3$ [r]
iii) $r = \sqrt{3b - 4ac}$ [a]
iv) $x = \sqrt{b^2 - 4ac}$ [b]
v) $am = 4(a - b)$ [a]
vi) $w = \frac{x^2 - 2}{3}$ [x]

Q5. Solve each of the following non-linear equations:

a) $3k^2 - 7 = 41$
b) $\frac{p^2 - 7}{3} = 14$
c) $2m^2 + 7 = 135$
d) $m^3 = -27$
e) $\frac{m^3 - 9}{5} = 11$
f) $4a^3 + 132 = -368$

Q6. A rectangular prism of height h has a base which is three times as long as it is wide. The surface area of the prism is 42 cm^2.

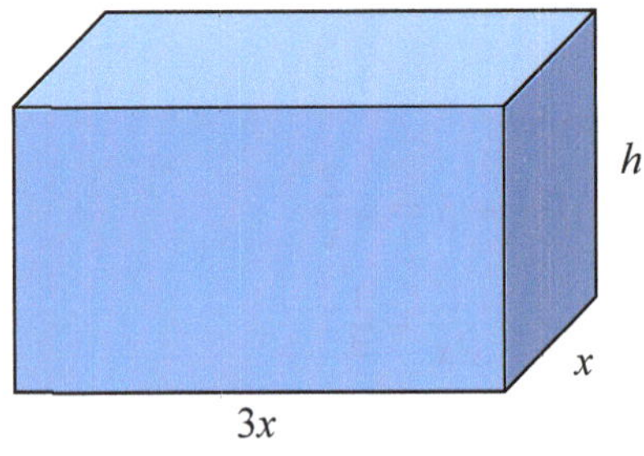

a) (i) Letting the width be x cm express the height h in terms of x.
(ii) If the height of the prism is equal to the width, find the exact value for x.

b) (i) Express the volume $V\text{ cm}^3$ in terms of x.
(ii) Show that for $x = 1$ and $x = 2$ the volume of the prisms are the same.

Q7. Solve for x:

a) $\frac{1}{2}(8x - 1) = \frac{1}{3}(6x + 9)$
b) $\frac{5x - 3}{6} = x + 2$
c) $\frac{x - 1}{x + 6} - 3 = 5$
d) $\frac{x + 2}{4 - x} + 2 = \frac{2}{3}$
e) $4 - \frac{2x + 3}{3 - 2x} = \frac{1}{2}$
f) $\frac{x - 2}{3} + \frac{x + 3}{2} = x$
g) $\frac{3(4x - 3)}{4} - \frac{2(3x + 1)}{3} = \frac{5x}{4}$
h) $\frac{3x - 1}{5} - \frac{2x + 3}{4} = x - 3$
i) $\frac{5}{x} + x = \frac{9}{x}$

Q1. Calculate where the following parabolas cut or intersect the x-axis:

a) $y = x^2 - 2x - 8$ b) $y = x^2 - 7x$ c) $y = x^2 - 8x + 15$
d) $y = 16 - x^2$ e) $y = x^2 - 6x + 9$ f) $y = 15 + 2x - x^2$

Q2. The line $y = 3x + 5$ intersects the parabola $y = x^2 + 5x - 3$ at 2 points. Solve the two equations simultaneously to find to the coordinates of the points of intersection.

Q3. A ball is thrown upwards and its height, h metres, after t seconds is given by the formula $h = 12t - t^2$.

a) What will be the height of the ball after 3 seconds?
b) At what times will the height be 32 m?
c) How long will it take for the ball to land on the ground?

Q4. Use the quadratic formula to solve the following for x.

a) $x^2 + 2x - k = 0$ b) $x^2 + 2x + 4k = 0$ c) $x^2 - 4x - 4k = 0$
d) $x^2 - 2kx - 1 = 0$ e) $x^2 + 4kx - 1 = 0$ f) $x^2 - 6kx - 4 = 0$
g) $kx^2 - 4x + 4 = 0$ h) $2kx^2 - kx - 4k = 0$ i) $2kx^2 - kx + 4k^2 = 0$

Q5. The discriminant (denoted by Δ) is the part under the square root section of the general quadratic formula: $\Delta = b^2 - 4ac$.

a) Explain how the discriminant can be used to give the number of solutions to any quadratic equation.
b) Find the discriminant for the following quadratic equations and hence state the number of solutions for each one.
(i) $x^2 - 4x - 3$ (ii) $2x^2 + 4x + 10$ (iii) $3x^2 + x + 8$ (iv) $x^2 - 4x + 4$
c) Find the value of m for $x^2 + mx + 4$ so that each equation has
(i) one solution (ii) no solution (iii) two solutions.
d) (i) For $x^2 + mx + (m - 1) = 0$ find the value of m for which there is one solution.
(ii) Show that it is impossible for $x^2 + mx + (m - 1) = 0$ to have no solutions.

Q6. In this question do NOT find the point(s) of intersection (if they exists). Determine whether the line $y = 18x - 24$ intersects twice, is a tangent to, or completely misses the parabola $y = 9x^2 - 30x + 40$.

Q7. Solve the following simultaneously for the indicated variables.
a) $a + b + c = 10,\ a + b - c = 0,\ 2a + b + c = 11$
b) $a + b + c = 10,\ a + b - c = 6,\ 2a + b + c = 13$
c) $a + b + c = 8,\ a + b - c = 0,\ 2a + b + c = 8$
d) $a + b + c = 34,\ b - 2a = c - b = 8$
e) $c - b = b - a = 2,\ a + b = 8$

Q8. What fraction of the square is shaded grey?

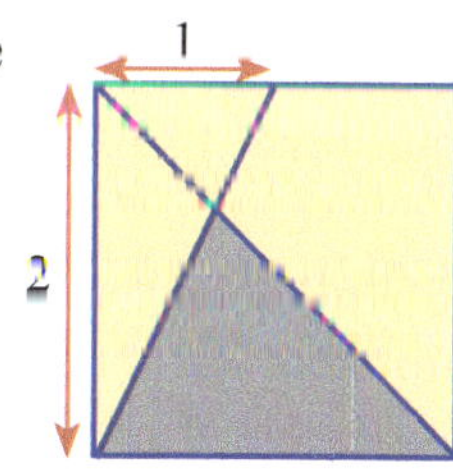

Q9. Find the green area.

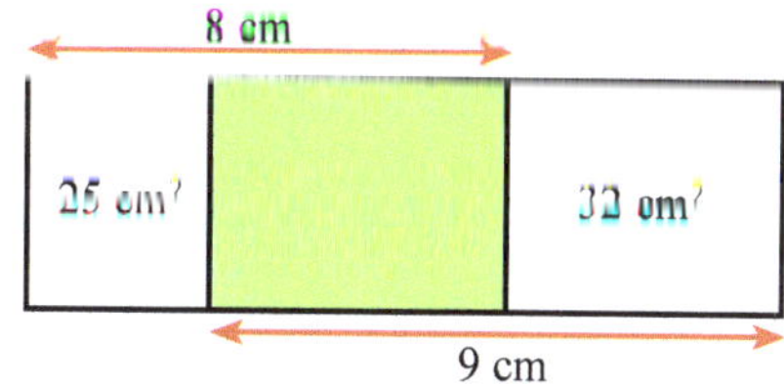

Q1. The equation for an ellipse in terms of the variables x and y is $\frac{x^2}{4} + y^2 = 1$.

a) (i) Express y in terms of x. (ii) Hence explain why $-2 \le x \le 2$.

(iii) Complete the table of values expressing the values for y correct to three decimal places where necessary.

x	–2	–1.5	–1	–0.5	0	0.5	1	1.5	2
y									

b) Express x in terms of y. (ii) Hence explain why $-1 \le y \le 1$.

(ii) Complete the table of values expressing the values for x correct to three decimal places where necessary.

y	–1	–0.75	–0.5	–0.25	0	0.25	0.5	0.75	1
x									

c) The equation for an ellipse in terms of the variables x and y is $\frac{x^2}{a^2} + \frac{y^2}{b^2} = 1$, where a and b are constants. Make the following the subject of the equation formula. (i) x (ii) y (iii) a (iv) b

Q2. The number of diagonals, d that can be drawn inside a polygon with n sides is given by the formula $d = \frac{1}{2}n(n - 3)$.

a) (i) Using algebra show that $n \ne 0, 1, 2$.

(ii) Using geometry show that $n \ne 0, 1, 2$.

b) (i) Using the general quadratic formula express n in terms of d.

(ii) Show that whole values for n result for $d = \{2, 5, 9, 14\}$

c) The values for d as $\{2, 5, 9, 14\}$ form a pattern.

(i) Explain the pattern and write down the next four terms in the pattern.

(ii) Using t for the term number write the quadratic rule that expresses the value of d in terms of t.

(iii) Use the quadratic rule to find the first eight terms in the pattern.

Q3. The formula to find the amount of interest earned (I) when a principal amount (P) is invested with an annual interest rate (R) while compounding annually for n years is: $I = P(1 + R/100)^n - P$

a) Find the amount of compound interest earned when \$2 000 is invested at 5% per annum for 4 years.

b) (i) Express the compound interest formula with P as the subject.

(ii) Hence find the principal, expressed to the nearest dollar that needs to invested so that the interest earned is \$1000 when it is invested with an annual interest rate of 8%, compounding for 4 years.

c) (i) Write a formula in terms of R and n that describes the situation of an investment doubling in value.

(ii) Use the formula to find the time to the nearest month that it will take for an investment to double when invested at 10% per annum.

Q4. Leonhard Euler is credited with first using function notation where $f(x) = x^2$, it is read as 'the function f with variable x is equal to x squared'. f is the function or rule and x is the variable. For $f(x) = x^2$, then $f(a) = a^2$, $f(3) = 3^2 = 9$, $f(a + 1) = (a + 1)^2 = a^2 + 2a + 1$.

a) For $f(x) = x + 1$, express the following in simplest form. (i) $f(a)$ (ii) $f(2)$ (iii) $f(a + 1)$

b) For $f(x) = 2x$, express the following in simplest form. (i) $f(a)$ (ii) $f(3)$ (iii) $f(a - 1)$

c) For $f(x) = x^2 + 1$, express the following in simplest form. (i) $f(a)$ (ii) $f(4)$ (iii) $f(a + 1)$

d) For $f(x) = 2x^2$, express the following in simplest form. (i) $f(a)$ (ii) $f(5)$ (iii) $f(a - 1)$

e) For $f(x) = x^2 + x$, express the following in simplest form. (i) $f(a)$ (ii) $f(6)$ (iii) $f(a + 1)$

COORDINATE GEOMETRY

The 'Australian Curriculum Mathematics' (ACM) references for this sub-strand of 'Numbers and Algebra' (NA) are given below. This chapter may contain additional extension work which the author feels will be beneficial to the student.

- *The distance between two points (ACMNA 214).*
- *Graphing linear equations (ACMNA 215).*
- *Parallel and perpendicular lines (ACMNA 238).*
- *The midpoint and gradient of an interval (ACMNA 294).*
- *The gradient - intercept equation (NSW).*
- *Sketching parabolas (ACMNA 296).*
- *The point-gradient form of a line (NSW).*

Some of the opening concepts are revision of Year 9 work. New concepts are explained in the second half of the chapter. Please read the note at the bottom of page (v).

Rene Descartes (1596 - 1650)

He was a famous French philosopher, mathematician and scientist and he was the person who derived the method of describing the position of a point on a flat number plane. So it is often called the CARTESIAN coordinate system after the last 6 letters of his name. His best known philosophical statement is.

'I think, therefore I am.' or 'Cogito, ergo sum.'

ORDERED PAIRS

One can plot a point on a number plane using a pair of numbers. This pair of numbers (written in brackets and separated by a comma) is called the **COORDINATES** of the point. Because the order in which these 2 numbers are written is important, it is also often called an **ORDERED PAIR**.

> The first number always gives the number of units along the horizontal x-axis.
> The second number always gives the number of units up or down the vertical y-axis.

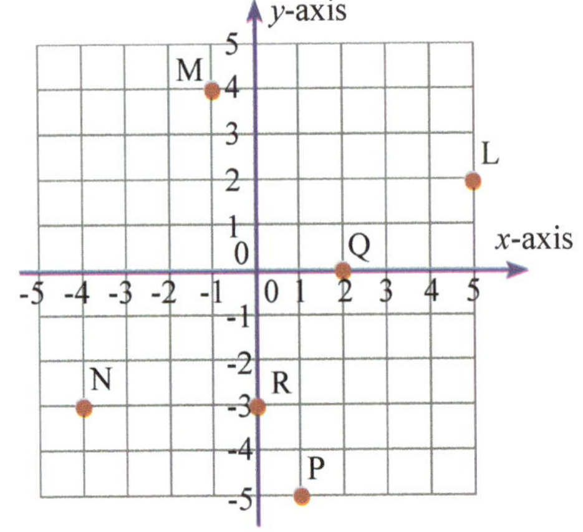

The point (0,0) is called the ORIGIN.

Example: Give the coordinates of the 6 points.

(i) L (ii) M (iii) N (iv) P (v) Q (vi) R

Solution:

(i) +5 units to the right (x-axis), +2 units up (y-axis)
$\therefore$ Coordinates of L are (5, 2)
(ii) –1 unit to left (x-axis), +4 units up (y-axis)
$\therefore$ Coordinates of M are (–1, 4)
(iii) –4 units to left (x-axis), –3 units down (y-axis)
$\therefore$ Coordinates of N are (–4, –3)
(iv) +1 unit to right (x-axis), –5 units down (y-axis)
$\therefore$ Coordinates of P are (1, –5)
(v) 2 units to right (x-axis), 0 units up or down (y-axis)
$\therefore$ Coordinates of Q are (2, 0)
(vi) 0 units to left or right (x-axis), –3 units down (y-axis)
$\therefore$ Coordinates of R are (0, –3)

MIDPOINT FORMULA

Consider a point A represented by coordinates (x_1, y_1) and another point B represented by the coordinates (x_2, y_2).

Then the coordinates of the middle point between A and B can be calculated using the formula:

$$\left(\frac{x_1 + x_2}{2}, \frac{y_1 + y_2}{2}\right)$$

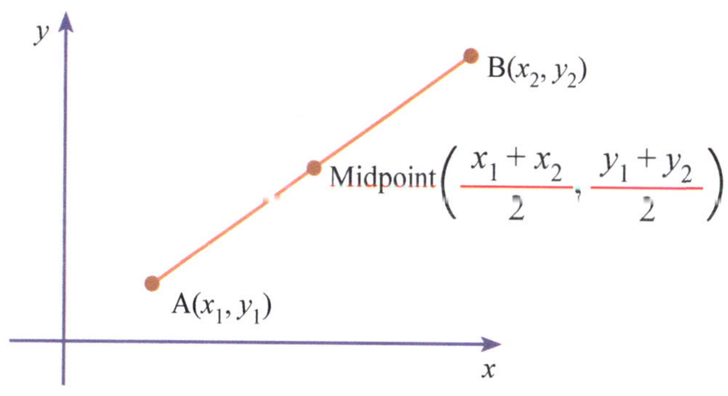

It is really the average of the x coordinates, and the average of y coordinates.

Example 1: Calculate the midpoint between A(1, –2) and B(5, 6).
Let $(x_1, y_1) = (1, -2)$ and $(x_2, y_2) = (5, 6)$

$$\therefore \quad \text{Midpoint of AB} = \left(\frac{x_1 + x_2}{2}, \frac{y_1 + y_2}{2}\right)$$

$$= \left(\frac{1 + 5}{2}, \frac{-2 + 6}{2}\right)$$

$$= (3, 2)$$

Example 2: The midpoint of an interval is (–2, 1). If the coordinates of one end of the interval are (–5, 3), find the coordinates of the other end.
Let $(x_1, y_1) = (-5, 3)$ and the coordinates of the unknown end (x_2, y_2).

$$\text{Midpoint} = \left(\frac{x_1 + x_2}{2}, \frac{y_1 + y_2}{2}\right)$$

$$\therefore \quad (-2, 1) = \left(\frac{-5 + x_2}{2}, \frac{3 + y_2}{2}\right)$$

$$\therefore \quad \frac{-5 + x_2}{2} = -2 \quad \text{and} \quad \frac{3 + y_2}{2} = 1$$

$$\therefore \quad -5 + x_2 = -4 \qquad \therefore \quad 3 + y_2 = 2$$

$$\therefore \quad x_2 = 1 \qquad \therefore \quad y_2 = -1$$

The coordinates of the other end are (1, –1).

DISTANCE BETWEEN TWO POINTS

If A has coordinates (x_1, y_1) and B has coordinates (x_2, y_2), then the straight line distance (d) between A and B is given by the formula:

$$d = \sqrt{(x_2 - x_1)^2 + (y_2 - y_1)^2}$$

It can also be thought of as the length of the line interval joining A and B. This formula can easily be proved using Pythagoras' Theorem.

Example 1: Find the distance between A(l, 5) and B(4, 9).

$$\therefore \quad d = \sqrt{(x_2 - x_1)^2 + (y_2 - y_1)^2}$$
$$= \sqrt{(4 - 1)^2 + (9 - 5)^2}$$
$$= \sqrt{25}$$
$$= 5$$

Always start by making sure you get your x_1, y_1 and x_2, y_2 values correct.

Example 2: Show that the point A (4, 5), B(–2,5) and C (1, –4) are the vertices of an isosceles triangle.

Solution: In these types of questions, it is always best to start by drawing a quick sketch of the 3 points.

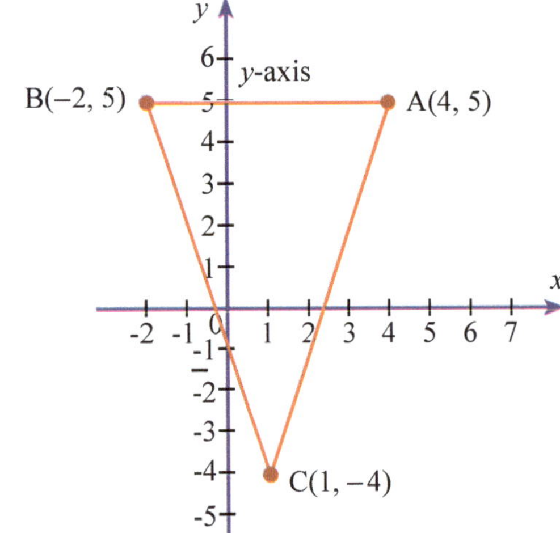

From the diagram, it is now a simple matter to see that we have to prove that BC = AC. Using the distance formula twice:

$$BC = \sqrt{(1 - -2)^2 + (-4 - 5)^2}$$
$$= \sqrt{3^2 + (-9)^2}$$
$$= \sqrt{90}$$
$$= 3\sqrt{10}$$

$$AC = \sqrt{(1 - 4)^2 + (-4 - 5)^2}$$
$$= \sqrt{(-3)^2 + (-9)^2}$$
$$= \sqrt{90}$$
$$= 3\sqrt{10}$$

Since BC = AC = $\sqrt{90}$, therefore triangle ABC is isosceles.

GRADIENT OF A LINE

Each straight line in the Cartesian (x, y) plane has a certain 'slope' or 'steepness'. In Mathematics, it is called the 'gradient' of the line and it is usually denoted by the small letter 'm'.

Pictorially, the gradient is defined as:

$$m = \frac{\text{vertical rise}}{\text{horizontal run}}$$

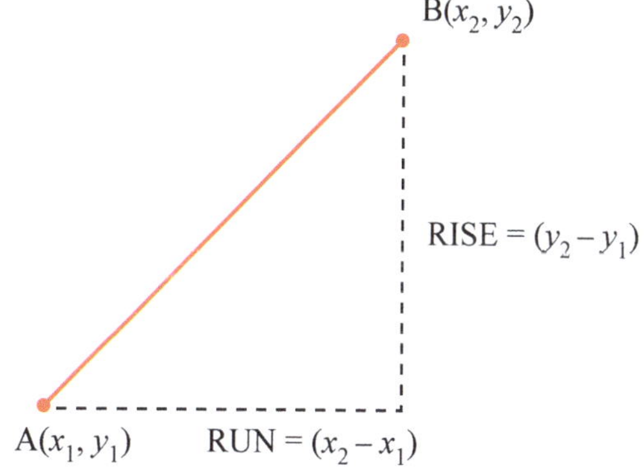

If A has coordinates (x_1, y_1) and B has coordinates (x_2, y_2) then the gradient (m) of the line joining AB can be calculated using the formula:

$$m = \frac{y_2 - y_1}{x_2 - x_1}$$

Example: Find the gradient of the line joining (3, –1) and (1, 5).

Let $(x_1, y_1) = (3, -1)$ and $(x_2, y_2) = (1, 5)$

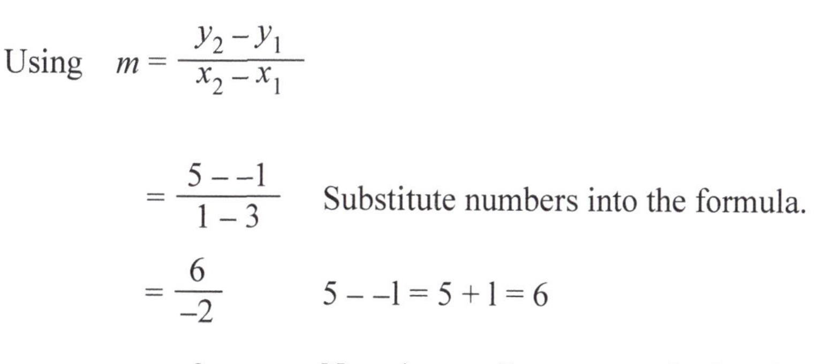

Using $m = \dfrac{y_2 - y_1}{x_2 - x_1}$

$= \dfrac{5 - -1}{1 - 3}$ Substitute numbers into the formula.

$= \dfrac{6}{-2}$ $5 - -1 = 5 + 1 = 6$

$= -3$ Negative gradient means sloping down to the right.

We simply substitute the given coordinates into the gradient formula.

Note: The midpoint, distance and gradient formulae are tested regularly in almost every exam which involves this topic. In most cases, diagrams are not necessary, because the formulae will calculate the answer for you providing you substitute in the correct values of x_1, x_2, y_1 and y_2.

GRAPHING STRAIGHT LINES

There are 4 different ways of expressing a relationship between the two variables x and y.

1. ALGEBRAIC

This is written as an equation with an 'equal sign'. It shows in algebra form the connection between x and y.

Example: $y = 2x + 1$

This means: y must always equal twice the value of x plus the number 1 added on.

In this particular relationship, if x has the value of 10, then y must have the value of twice ten plus one, i.e. $y = 21$

2. TABLE FORM

A table consisting of at least three x values is written down. The y values are then obtained by substituting the three x values into the algebraic equation.

x	0	1	2
y	1	3	5

A table of values for $y = 2x + 1$

When $x = 0$, $y = 2 \times 0 + 1 = 1$
$x = 1$, $y = 2 \times 1 + 1 = 3$
$x = 2$, $y = 2 \times 2 + 1 = 5$

We actually need only a minimum of 2 points to determine a line.

The x values chosen are usually 0, 1 and 2 because they make the calculations easier. However, any 3 numbers could have been chosen.

3. ORDERED PAIRS

This is done by simply changing the 3 pairs of numbers from the table into ordered pair format. From the table above, the ordered pairs are (0, 1), (1, 3) and 2, 5).

4. GRAPH

The 3 ordered pairs are now plotted on the number plane. A straight line is then drawn through the points. The line shouldn't stop at the end of the 2 outer most points. It must extend beyond them, with arrows attached at the end points, as shown in the diagram.

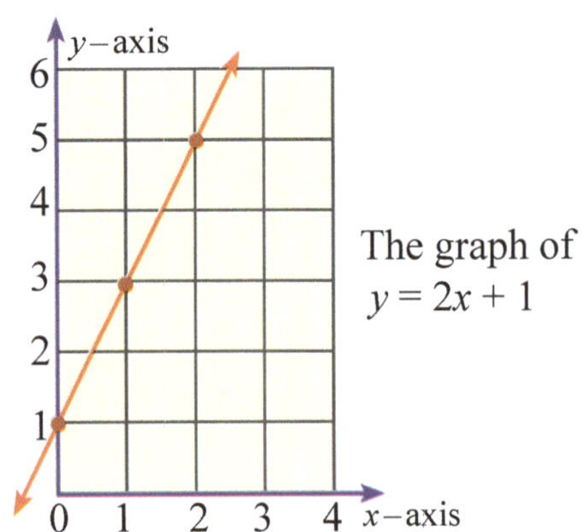

The graph of $y = 2x + 1$

Notes:
(i) Only 2 points are needed to determine a straight line, but 3 points are usually chosen as a check against errors.
(ii) If the 3 points don't connect in a line, then you know at least one of the answers is wrong.
(iii) The gradient of the line $y = 2x + 1$ is constant and equal to 2. This can be found by drawing a right angled triangle anywhere along the line.
(iv) The line $y = 2x + 1$ cuts the y-axis at +1.

FURTHER EXAMPLES

It is not necessary to go through every step shown in the examples in the previous section. These steps were done so that you could fully understand the 4 main ways of expressing a relation between 'y' and 'x'.
However, as you gain confidence, it should be a very fast and simple process to draw up a table of values, and then plot these directly onto the number plane.

Example 1: Graph $y = \frac{1}{2}x$ on the number plane.

Solution: When $x = 0, \quad y = \frac{1}{2} \times 0 = 0$

$x = 2, \quad y = \frac{1}{2} \times 2 = 1$

$x = 4, \quad y = \frac{1}{2} \times 4 = 2$

i.e.

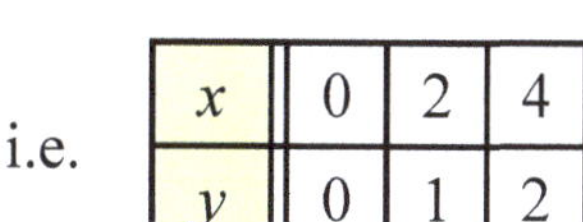

x	0	2	4
y	0	1	2

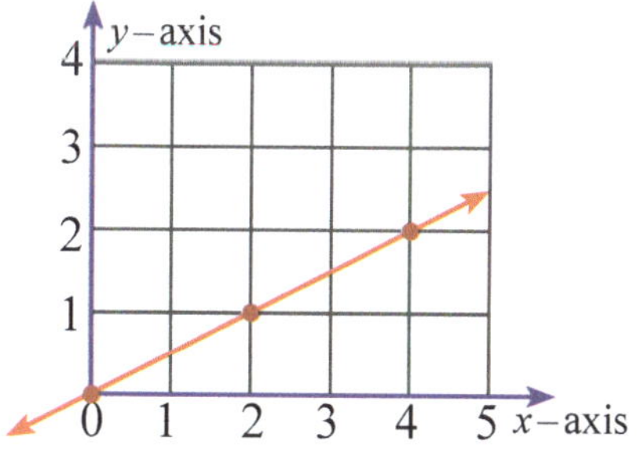

Note: Remember that we can use any values we like for x.
The 3 values chosen for x in the above example are 0, 2 and 4 to make the calculations easier.
Otherwise, y would have had fractional values.

Example 2: Graph this pair of equations and find their point of intersection on the number plane:

$x + y = 3$ and $y = 2x$

Solution:

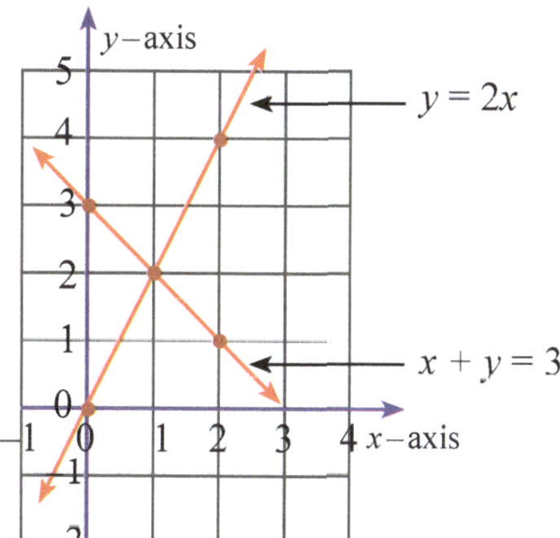

$x + y = 3$

x	0	1	2
y	3	2	1

$y = 2x$

x	0	1	2
y	0	2	4

It can be seen that the 2 lines intersect (cut each other) at the point (1, 2).

Note: If the 3 points are not in line, then a mistake has been made and you should then carefully check all your working.

TWO SPECIAL TYPES OF LINE GRAPHS

So far you have seen how to graph linear relationships involving y in terms of x. However, there are 2 other types which involve only one of the letters (either x or y) and a number.

For example: $y = 3$, $y = -2$, $x = 5$, $x = -3$

FOR $y = c$ WHERE c IS A NUMBER:

> Draw a horizontal line parallel to the x-axis, which cuts the y-axis at the number c.

Example 1: Graph $y = 3$.
A horizontal line is drawn parallel to the x-axis which cuts the y-axis at 3.

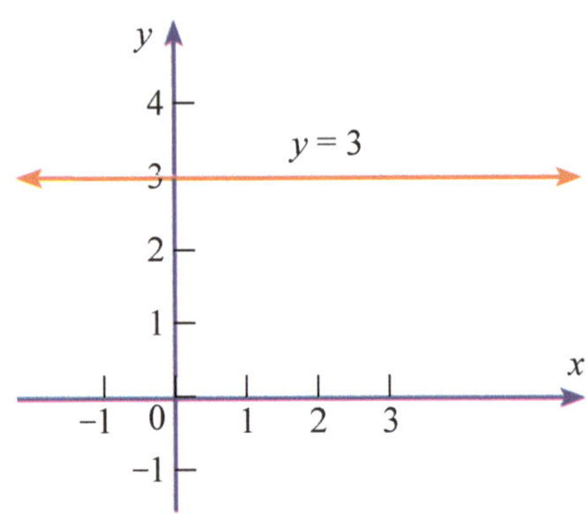

FOR $x = c$ WHERE c IS A NUMBER:

> Draw a vertical line parallel to the y-axis, which cuts the x-axis at the number c.

Example 2: Graph $x = 5$.
A vertical line is drawn parallel to the y-axis, which cuts the x-axis at 5.

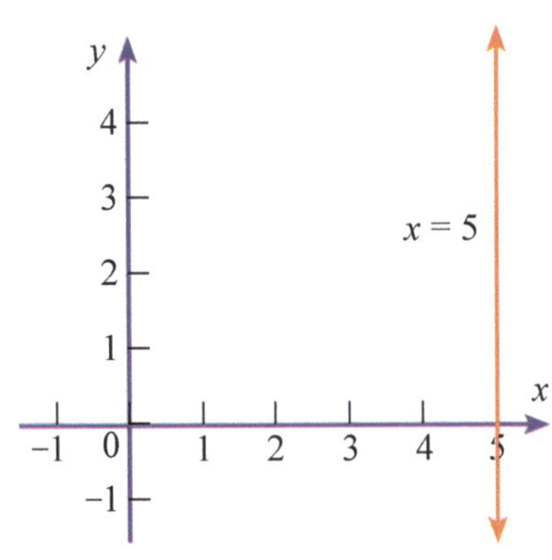

THE GRADIENT — INTERCEPT FORM OF A LINE

Each different straight line on the Cartesian number plane has a unique name or equation associated with it — depending on the gradient (slope) and where the line cuts the y-axis.
Each equation can be written in the following format or pattern:

$$y = mx + b$$

The fraction or whole number 'm' represents the gradient of the line. The number 'b' gives the y intercept — that is, where the line cuts the y-axis.

If 'm' is positive the line slopes up to the right.

If 'm' is negative the line slopes down to the right.

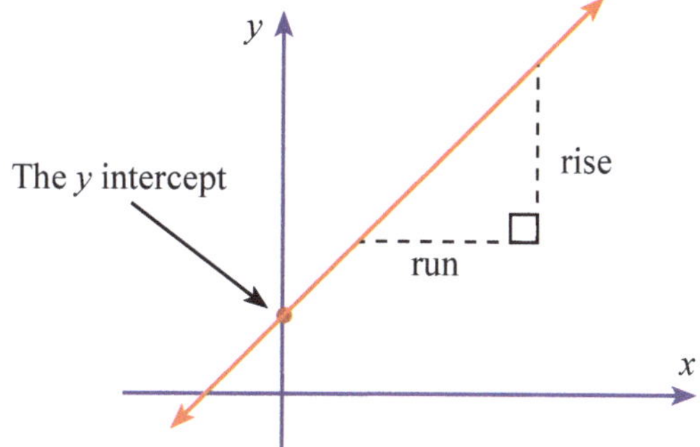

$$m = \frac{\text{vertical rise}}{\text{horizontal run}}$$

Example 1: If $y = \frac{3}{4}x - 5$ find the gradient and the y intercept.

Compare $y = \frac{3}{4}x - 5$ to $y = mx + b$.

It can be seen that $m = \frac{3}{4}$ and $b = -5$

Therefore the gradient $= \frac{3}{4}$ and y intercept $= -5$.

Example 2: Find the equation of a line with gradient of -2 and y intercept 4.
Substitute $m = -2$ and $b = 4$ into $y = mx + b$
Therefore the equation is $y = -2x + 4$

Example 3: If $2y = 5x - 6$ find the gradient and y intercept.
Dividing both sides by 2

$\therefore \quad y = \frac{5}{2}x - 3$

$\therefore \quad m = \frac{5}{2}$ and $b = -3$

Must always have 1y or y by itself on the left hand side in order to determine 'm' and 'b'.

SKETCHING A LINE USING $y = mx + b$

In a previous section, you learned how to graph a line by plotting 3 different points on the number plane, and drawing a line through them. In this section, you will learn a different quicker method of sketching a line $y = mx + b$ using the **GRADIENT** (m) and the y **INTERCEPT** (b).

THE 3 STEPS:

(i) Mark point 'b' on the y-axis (i.e. the y intercept).

(ii) From point b, construct a gradient $m = \frac{\text{rise}}{\text{run}}$.

(iii) Draw a line through the 2 points.

Example: Sketch $y = \frac{3}{4}x + 1$

By comparing $y = \frac{3}{4}x + 1$ to $y = mx + b$,

we see that $m = \frac{3}{4}$ and $b = +1$

Step (i): Mark '+1' on the y-axis

Step (ii): $m = \frac{\text{rise}}{\text{run}}$ From '+1' on the y-axis, construct a gradient of $\frac{3}{4}$ by going 4 units to the right, and 3 units up.

Step (iii): Draw a line through the two points.

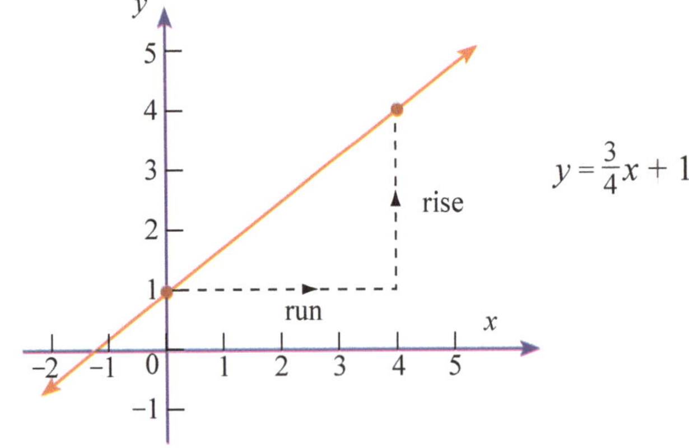

Note: (i) If the gradient had been $m = -\frac{3}{4}$, then go 4 units to the right and 3 units down. This would produce a line which slopes down to the right.

(ii) If the gradient is a whole number, then rewrite it as a fraction over the number 1.

If $y = 3x - 2$ then $m = 3 = \frac{3}{1}$.

FINDING THE EQUATION OF A LINE

In this section, we are really applying the reverse logic to the previous page. You will be given a diagram of the line and asked to find its equation.

> Step (i) Determine the intercept (b) by reading off where the line cuts the y-axis.
>
> Step (ii) Determine the gradient (m) of the line by finding a suitable right angled triangle and using
>
> $$m = \frac{\text{vertical rise}}{\text{horizontal run}}.$$
>
> Step (iii) Substitute the values of 'b' and 'm' into $y = mx + b$.

Example 1: Find the equation of the line shown in the graph.

Solution:

y intercept is -1

$\therefore \quad b = -1$

$\text{gradient} = \frac{\text{rise}}{\text{run}} = \frac{1}{2}$

$\therefore \quad m = \frac{1}{2}$

Substitute $m =$ and $b = -1$ into $y = mx + b$

$\therefore$ Equation of line is $y = \frac{1}{2}x - 1$

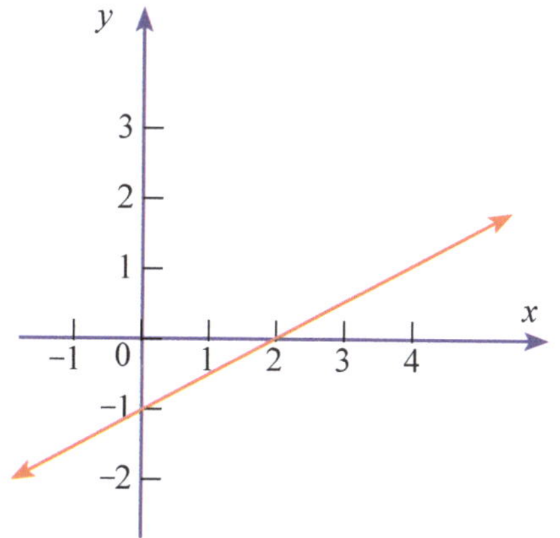

Example 2: Find the equation of the line shown in the graph.

Solution:

y intercept is 2

$\therefore \quad b = 2$

$\text{gradient} = \frac{\text{rise}}{\text{run}} = -\frac{2}{3}$

$\therefore \quad m = -\frac{2}{3}$

Substitute $m = -\frac{2}{3}$ and $b = 2$ into $y = mx + b$

$\therefore$ Equation of line is $y = -\frac{2}{3}x + 2$

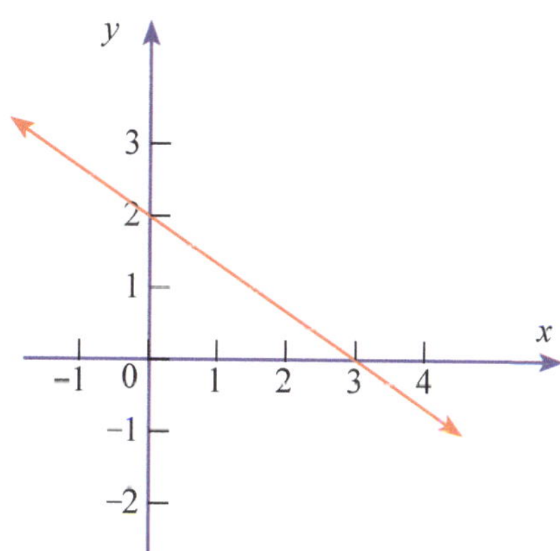

Note: It is automatically a negative gradient because it can be seen that the line is sloping down to the left.

THE GENERAL FORM OF A STRAIGHT LINE

It is sometimes more convenient and neater to express the equation of a line in a different format known as 'general form'.

$$ax + by + c = 0$$

where a. b, c are WHOLE NUMBERS.

It is neater, because there are no fractions involved in this format. The student will find that many of the solutions at the end of the text book are expressed in 'general form'.

It is important to be able to transpose 'general form' to 'gradient-intercept form' and vice versa.

Example 1: The equation $y = \frac{2}{3}x - 5$ is in the form $y = mx + b$. Rewrite this in general form.

Solution:

$3y = 2x - 15$	Multiply both sides by 3.
$0 = 2x - 3y - 15$	Subtract $3y$ from both sides.
$\therefore\ 2x - 3y - 15 = 0$	It is now in general form $ax + by + c = 0$.

Example 2: If the equation of a line is given in general form as $3x + 4y - 8 = 0$, determine the gradient and y intercept.

Solution: Before we can do this, it is necessary to transpose the equation to the form $y = mx + b$.

$3x + 4y - 8 = 0$

$\therefore\ 4y = -3x + 8$	Subtract $3x$ and add 8 to both sides.
$\therefore\ y = -\frac{3}{4}x + 2$	Divide both sides by 4.
$\therefore$ Gradient $= m = -\frac{3}{4}$	Compare it to $y = mx + b$ to find values of 'm' and 'b'.
y intercept $= b = +2$	

THE POINT – GRADIENT FORMULA

If a line passes through a given point (x_1, y_1) and has a known gradient (m), then the **EQUATION** of the line can be determined using the formula:

$$y - y_1 = m(x - x_1)$$

In some textbooks, it hasn't been transposed and the student might recognise it in this format:

$$\frac{y - y_1}{x - x_1} = m$$

This is the same formula as above!

Note: It is very important to understand and be able to use this formula, as its use occurs many times in Years 9 and 10 — and even more frequently in Years 11 and 12.

Example: a) Find the equation of a line which has a gradient of $\frac{3}{5}$ passing through the point (–2, 4).

b) Also give the final answer in general form.

Solutions: The given point $(x_1, y_1) = (-2, 4)$ and the gradient $m = \frac{3}{5}$

a)	$y - y_1 = m(x - x_1)$	Point - gradient formula
$\therefore$	$y - 4 = \frac{3}{5}(x - -2)$	Substitute coordinate values in
$\therefore$	$y - 4 = \frac{3}{5}x + \frac{6}{5}$	
$\therefore$	$y = \frac{3}{5}x + 5\frac{1}{5}$	This is in $y = mx + b$ form
b)	$5y = 3x + 26$	Multiply through by 5 to clear fractions
$\therefore$	$0 = 3x - 5y + 26$	
$\therefore$	$3x - 5y + 26 = 0$	$ax + by + c = 0$

THE TWO POINT FORMULA

If a line passes through two given points (x_1, y_1) and (x_2, y_2) then the **EQUATION** of the line can be determined using this one formula:

$$\frac{y - y_1}{x - x_1} = \frac{y_2 - y_1}{x_2 - x_1}$$

ALTERNATIVELY the equation of the line can be determined by using two ideas already covered:

(i) Find the gradient by using $m = \dfrac{y_2 - y_1}{x_2 - x_1}$

(ii) Determine the equation by using $y - y_1 = m(x - x_1)$

Examples: Find the equation of the line passing through (2, –3) and (4, 1).
Let $(x_1, y_1) = (2, -3)$ and $(x_2, y_2) = (4, 1)$.

$$\frac{y - y_1}{x - x_1} = \frac{y_2 - y_1}{x_2 - x_1} \qquad \text{Two point formula}$$

$$\frac{y - -3}{x - 2} = \frac{1 - -3}{4 - 2} \qquad \text{Simply substitute in values}$$

$$\therefore \quad \frac{y + 3}{x - 2} = 2$$

$$\therefore \quad y + 3 = 2(x - 2) \qquad \text{Cross multiply}$$

$$\therefore \quad y = 2x - 7 \qquad \text{Written in form } y = mx + b$$

Alternative Method:

$$m = \frac{y_2 - y_1}{x_2 - x_1} \qquad \text{Gradient formula}$$

$$\therefore \quad m = \frac{1 - -3}{4 - 2}$$

$$= 2$$

$$\therefore \quad y - y_1 = m(x - x_1) \qquad \text{Point - gradient formula}$$

$$\therefore \quad y + 3 = 2(x - 2)$$

$$\therefore \quad y = 2x - 7$$

This method uses 2 steps instead of 1 step.

PARALLEL LINES

If two lines of gradient m_1 and m_2 are known to be parallel, then their gradients must be equal. It should make sense that if lines are parallel, then they must have the same slope.

i.e. $$m_1 = m_2$$

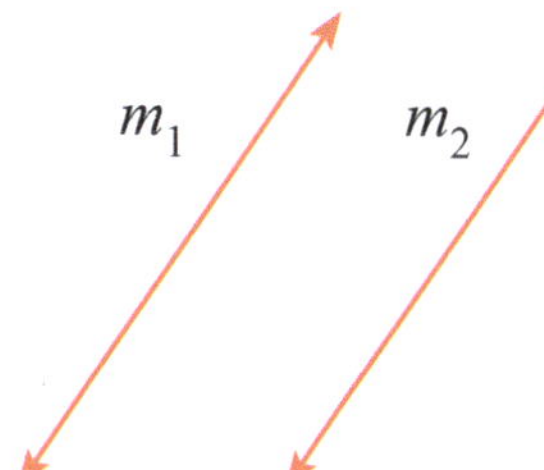

These two lines are parallel.

They have equal gradients.

Although this appears to be a fairly simple and obvious idea, it is nevertheless important to understand — particularly when studying calculus and tangents in Years 11 and 12.

Example 1: Which of the following lines are parallel?

$y = -2x + 5$, $\quad y = \frac{2}{3}x - 1$, $\quad y = 1 - 2x$, $\quad 3y = -6x + 7$

Solution:

$y = -2x + 5$
$m = -2$

$y = \frac{2}{3}x + 1$,
$m = \frac{2}{3}$

$y = 1 - 2x$
$\therefore \; y = -2x + 1$
$m = -2$

$3y = -6x + 7$
$\therefore \; y = -2x + \frac{7}{3}$
$m = -2$

Three of the lines are parallel because they each have a gradient of –2 when they are written in the form $y = mx + b$.

The second line $y = \frac{2}{3}x - 1$ has a gradient of $\frac{2}{3}$ and it is NOT parallel to the other 3 lines.

Example 2: Find the equation of the line which is parallel to $y = -3x + 7$ and passing through the point (4, –1).

Solution: Since it is parallel to $y = -3x + 7$, it must have the gradient of –3.

Using $y - y_1 = m(x - x_1)$ — Point-gradient formula

$\therefore \; y - -1 = -3(x - 4)$ — $(x_1, y_1) = (4, -1)$ and $m = -3$

$\therefore \; y + 1 = -3x + 12$

$\therefore \; y = -3x + 11$

PERPENDICULAR LINES

If two lines of gradient m_1 and m_2 are perpendicular, then the product of their respective gradients must equal -1.

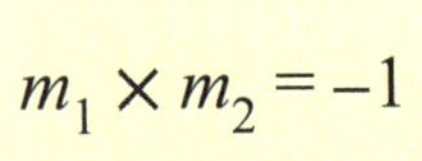

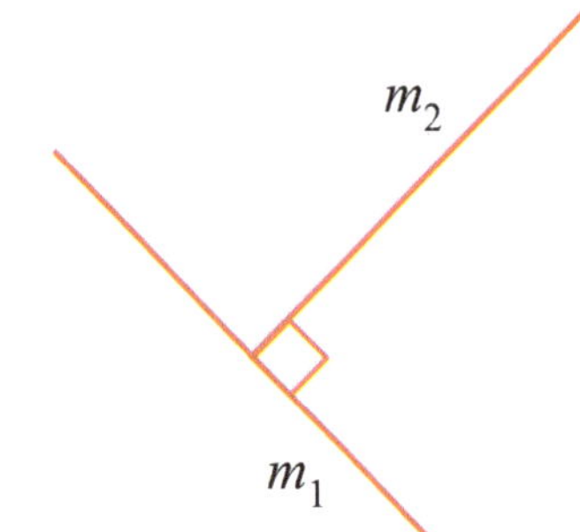

These two lines are at right angles to each other.

$\therefore \quad m_1 \times m_2 = -1$

OR

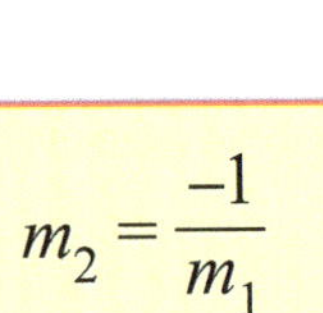

ALTERNATIVELY the gradient of one line must equal the negative reciprocal of the other gradient.

Example 1: Find the gradient of the line which is perpendicular to the line $y = \frac{3}{4}x - 5$

$m_1 \times m_2 = -1$

$\frac{3}{4}m_2 = -1$

$\therefore \quad m_2 = -\frac{4}{3}$

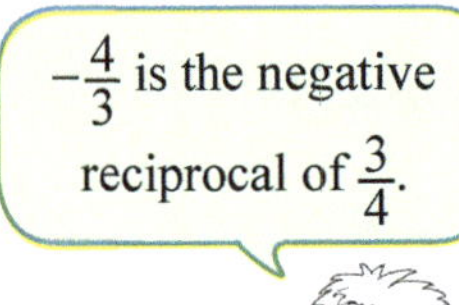

Example 2: Find the equation of the line which is perpendicular to $y = -\frac{1}{2}x + 3$ and passing through the point $(-1, 4)$.

Solution: Since this line is perpendicular to $y = -\frac{1}{2}x + 3$, it must have a gradient of 2 (i.e. negative reciprocal of $-\frac{1}{2}$).

Using $y - y_1 = m(x - x_1)$ — Point-gradient formula

$\therefore \quad y - 4 = 2(x - -1)$ — $(x_1 - y_1) = (-1, 4)$ and $m = 2$

$\therefore \quad y - 4 = 2x + 2$ — $(x - -1) = (x + 1)$

$\therefore \quad y = 2x + 6$ — Add 4 to both sides.

TO DETERMINE IF A POINT LIES ON A LINE

To prove mathematically, without the aid of diagrams, whether a point lies on a given line (or whether a line passes through a given point), carry out the following steps:

(i) Substitute the coordinates of the point into the equation of the line.
(ii) If the point satisfies the equation, then it lies on the line.
(iii) If the point doesn't satisfy the equation, then it doesn't lie on the line.

Note: To satisfy an equation means that after substituting in the x and y values, the numerical value of the left hand side of the equation is equal to the numerical value of the right hand side of the equation.

In other words, it means that the equation is balanced.

Example 1: Does the point $(-2, 3)$ lie on the line $y = 3x + 8$?

Substitute $x = -2$ and $y = 3$ into $y = 3x + 8$

$\therefore \quad 3 \neq 3 \times -2 + 8$ Left hand side $\neq$ Right hand side

$\therefore \quad$ Point $(-2, 3)$ does NOT lie on the line $y = 3x + 8$.

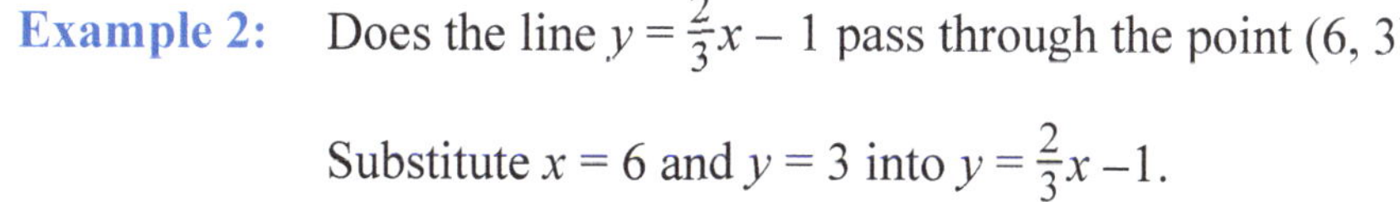

Example 2: Does the line $y = \frac{2}{3}x - 1$ pass through the point $(6, 3)$.

Substitute $x = 6$ and $y = 3$ into $y = \frac{2}{3}x - 1$.

Asking 'if a point lies on the line' or 'if the line passes through the point' is another way of asking the same question.

$\therefore \quad 3 = \frac{2}{3} \times 6 - 1$ Left hand side = Right hand side

$\therefore \quad$ The line $y = \frac{2}{3}x - 1$ does pass through the point $(6, 3)$.

COLLINEAR POINTS

Three or more points are said to be 'collinear' if they lie on a straight line.

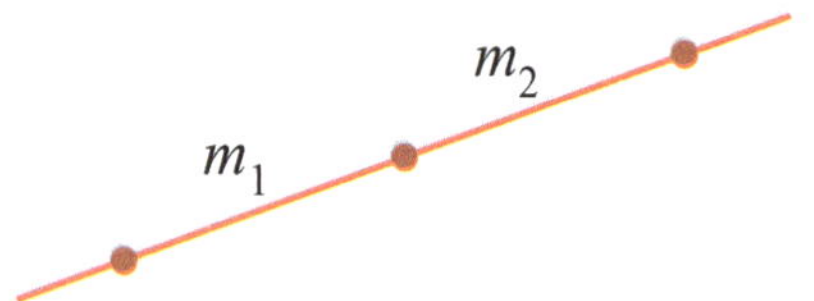

'Collinear points' $m_1 = m_2$

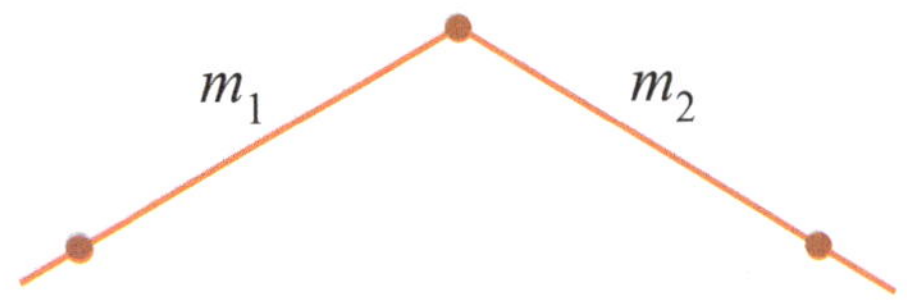

'Non collinear points' $m_1 \neq m_2$

To determine whether or not 3 points are collinear, the following procedure is used:

(i) Find the gradient between the first and second point.
(ii) Find the gradient between the second and third point.
(iii) If the two gradients are equal (i.e. $= m_2 = m_2$) then the 3 points are collinear.

Example 1: Determine whether or not the points A(–1, –7), B(2, 5) and C(3, 9) are collinear.

Solution:

$$\text{Gradient of AB} = m_1 = \frac{y_2 - y_1}{x_2 - x_1}$$

$$= \frac{5 - -7}{2 - -1}$$

$$= 4$$

Watch the negative signs carefully. Subtracting a negative number is the same as adding a positive number.

$$\text{Gradient of BC} = m_2 = \frac{y_2 - y_1}{x_2 - x_1}$$

$$= \frac{9 - 5}{3 - 2}$$

$$= 4$$

Since $m_1 = m_2$, the three points are collinear and each line segment includes the point B.

ALTERNATIVE METHOD:

Find the equation of the line through AB using the point-gradient formula $y - y_1 = m(x - x_1)$. Determine whether or not the third point satisfies this equation (i.e. lies on the line) using the method shown on the opposite page.

SKETCHING A PARABOLA

Graphing a parabola, as explained in one of the earlier chapters using a table of values, is a long process. In addition, the maximum or minimum, and roots might not have been easy whole number values as shown in the examples.

Parabolas have the format or pattern '$y = ax^2 + bx + c$'. To find the x and y intercepts, exactly the same method is used as with straight lines (see p. 83). The only difference is that the solution of the quadratic equation results in two x values or roots. Furthermore, the 'axis of symmetry' and the coordinates of the 'maximum' or 'minimum' can easily be calculated.

To sketch a parabola $y = ax^2 + bx + c$ requires 5 easy steps:

(i) If a > 0, the shape is concave up ∪ i.e. minimum.
If a < 0, the shape is concave down ∩ i.e. maximum.

(ii) To find the y intercept, let $x = 0$.

(iii) To find the x intercepts, let $y = 0$ and solve the resulting quadratic equation.

(iv) The axis of symmetry is given by the formula:

$$x = \frac{-b}{2a}$$

(v) Coordinates of maximum or minimum are:

$$\left[\frac{-b}{2a}, f\left(\frac{-b}{2a}\right)\right]$$

Note: The x coordinate of the maximum or minimum must lie on the axis of symmetry, i.e. $x = \frac{-b}{2a}$

The y coordinate of the maximum or minimum is found by substituting this value of x into the original equation $y = ax^2 + bx + c$.

We can also write this process in Maths as $f\left(\frac{-b}{2a}\right)$.

Therefore coordinates are $\left[\frac{-b}{2a}, f\left(\frac{-b}{2a}\right)\right]$.

Using $f(x)$ instead of the letter y is called function notation.

EXAMPLES OF SKETCHING PARABOLAS

Example 1: Sketch $y = x^2 - 2x - 8$ showing clearly all the main features.

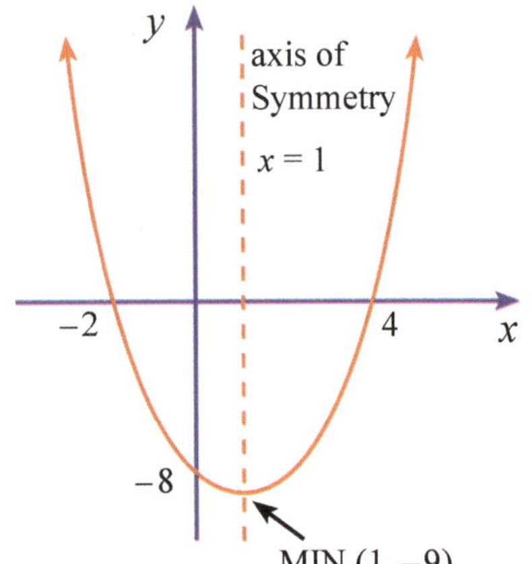

(i) $a = 1 \therefore$ concave up shape $\cup$

(ii) Let $x = 0$ ⇨ $y = -8$

(iii) Let $y = 0$ ⇨ $x^2 - 2x - 8 = 0$

$\therefore \quad (x - 4)(x + 2) = 0$

$\therefore$ Roots are $x = 4$ or $x = -2$

(iv) Compare $x^2 - 2x - 8$ to $ax^2 + bx + c$

$\therefore \quad a = 1, b = -2$ and $c = -8$

Axis of symmetry is $x = \dfrac{-b}{2a}$

$\therefore \quad x = \dfrac{2}{2} = 1$

(v) Substitute $x = 1$ back into $y = x^2 - 2x - 8$

$\therefore$ Minimum $= (1, -9)$

You should recognise the above same example from page 84.

Example 2: Sketch $y = 6x - x^2$ showing clearly all the main features.

(i) $y = -x^2 + 6x$

Since $a = -1 \therefore$ curve has concave down shape $\cap$

(ii) Let $x = 0$ ⇨ $y = 0$

(iii) Let $y = 0$ ⇨ $6x - x^2 = 0$

$\therefore \quad x(6 - x) = 0$

$\therefore$ Roots are $x = 0$ or $x = 6$

(iv) Compare $-x^2 + 6x + 0$ to $ax^2 + bx + c$

$\therefore \quad a = -1, b = 6$ and $c = 0$

Axis of symmetry is $x = \dfrac{-b}{2a}$

$\therefore \quad x = \dfrac{-b}{2a} = 3$

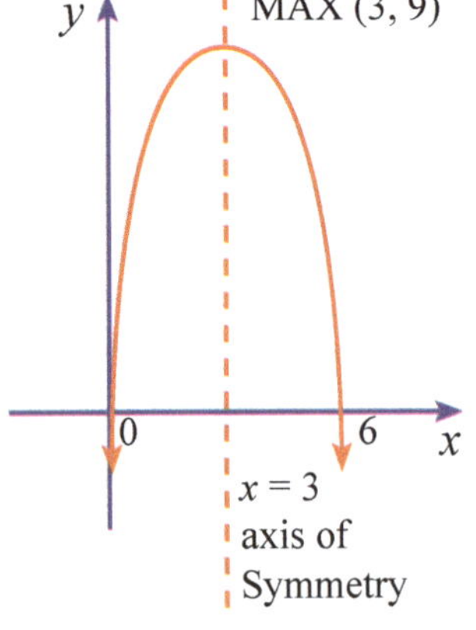

(v) Substitute $x = 3$ back into $y = -x^2 + 6x$

$\therefore$ Maximum $= (3, 9)$

OR $f(x) = -x^2 + 6x$

$\therefore f(3) = -3^2 + 6 \times 3 = 9$

CHAPTER SUMMARY

PLOTTING POINTS

Any point $P\ (x, y)$ can be plotted on the number plane. The x coordinate gives the distance from the y-axis, and the y coordinate gives the distance from the x-axis. The x value is sometimes called the **ABSCISSAE** and the y value is sometimes called the **ORDINATE**. Given any two points on the number plane, $A\ (x_1, y_1)$ and $B\ (x_2, y_2)$, then the following 3 important formulae can be applied:

Midpoint of AB: $(x, y) = \left(\dfrac{x_1 + x_2}{2}, \dfrac{y_1 + y_2}{2}\right)$

Length of AB: $d = \sqrt{(x_2 - x_1)^2 + (y_2 - y_1)^2}$

Gradient of AB: $m = \dfrac{\text{vertical rise}}{\text{horizontal run}} = \dfrac{y_2 - y_1}{x_2 - x_1}$

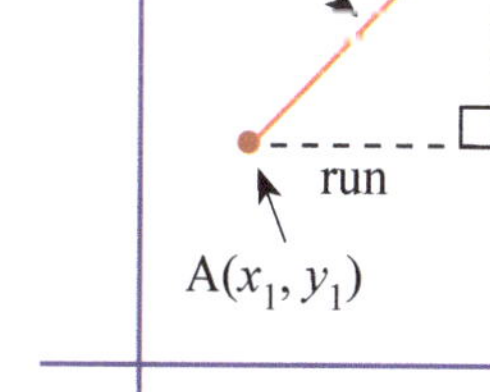

Note: If a line leans upwards to the right it has a **positive** (+) slope.
If a line leans downwards to the right it has a **negative** (–) slope.

GRADIENT-INTERCEPT FORM OF A LINE

Each different line on the Cartesian number plane has a unique name or equation associated with it – depending on the gradient (slope) and where the line cuts the y-axis. Each equation can be written in the following format or pattern:

$\boldsymbol{y = mx + b}$ where m = the gradient
$b = y$ intercept (where it cuts the y-axis)

GENERAL FORM OF A LINE

It is sometimes more convenient and neater to express the equation of a line in a different format known as 'general form'.

$\boldsymbol{Ax + By + C = 0}$ where A, B and C are whole numbers

Example: If $3x - 4y + 8 = 0$ find the gradient and y intercept of the line.
Solution: gradient $= \frac{3}{4}$, y intercept $= 2$

SKETCHING LINES

There are several methods, but the easiest and fastest is:

1. Let $x = 0$ in the given equation, then solve it to find the y intercept.

2. Let $y = 0$ in the given equation, and solve it to find the x intercept.

3. Draw a line which passes through the 2 known intercepts.

POINT GRADIENT FORMULA: $y - y_1 = m(x - x_1)$

Used to find the equation of a line passing through a given point (x_1, y_1) with given gradient m. This formula is one of the most important and frequently used formulas from Year 9 all the way through to Year 12.

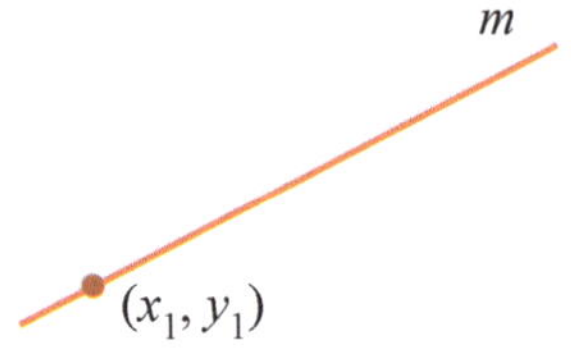

TWO POINT FORMULA: $\dfrac{y - y_1}{x - x_1} = \dfrac{y_2 - y_1}{x_2 - x_1}$

Used to find the equation of a line passing through 2 given points (x_1, y_1) and (x_2, y_2).

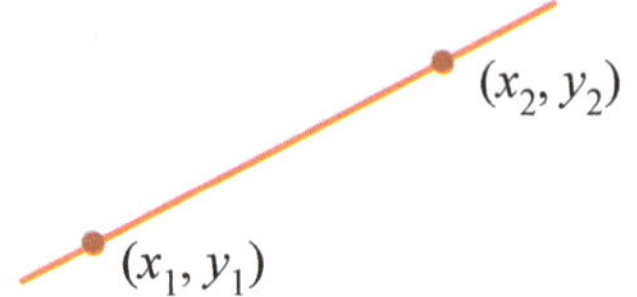

PARALLEL LINES: $m_1 = m_2$

If 2 lines are parallel, then their gradients are equal.

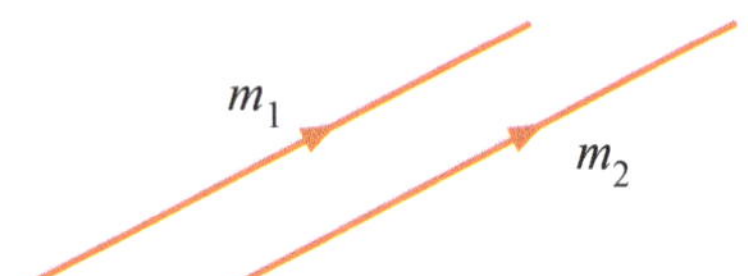

PERPENDICULAR LINES: $m_1 \times m_2 = -1$ or $m_2 = \dfrac{-1}{m_1}$

If 2 lines are perpendicular, then the gradient of one line is equal to the negative reciprocal of the gradient of the other line.

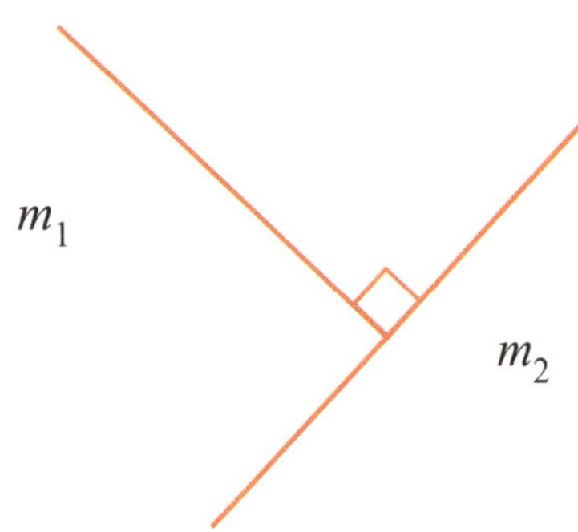

Example: Find the equation of the line which passes through the point (–5, 4) and which is perpendicular to the line $y = -\frac{3}{4}x + 1$.

Express your final answer in general form.

Solution: $3y - 4x + 8 = 0$

SKETCHING THE PARABOLA

The most basic way of graphing a parabola is by using a table of values. But this method is very time consuming and inefficient because at least 10 ordered pairs (x and y values) have to be calculated and plotted.

To sketch a parabola $y = ax^2 + bx + c$ requires 5 easy steps:

Step 1: If $a > 0$. the shape is concave up. i.e. minimum

If $a < 0$, the shape is concave down. i.e. maximum

Step 2: To find the y intercept, let $x = 0$.

Step 3: To find the x intercepts, let $y = 0$ and solve the resulting quadratic equation.

Step 4: The axis of symmetry is given by the formula:

$$x = \frac{-b}{2a}$$

Step 5: Coordinates of the maximum or minimum are:

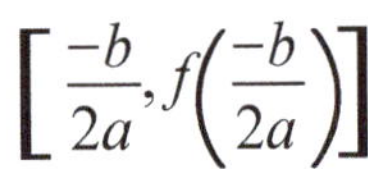

$$\left[\frac{-b}{2a}, f\left(\frac{-b}{2a}\right)\right]$$

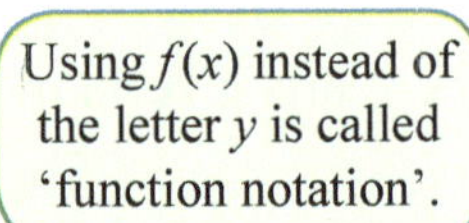

LEVEL 1 — COORDINATE GEOMETRY

EASIER QUESTIONS

Note: Only turn back to page number shown if you have difficulty.

Page

Q1. — 196

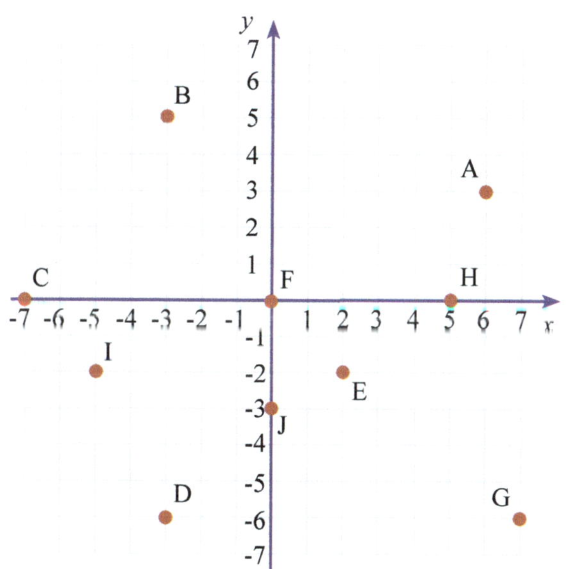

Name the coordinates of the points:

a) A f) F
b) B g) G
c) C h) H
d) D i) I
e) E j) J

The image shown is called a Cartesian plane, named after the mathematician Descartes.

Q2. Find the midpoint of the interval joining: — 197

a) (2, 3) and (6, 9) b) (4, 5) and (2, 1)
c) (7, –3) and (0, 7) d) (–5, 2) and (9, –4)
e) (–6, –1) and (5, 3) f) (1, –8) and (–4, –3)

Q3. Find the distance between the points: — 198

a) (4, 3) and (2, 9) b) (6, 1) and (–3, 5)
c) (–1, 4) and (2, –8) d) (2, –3) and (–1, –4)
e) (5, –8) and (–7, 2) f) (0, –3) and (8, 0)

Q4. Find the gradient of the line passing through: — 199

a) (0, 0) and (6, 3) b) (5, 1) and (2, 6)
c) (–2, 0) and (7, 1) d) (–4, 1) and (–2, –4)
e) (6, –6) and (–1, 3) f) (8, –2) and (–5, 2)

Q5. Graph these lines: — 200, 201

a) $y = 2x + 1$ b) $x = 5$ c) $y = x$
d) $y = 3 - x$ e) $y = 4x - 5$ f) $y = -3$

Q6. Find the gradient and y-intercept of the following lines, then sketch each line: — 83, 203

a) $y = 5x - 2$ b) $y = -2x + 3$ c) $y = \frac{3}{2}x - 4$
d) $y = 1 - \frac{1}{2}x$ e) $y = -x$ f) $y = \frac{1}{2} - x$

Note: Look at bottom of page 215 or on page 83 for an easy quick method to sketch straight lines.

 AVERAGE QUESTIONS

Note: Only turn back to page number shown if you have difficulty. | Page

Q1. Write down a relationship between x and y which satisfies each table of values: below: 200, 201

a)

x	2	3	4	8
y	5	4	3	–1

b)

x	–3	1	4	8
y	–10	2	11	23

c)

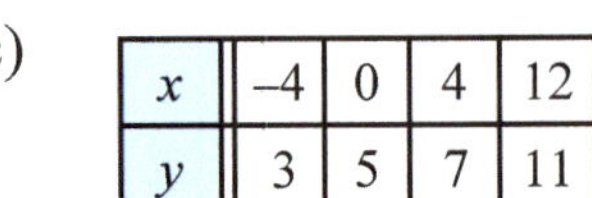

x	–4	0	4	12
y	3	5	7	11

d)

x	0	3	5	7
y	–4	5	21	45

e)

x	–5	3	7	13
y	–4	0	2	5

f)

x	–2	3	5	7
y	–21	–11	–7	–3

Q2. Match each equation below with the graph on the number plane. 204, 205

a) $y = \frac{1}{2}x + 1$
b) $y = 2x - 3$
c) $y = 3 - 5x$
d) $x + y = 4$
e) $y = 5 - \frac{1}{2}x$
f) $y = 3x$
g) $x = 4$
h) $y = -3$

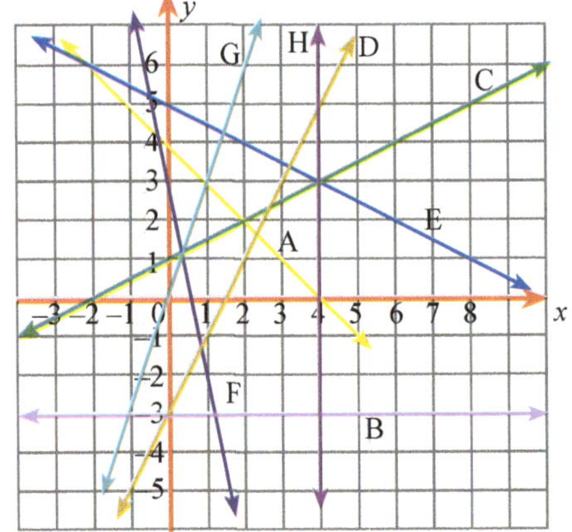

Q3. Graph the equations $y = 2x$ and $y = 6 - x$. What is the area enclosed by the two lines and the x-axis? 200, 201

Q4. a) For the parabolic relationship $y = x^2 - 2x - 8$, draw up a suitable table of x and y values. 84

b) Draw the graph on a number plane and give the coordinates of the lowest point (called the 'minimum') on the parabola.

c) Also graph the relationship $y = 2x - 3$.

d) Write down the coordinates of the 2 points of intersection of the parabola and the straight line.

Q5. Repeat all the above steps in Q4. to the following pairs of equations: 84

a) $y = x^2 - 4$, $y = 2x - 1$
b) $y = x^2 - x - 6$, $y = 3 - x$
c) $y = 4x - x^2$, $y = 2x - 3$

Q6. From the previous chapter, use your knowledge of solving simultaneous equations to find the 2 points of intersection in Q5. **algebraically**. They should be the same answers that you obtained by using the longer method of graphing the lines. 173

Q7. **Sketch** the parabolas showing all important features 213, 214

a) $y = x^2 - 2x - 8$
b) $y = x^2 - 8x + 12$
c) $y = -x^2 + 7x - 10$
d) $y = -x^2 + x + 12$
e) $y = 2x^2 - 3x - 2$
f) $y = -2x^2 + 5x - 3$

LEVEL 3 — COORDINATE GEOMETRY

AVERAGE QUESTIONS

Note: Only turn back to page number shown if you have difficulty.

	Page
Q1. Write the following equations in the form $y = mx + b$, and hence find the gradient and y-intercept of each one: a) $2x - y + 3 = 0$ b) $4x + 2y - 6 = 0$ c) $3x + 3y + 5 = 0$ d) $x + 5y = 0$ e) $2x - 3y + 1 = 0$ f) $x - 4y = 4$	203
Q2. Find the equation of a straight line with: a) a gradient of 3 and passing through the point (4, 3) b) a gradient of –2 and passing through the point (2, 7) c) a gradient of 5 and passing through the point (–3, 5) d) a gradient of $-\frac{1}{2}$ and passing through the point (5, –2) e) a gradient of $\frac{2}{3}$ and passing through the point (–4, 3)	207
Q3. Find the equation of the straight line which passes through the points: a) (2, 1) and (5, 7) b) (1, 3) and (4, –3) c) (4, –8) and (0, 8) d) (–1, –1) and (1, 5) e) (–2, 7) and (6, –5) f) (–3, –4) and (9, 2)	208
Q4. Which of the following lines are parallel to $y = 2x - 5$: $2x + y - 2 = 0$, $2x - 3y + 2 = 0$, $2x - y + 1 = 0$, $4x - 2y + 3 = 0$	209
Q5. Find the equation of the line which cuts the y-axis at – 3 and is parallel to $4x - y + 1 = 0$.	209
Q6. Which of the following lines are perpendicular to $y = -3x + 2$: $3y - x + 3 = 0$, $3x + y - 2 = 0$, $6y - 2x + 4 = 0$, $3x - 3y + 3 = 0$	210
Q7. Find the equation of the line which is perpendicular to $y = \frac{2}{3}x - 1$ and has y-intercept 5.	210
Q8. Which of the following points lie on the line $4x - y + 3 = 0$ (3, 15) (–5, –15) (8, 35) (–8, –35) (4, 19)	212
Q9. Show that the points (–4, –5), (2, –2) and (8, 1) are collinear.	212
Q10. Write the following equations in general format $ax + by + c = 0$: a) $y = 5x - 8$ b) $y = \frac{2}{3}x + 7$ c) $y = \frac{4}{5}x - \frac{1}{3}$	206
Q11. Sketch these parabolas showing all important features: a) $y = x^2 - 4$ b) $y = x - x^2$ c) $y = 10x^2 - x - 3$ d) $y = 6x^2 - 8x$ e) $y = 4x^2 - 64$ f) $y = 13x - 6 - 6x^2$ g) $y = x^2 - 2x + 1$ h) $y = 4x^2 - 12x + 9$ i) $y = 4 + 12x - 18x^2$	213, 214

HARDER QUESTIONS

Q1. In the following 3 lines you are only given the x and y intercepts. Give the equation of each of the lines in general form.

a)

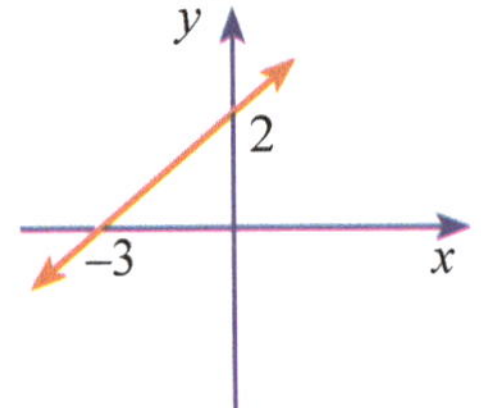

b)

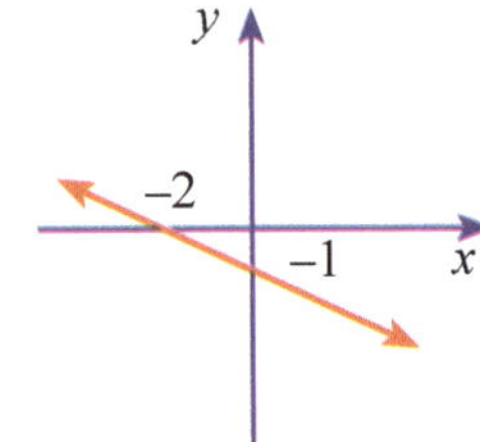

c)

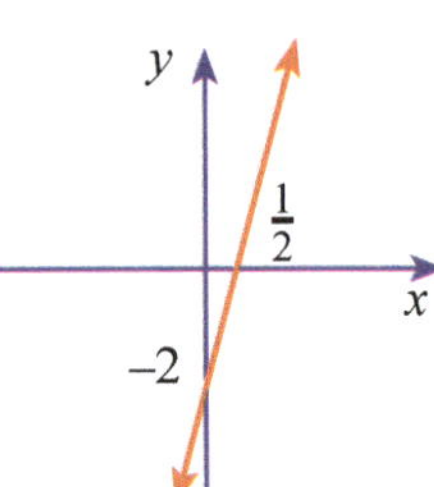

Q2. Find the equation of the straight line passing through the midpoint of the interval joining (6, 2) and (–2, 4), and the origin.

Q3. The midpoint of an interval is (2, 4). If the coordinates at the ends of the interval are $(m, 5)$ and $(8, n)$, find m and n.

Q4. A and B are the points (–3, 4) and (6, –2) respectively. If M is the midpoint of AB, find the distance BM.

Q5. Find the equation of the line which passes through the point (6, –4) and is parallel to the line $2x - 3y + 3 = 0$.

Q6. Find the equation of the line which passes through the points $P(0, a)$ and $Q(a, 2a)$.

Q7. Find the equation of the line perpendicular to $5x + 2y - 1 = 0$, and passing through the point (–2, 7).

Q8. Show that the triangle whose vertices are A(3, 6), B(–2, –4) and C(–5, 2) is a right angled triangle. Find the area of the triangle.

Q9. Prove that a quadrilateral ABCD with vertices A(–1, 4), B(–3, 3), C(1, 2) and D(3, 3) is a parallelogram.

Q10. The vertices of a quadrilateral are A(–2, 3), B(5, 4), C(4, –3) and D(–3, –4). Prove that ABCD is a rhombus. Find the area of ABCD.

Q11. Find the equation of the straight line passing through the point (–2, 1) and through the point of intersection of $4x - y - 1 = 0$ and $2x - y + 5 = 0$.

Q12. A, B and C are collinear points and AB = BC. If A is the point (4, 5) and B is the point (1, –1), find the coordinates of C.

Q13. Find the equation of the line which passes through the point of intersection of $5x - 3y + 2 = 0$ and $x + 3y + 1 = 0$, and is parallel to $2x + 3y + 3 = 0$.

Q1. A groundsman was mapping out the shape of a particular quadrilateral on the oval. He modelled it on paper first. In the quadrilateral ABCD, the points A, B, and D are at (3, 3), (0, 1) and (6, 2) respectively. The line BD intersects the line AC at right angles at the point M.

a) Plot and label the points A,B, and D on the graph and rule the lines AB, AD and BD.

b) Find the gradients of (i) AB (ii) AD (iii) BD

c) Find the equations of (i) AB (ii) AD (iii) BD

d) AC is perpendicular to BD.
   (i) Write the gradient of AC
   (ii) Find the equation of AC

e) Find the coordinates of M, the point of intersection of AC and BD. (Answers will include fractions.)

f) On the model, the groundsman thinks that C appears to be at the point (4, –3). Show that this could be the point C and draw the line AC on the graph.

g) Find the exact length of AC.

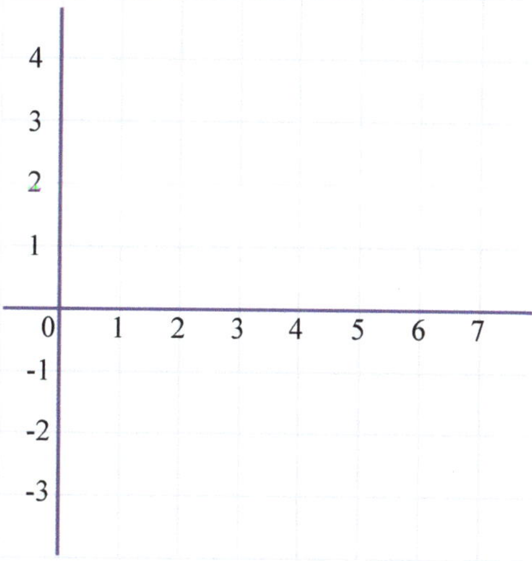

Q2. a) Plot the points on the following graph:
A(0, 3) B(3, 10) C(10, 7) D(7, 0).

b) Find the lengths: AB, BC, CD, AD

c) What type of quadrilateral is ABCD?

d) (i) Find the gradient of the lines AB and BC.
   (ii) What mathematical fact is indicated between the gradients of the lines AB and BC?

e) Find the area of ABCD.

f) Find the coordinates of the points A', B', C' and D' which are the reflected image points of A, B, C and D in the *x*-axis.

g) Find the coordinates of the points A'', B'', C'' and D'' which are the reflected image points of A, B, C and D in the *y*-axis.

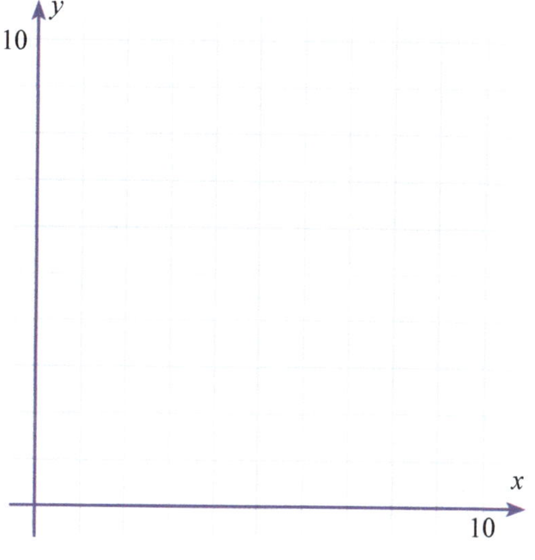

Q3. a) Draw the line between the points A (6, 8) and B (3, 2).

b) Find the coordinates of the mid-point of the line AB and mark it C.

c) Find the coordinates of the point D which divides the line AB in the ratio 1:2 closer to A.

d) Find the length of the line CD.

e) Show that the ratio of the distances of the lines AB to CD is 6:1.

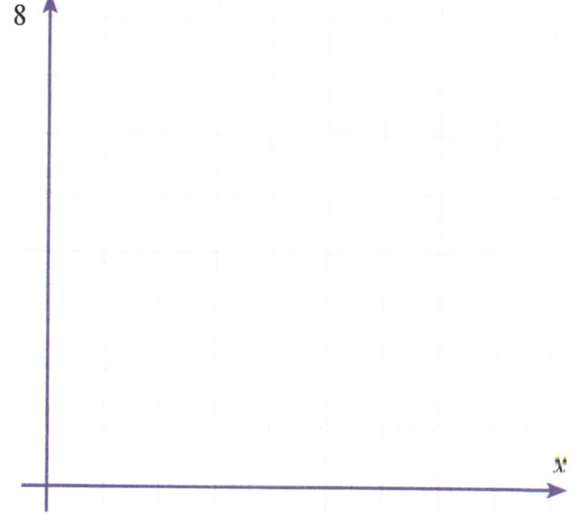

EXTENSION QUESTIONS

Q1. A straight line is drawn between two points, $A(-2, 8)$ and $B(6, -12)$.
a) Find (i) the gradient between the points (ii) the equation of the line AB.
b) Find the exact distance between the points.
c) Find the midpoint, M of AB and show that the distance of AM is the same as the distance BM.
d) Find the equation of a line which is parallel to AB which passes through the point (2, 5).

Q2. If a line with gradient m_1 is perpendicular to a line with gradient m_2 then $m_1 \times m_2 = -1$.
a) Find the gradient of the line A with equation $2x - 6y + 2 = 0$.
b) Find the gradient of a line, B which is perpendicular to the line A.
c) Find the equation of B if it passes through the point (2, 8).
d) Find the coordinates of the point where line A crosses line B.

Q3. Four points, $A(-1, 0)$ $B(2, 4)$ $C(6, 1)$ $D(3, -3)$ are joined to make the quadrilateral $ABCD$.
a) Plot the points on the set of axes.
b) Find the lengths (i) AB (ii) BC (iii) CD (iv) AD and state possible names for quadrilateral $ABCD$.
c) Find the gradients of (i) AB (ii) BC (iii) CD (iv) AD and name quadrilateral $ABCD$.
d) Find the area of $ABCD$.
e) Find the equations of the lines
(i) AB (ii) BC (iii) CD (iv) AD
f) Show that when the midpoints of AB, BC, CD and AD are joined then this quadrilateral is a square.

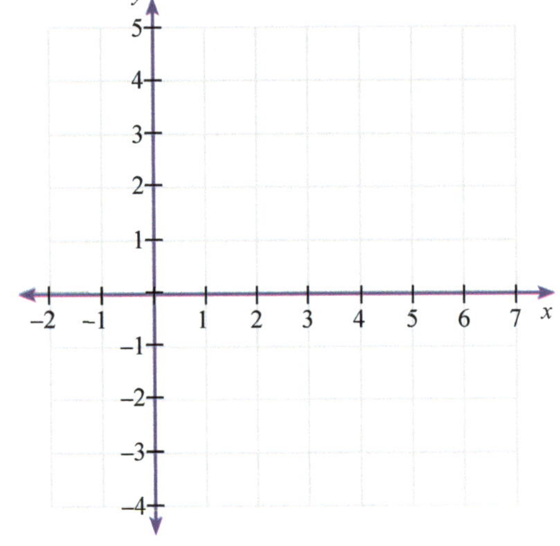

Q4. Line A passes through the points (–2, 0) and (4, 10) while Line B passes through the points (0, 10) and (10, 0).
a) Plot the points for both lines on the set of axes.
b) Find the equation of (i) line A (ii) line B.
c) State the coordinates of the intersection of both lines.
d) Find the area between the lines and the x-axis.
e) If lines A and B were both translated up 2 units, find (i) the new equation for each line
(ii) the point of intersection of both lines
(iii) the area between the lines and the x-axis.

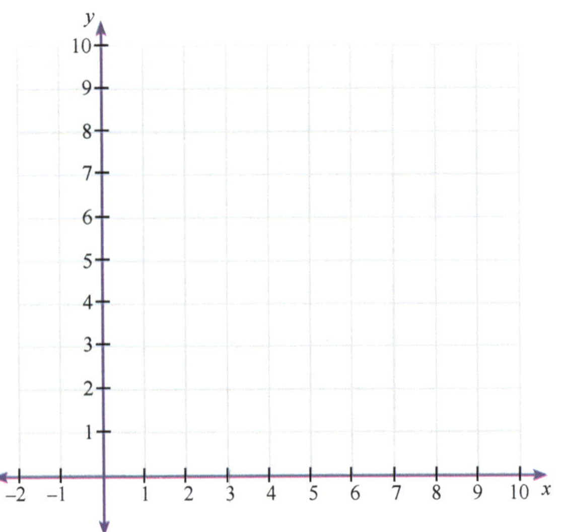

FURTHER TRIGONOMETRY

The 'Australian Curriculum Mathematics' (ACM) references for this sub-strand of 'Measurement and Geometry' (MG) are given below. This chapter may contain additional extension work which the author feels will be beneficial to the student.

- *Right-angled trigonometry (ACMMG 223, 224).*
- *Angles of elevation, depression and bearings (ACMMG 225, 245).*
- *Sine rule, cosine rule and area of triangle (ACMMG 273).*
- *Trigonometry in 3D and Pythagoras' theorem (ACMMG 276).*
- *Special angles and angles greater than 90º (NSW).*

Johann Gauss (1777 - 1855)

Gauss was a German mathematician and physicist who made significant contributions to many areas of Maths (particularly Number Theory) and physics. He is sometimes referred to as the Princeps Mathematicorum (latin : the foremost of the mathematicians) and the 'greatest mathematician since antiquity'. Like many other famous mathematicians and scholars he was a child prodigy, and at the age of only three years old he was already correcting his fathers mathematical errors. One famous story is how, at about the age of ten, a teacher tried to slow him down in the classroom because he finished all the usual assignments so quickly. The teacher asked him to add up all the counting numbers from 1 to 100, and was completely surprised when Gauss gave him the correct answer in a matter of seconds using a clever technique involving a pattern.

REVIEW OF TRIGONOMETRY

Most of the work in the first half of this chapter relating to right-angled triangles should be important revision of concepts already explained in Year 9.

The 3 formulas summarized below will enable you to calculate unknown sides and angles in right angled triangles. In the last part of the chapter new formulas will be explained to you – and these will enable you to calculate unknown sides and angles in non right angled triangles.

Consider a right-angled triangle with reference angle θ as shown below:

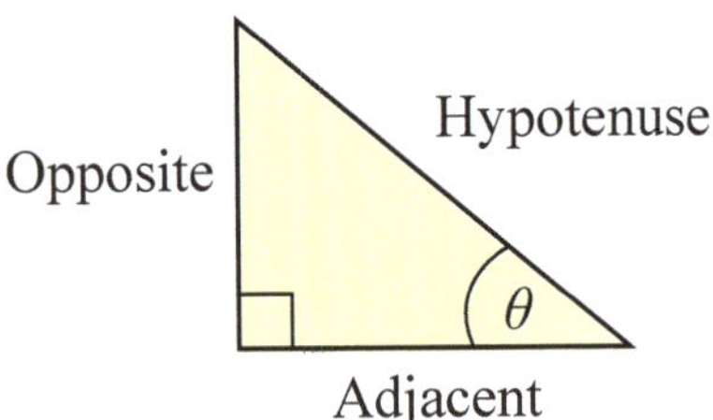

The first step is to determine the **opposite** side, the **hypotenuse** and the **adjacent** side. The hypotenuse is fixed because it is always the side which is directly facing the right angle. The other two sides depend on where the given reference angle is located.

To calculate the unknown measure of a side or an angle in a right-angled triangle, it is necessary to use one of the following formulae:

$$\sin\theta = \frac{\text{opposite}}{\text{hypotenuse}} = \frac{\text{O}}{\text{H}}$$

$$\cos\theta = \frac{\text{adjacent}}{\text{hypotenuse}} = \frac{\text{A}}{\text{H}}$$

$$\tan\theta = \frac{\text{opposite}}{\text{adjacent}} = \frac{\text{O}}{\text{A}}$$

An easy way of remembering these important formulae is:

S	**O**	**H**	**C**	**A**	**H**	**T**	**O**	**A**
Some	old	houses	can	always	hide	their	old	age.

Note: The formulas above are essential to remember, as every exam will contain questions relating to their application.
Which one of the three formulae to use will become obvious, once you have worked out the hypotenuse, adjacent and opposite sides.

THE CALCULATOR

The 3 main trigonometrical functions are called sine θ, cosine θ and tangent θ where θ represents an angle. These are usually shortened to the following:

sine θ = sin θ cosine θ = cos θ tangent θ = tan θ

The decimal value of these 3 trig functions can be found from special tables. However, in this day and age, we use the calculator because it is very much faster and more accurate.

As you know, 1 metre can be divided into 100 smaller units called centimetres, and each centimetre can be further subdivided into 10 smaller units called millimetres. So also 1 degree can be divided into 60 smaller units called 'minutes', and each minute can be further divided into 60 smaller units called 'seconds'.

1 degree = 60 minutes OR $1° = 60'$
1 minute = 60 seconds OR $1' = 60''$

The examples below illustrate how to use your calculator to find the decimal equivalent of angles relating to the 3 trig functions.

Example 1: Find tan 24° to 2 d.p.

Calculator steps: tan  24)  = 0.445 228 685   i.e 0.45 to 2 d.p.

You can round off by inspection or by using SHIFT FIX 2

Example 2: Find cos 36° 42' to 3 d.p.

Calculator steps: cos  36 °,,, 42 °,,, 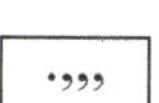) = 0.801 774 644

i.e. 0.802 to 3 d.p.

Example 3: If sin θ = 0.654 find the value of θ to the nearest minute.
In this example, we are given the decimal value and have to work in reverse to find the angle.

Calculator steps: SHIFT sin 0.654) = °,,, 40° 50' 37.93

The final answer on your screen means 40° 50' 37.93".
In most problems, we don't require the accuracy of seconds, and therefore we round up (i.e. add 1 to the minutes) if it is 30 seconds or more.
Therefore the final answer is: θ = 40° 51'

FINDING THE LENGTHS OF UNKNOWN SIDES

Example 1: A ladder leans against a wall 5 m high. If the base of the ladder is 4 m from the wall, find the angle (to the nearest minute) which the ladder makes with the ground.

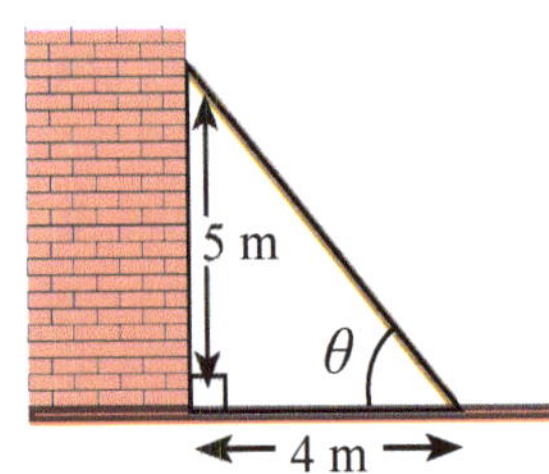

$\tan\theta = \dfrac{\text{opposite}}{\text{adjacent}}$ Formula

$\therefore \quad \tan\theta = \dfrac{5}{4}$ Substitution

$\therefore \quad \tan\theta = 1.25$ Convert to a decimal

$\therefore \quad \theta = 51^\circ\ 20'$

Calculator steps: [SHIFT] [tan] 5 [÷] 4 [)] [=] [°'''] [51° 20' 24.69]

Example 2: A boy is flying a kite on 300 m of string, which makes an angle of 37° with the horizontal. How much higher is the kite than the boy's hand?

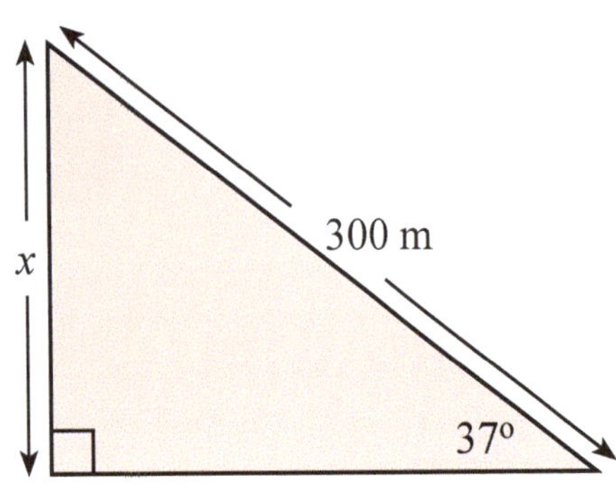

The first step is to draw a diagram from the description given in the question.

From the reference angle of 37°, it can be seen that opposite = x and hypotenuse = 300.

$\sin\theta = \dfrac{\text{opposite}}{\text{hypotenuse}}$

$\therefore \quad \sin 37^\circ = \dfrac{x}{300}$

$\therefore \quad x = 300 \times \sin 37^\circ$

$\therefore \quad x = 180.5$ m (correct to 1 d.p.)

Calculator steps: 300 [×] [sin] 37 [)] [=] [180.544 507]

Example 3: Calculate the length of x to 1 decimal place.

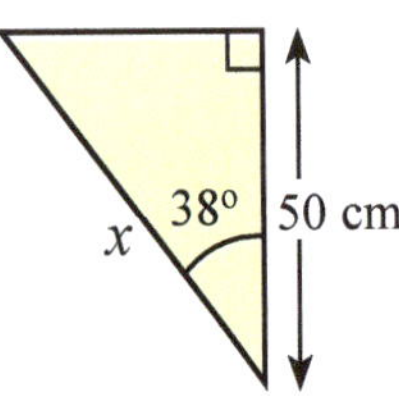

Sometimes the unknown length will be in the denominator and this makes the calculation slightly harder

Solution: This is an adjacent/hypotenuse problem which means the cosine ratio must be used.

$\cos\theta = \dfrac{\text{adjacent}}{\text{hypotenuse}}$

$\therefore \quad \cos 38^\circ = \dfrac{50}{x}$

$\therefore \quad x = \dfrac{50}{\cos 38^\circ}$ Divide both sides by cos 38°

$\therefore \quad x = 63.5$ cm

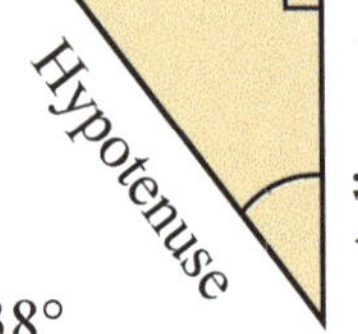

Calculator steps: 50 [÷] [cos] 38 [)] [=] [63.450 091 175]

FINDING THE SIZE OF UNKNOWN ANGLES

We will use one of the formulas summarized on page 224 . With the information given, it will be always obvious which of the 3 formulas that you will have to use.

Example 1: Calculate the size of angle θ, to the nearest minute.

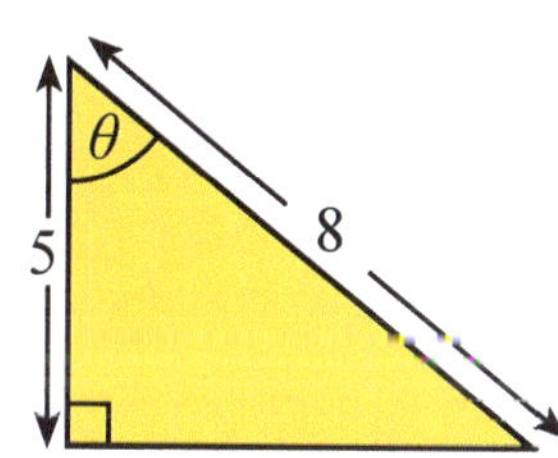

$\cos\theta = \dfrac{\text{adjacent}}{\text{hypotenuse}}$ Formula

$\therefore \quad \cos\theta = \dfrac{5}{8}$ Substitution

$\therefore \quad \cos\theta = 0.625$ Convert $\dfrac{5}{8}$ to a decimal

$\therefore \quad \theta = 51^\circ\ 19'$ To the nearest minute

Calculator steps: SHIFT | cos | 5 | ÷ | 8 |) | = | °'" | 51° 19' 4.13"

Example 2: In the diagram shown, find the size of 6 to the nearest minute.

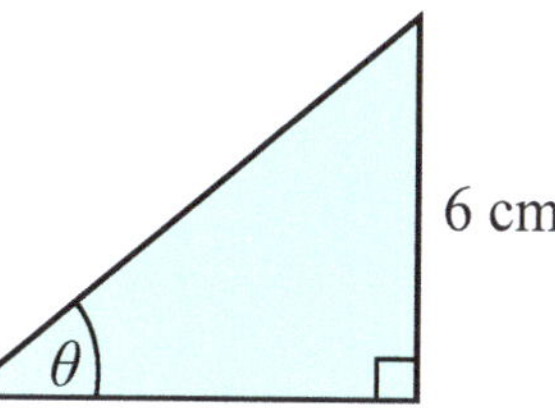

Solution: Must use tan formula to find θ.

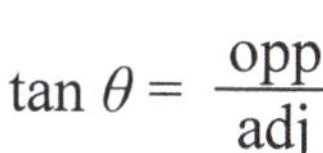

$\tan\theta = \dfrac{\text{opp}}{\text{adj}}$

$\therefore \quad \tan\theta = \dfrac{6}{8}$

$\therefore \quad \tan\theta = 0.75$

$\therefore \quad \theta = 36^\circ\ 52'$

A problem involving opp and adj is always a tan formula problem.

Calculator steps: SHIFT | tan | 6 | ÷ | 8 |) | = | °'" | 36° 52' 11.63"

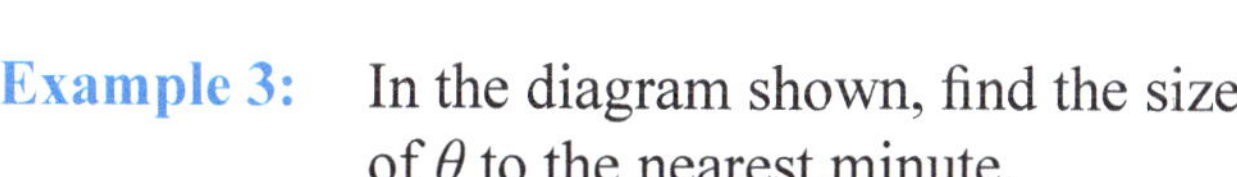

Example 3: In the diagram shown, find the size of θ to the nearest minute.

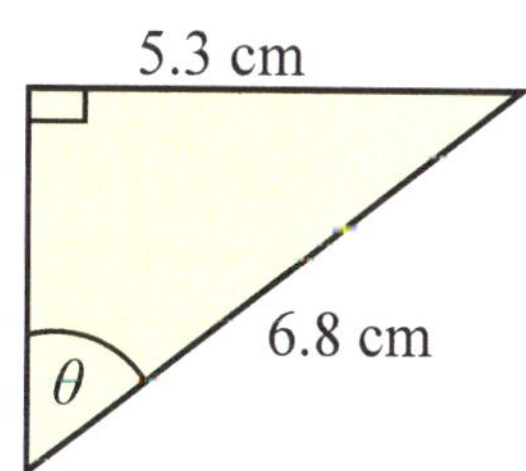

Solution: Must use sine formula to find θ.

$\sin\theta = \dfrac{\text{opp}}{\text{hyp}}$

$\therefore \quad \sin\theta = \dfrac{5.3}{6.8}$

$\therefore \quad \sin\theta = 0.779\,4117$

$\therefore \quad \theta = 51^\circ\ 12'$

A problem involving opp and hyp is always a sin formula problem.

Calculator steps: SHIFT | sin | 5.3 | ÷ | 6.8 |) | = | °'" | 51° 12' 24.3"

ANGLES OF ELEVATION AND DEPRESSION

ANGLE OF ELEVATION

This is the angle through which one has to **raise** or **elevate** one's eyes from the horizontal to **look up** at an object.

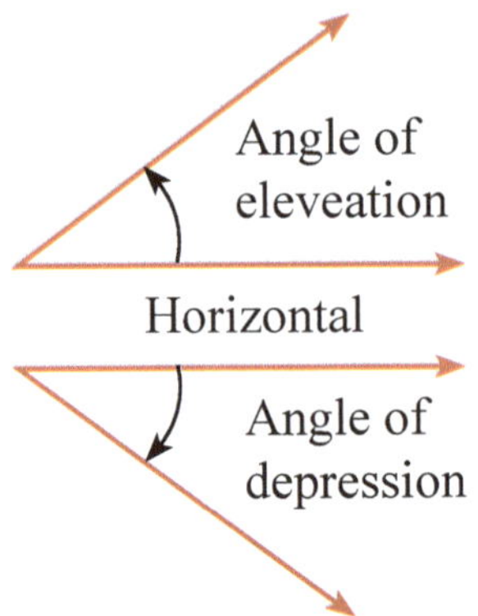

ANGLE OF DEPRESSION

This is the angle through which one has to **lower** or **depress** one's eyes from the horizontal to **look down** at an object.

Example: A ship is sighted from a plane at an angle of depression of 34°. If the distance from the plane to the ship is 5 000 m, calculate the height of the plane.

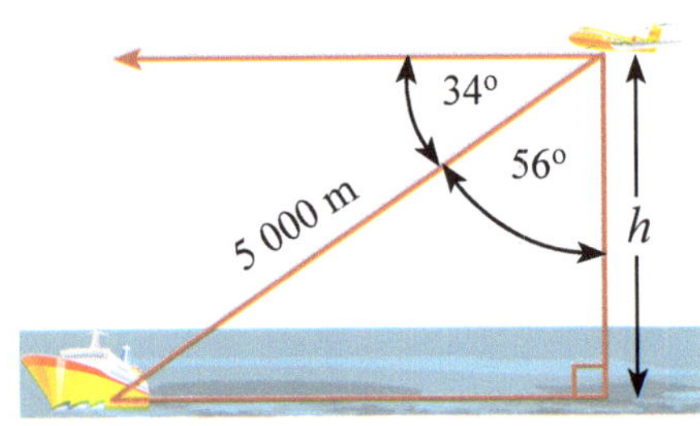

Solution: If the angle of depression is 34°, then the angle at the top of the triangle must be 90° – 34° = 56°

$$\text{Using } \cos\theta = \frac{\text{adjacent}}{\text{hypotenuse}}$$

$$\therefore \quad \cos 56^\circ = \frac{h}{5\ 000}$$

$$\therefore \quad h = 5\ 000 \times \cos 56^\circ$$

$$\therefore \quad h = 2\ 796 \text{ (to nearest metre)}$$

Note: The problem could have been done by calculating the other acute angle at the base of the triangle (i.e. 34°) and using the sine ratio.

$$\sin\theta = \frac{\text{opposite}}{\text{hypotenuse}}$$

$$\therefore \quad \sin 34^\circ = \frac{h}{5\ 000}$$

$$\therefore \quad h = 5\ 000 \times \sin 34^\circ$$

$$\therefore \quad h = 2\ 796 \text{ (to nearest metre)}$$

You will note that cos 56° gives the same decimal as sin 34°. Is there a connection?

THREE FIGURE BEARINGS

Bearings are used extensively in all types of navigation. They are angles that are used to give the exact direction of one position from a given fixed point.

> Bearings are angles written with 3 digits and they show the amount of turning in a clockwise direction from TRUE NORTH.

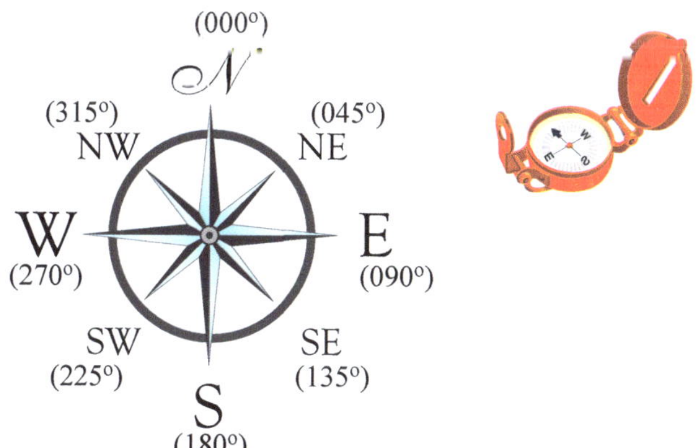

The compass points on the left show the three-figure bearings of 8 important directions of the compass. A bearing of NE is 045°, and a bearing of NW is 315°.

There is another alternative less used method of giving the exact direction from one location to another.

> We can use an angle to the East or West of the main North/South axis.

By studying the 3 examples below, you should be able to understand both methods:

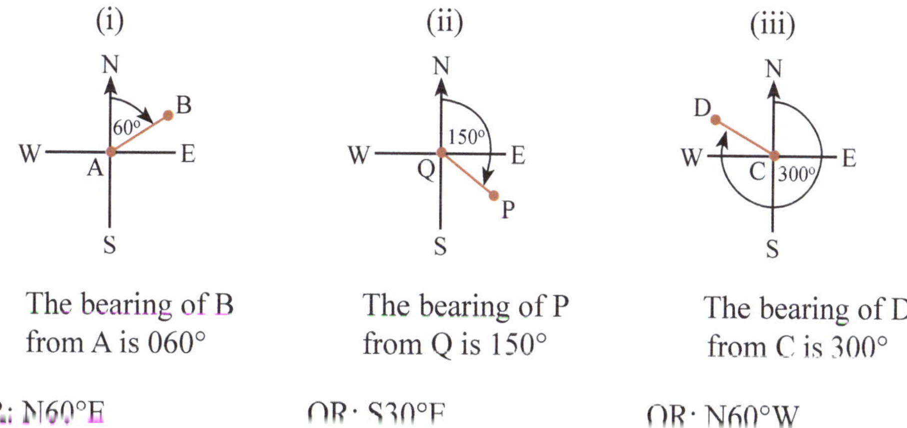

The bearing of B from A is 060°

The bearing of P from Q is 150°

The bearing of D from C is 300°

OR: N60°E

OR: S30°E

OR: N60°W

It is particularly important to read the question carefully in order to determine at which point the bearing is **being taken from**.

EXAMPLES USING BEARINGS

Example 1: Write the bearing of Q from P on each diagram using both methods.

a) b) c)

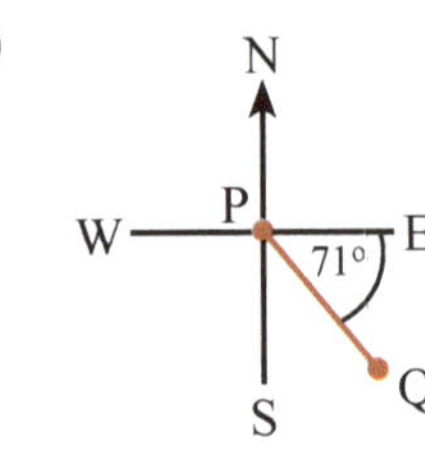

Solutions:

a) The bearing of Q from P is 328°
OR: N32°W

b) The bearing of Q from P is 212°
OR: S32°W

c) The bearing of Q from P is 161°
OR: S19°E

Example 2: If the bearing of B from A is 142°, what is the bearing of A from B?

It is important to note at which point the bearing is being taken from.

Solutions:

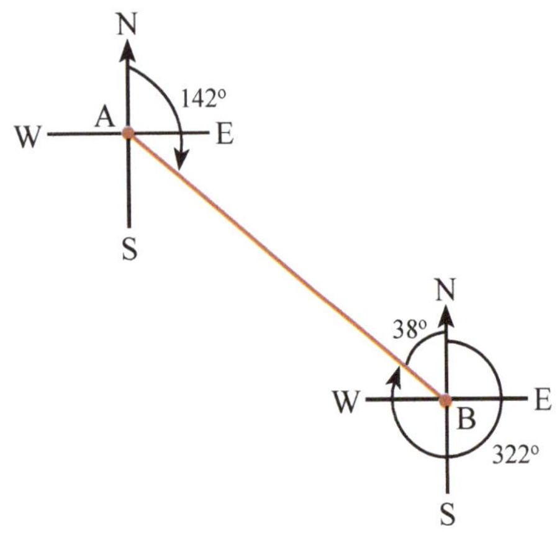

One of the best ways of solving these types of problems is to start with a sketch. The angle ∠NBY is cointerior to the angle at A because NA || NB.

The bearing of A from B can now easily be calculated.
Answer = 360° – 38° = 322°

Example 2: A ship sails 170 km on a bearing of 210° from a port P. How far directly South of the port is the ship?

Solutions:

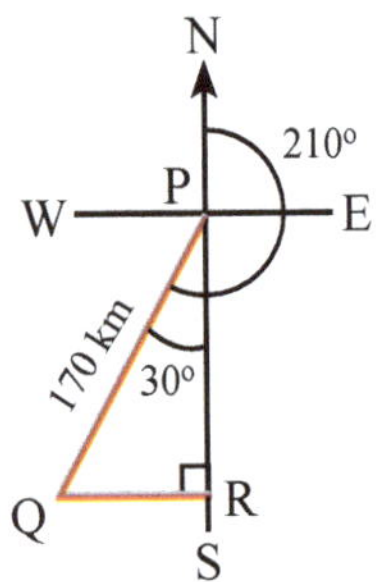

Step (i) Draw a diagram

Step (ii) Calculate ∠QPR = 30°

Step (iii) We need to find the distance PR.
This is an adjacent/hypotenuse problem.
∴ Must use cos ratio.

Step (iv) $\cos\theta = \dfrac{\text{adj}}{\text{hyp}}$

$\therefore \quad \cos 30^\circ = \dfrac{PR}{170}$

$\therefore \quad PR = 170 \times \cos 30^\circ$

$\therefore \quad PR = 147.22$

∴ The ship is 147.22 km due South of the port.

MORE DIFFICULT APPLICATIONS

USING TWO TRIANGLES

Some problems can only be solved by applying trig ratios to 2 different right angled triangles.

Example: A person finds the angle of elevation to the top of a tower 52 m tall to be 30°. He walks x metres closer to the tower and finds the new angle of elevation to be 40°. How far did he walk? (i.e. find x).

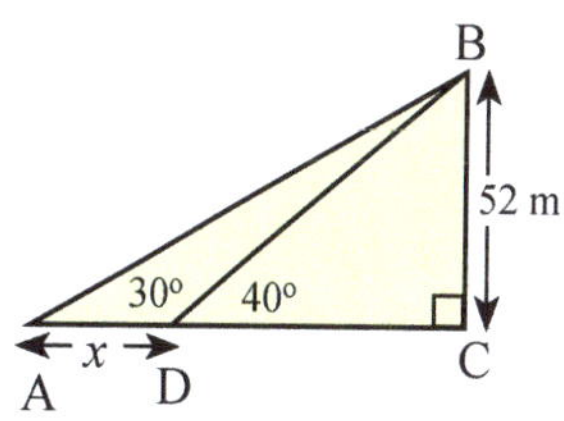

Using ΔABC:

$$\tan 30^\circ = \frac{52}{AC}$$

$$\therefore \quad \frac{1}{\tan 30^\circ} = \frac{AC}{52}$$

$$\therefore \quad AC = \frac{52}{\tan 30^\circ} = 90.07$$

Using ΔBCD:

$$\tan 40^\circ = \frac{52}{DC}$$

$$\therefore \quad \frac{1}{\tan 40^\circ} = \frac{DC}{52}$$

$$\therefore \quad DC = \frac{52}{\tan 40^\circ} = 61.97$$

$$\therefore \quad x = AC - DC = 90.07 - 61.97$$
$$= 28.1 \text{ metres.}$$

3 DIMENSIONAL PROBLEMS

It is important to look for right angled triangles, even though they may not look right angled from the 3 dimensional effect which the picture creates. Pythagoras' Theorem is also used regularly in the solution of these problems.

Example: From the diagram shown, find (i) $\angle CDH$ (ii) $\angle EDH$.

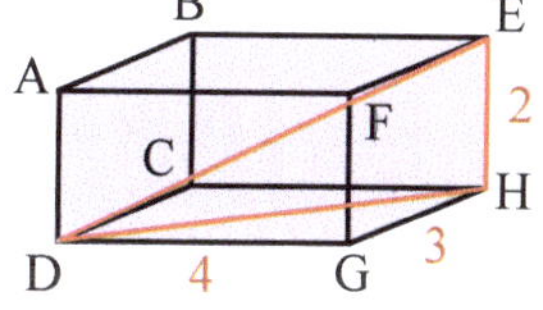

Solution: (i) ΔCDH has a right angle at C

From directly above it looks like:

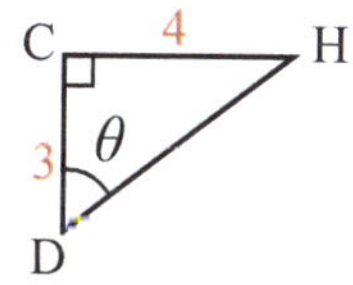

$$\tan \theta = \frac{4}{3}$$

$$\therefore \quad \angle CDH = \theta = 52^\circ \ 8'$$

(ii) Firstly applying Pythagoras to ΔDGH :

$$DH^2 = 4^2 + 3^2$$

$$\therefore \quad DH = 5$$

$$\tan \theta = \frac{2}{5}$$

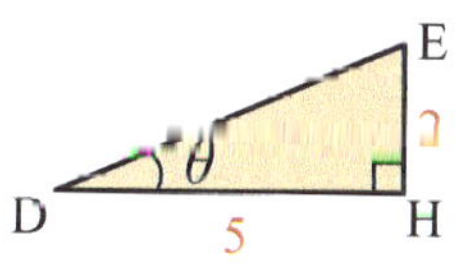

$$\therefore \quad \angle EDH = \theta = 21^\circ \ 48'$$

SPECIAL ANGLES

There are certain special angles in trigonometry where the values can be calculated exactly in surd form with the aid of the two triangles shown below:

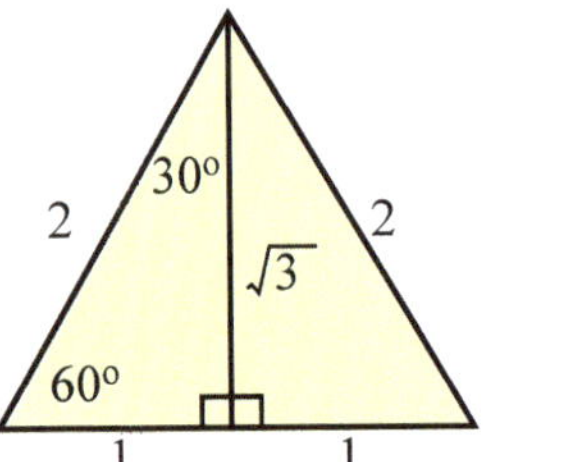

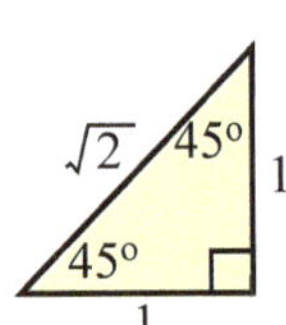

On the left, a right angled isosceles Δ will provide the trig ratios of 45º. On the far left, an equilateral Δ of side length 2 is split into 2 Δ's as shown.

Note: These two triangles, with the corresponding sides and angles, should be memorised. They will be used more frequently in Years 11 and 12.
Since they are right-angled triangles, we can apply SOH CAH TOA directly to find the exact values of the different trigonometric ratios.

Example: Find the exact values of the following:

(i) sin 60º (ii) cos 30º (iii) tan 30º
(iv) tan 60º (v) sin 45º (vi) tan 45º

Solution:

(i) $\sin 60^\circ = \dfrac{\text{opp}}{\text{hyp}} = \dfrac{\sqrt{3}}{2}$

(ii) $\cos 30^\circ = \dfrac{\text{adj}}{\text{hyp}} = \dfrac{\sqrt{3}}{2}$

(iii) $\tan 30^\circ = \dfrac{\text{opp}}{\text{adj}} = \dfrac{1}{\sqrt{3}}$

(iv) $\tan 60^\circ = \dfrac{\text{opp}}{\text{adj}} = \sqrt{3}$

(v) $\sin 45^\circ = \dfrac{\text{opp}}{\text{hyp}} = \dfrac{1}{\sqrt{2}}$

(vi) $\tan 45^\circ = \dfrac{\text{opp}}{\text{adj}} = 1$

RECIPROCAL RATIOS

$$\operatorname{cosec}\theta = \frac{1}{\sin\theta} \qquad \sec\theta = \frac{1}{\cos\theta} \qquad \cot\theta = \frac{1}{\tan\theta}$$

Example: Find the exact value of sec 30º.

$$\sec\theta = \frac{1}{\cos\theta} = \frac{\text{hyp}}{\text{adj}}$$

$$\therefore \quad \sec 30^\circ = \frac{2}{\sqrt{3}}$$

ANGLES LARGER THAN 90°

For the angles greater than 90° the three trigonometric ratios of sin, cos, and tan (and similarly for cosec, sec, and cot) are either positive or negative depending in which of the four quadrants the angle lies.

The rules shown below relating to different quadrants will be proved to you mathematically in later years. For the present time, just accept these rules and test them out using your calculator.

Example:

$\sin 150^\circ = 0.5$	$\cos 150^\circ = -0.87$	$\tan 150^\circ = -0.58$
$\sin 210^\circ = -0.5$	$\cos 210^\circ = -0.87$	$\tan 210^\circ = 0.58$
$\sin 300^\circ = -0.87$	$\cos 300^\circ = 0.5$	$\tan 300^\circ = -1.73$

From the above examples, it can be seen that sin is the only trigonometric function that is positive in the second quadrant ($90^\circ \leq \theta = 180^\circ$).

Tan is the only trigonometric function that is positive in the third quadrant ($180^\circ \leq \theta \leq 270^\circ$).

Cos is the only trigonometric function that is positive in the fourth quadrant ($270^\circ \leq \theta \leq 360^\circ$).

Depending on the size of the angle, you must remember the following rules:

QUADRANT I	All trig. ratios are positive.
QUADRANT II	$\sin\theta$ is ⊕, $\cos\theta$ and $\tan\theta$ are negative.
QUADRANT III	$\tan\theta$ is ⊕, $\sin\theta$ and $\cos\theta$ are negative.
QUADRANT IV	$\cos\theta$ is ⊕, $\sin\theta$ and $\tan\theta$ are negative.

An easy way to remember these rules is with the jingle:

ALL STATIONS TO CENTRAL

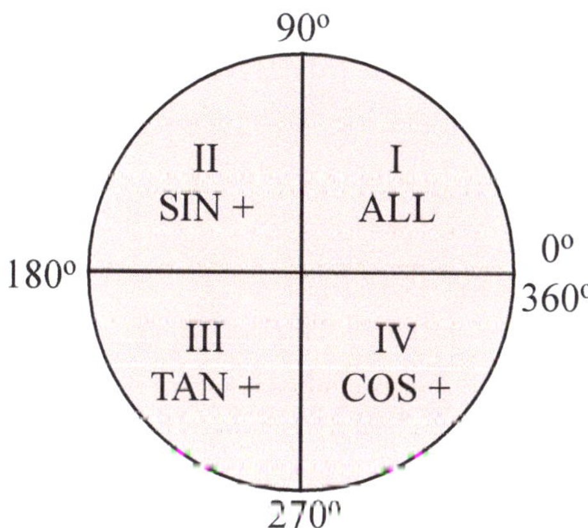

Note: The reciprocal functions of cosec, sec, and cot are also positive in the same quadrants as the original functions; that is, $\cot\theta$ is positive in Quadrant I and Quadrant III.

CALCULATIONS OF ANGLES OF ANY MAGNITUDE

To calculate the exact value of angles > 90°, it is necessary to follow the following steps:

STEP 1: Calculate the acute angle which the ray makes with the x-axis.

STEP 2: Determine the correct sign, using 'ALL STATIONS TO CENTRAL'

Note: The acute angle must always be calculated relative to the x-axis, and never the y-axis.

For example, an angle of 150° makes an acute angle of 30° with the x-axis.

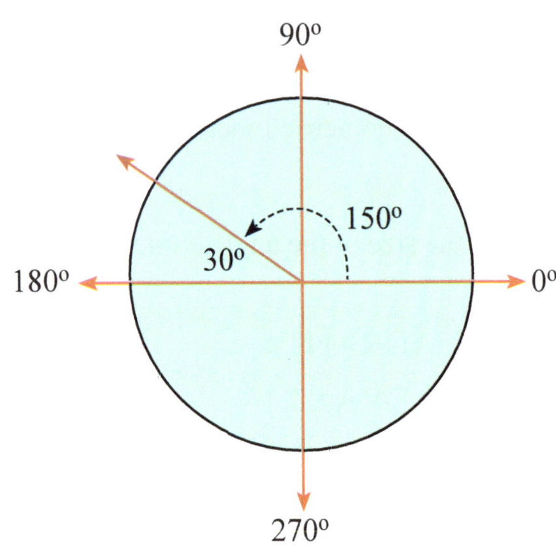

Example 1: tan 150°

tan 150° = tan (180° – 30°)

Acute angle = 30°.

Tan is negative in Quadrant II.

$\therefore$ tan 150° = –tan 30°

$= -\frac{1}{\sqrt{3}}$

Example 2: cos 225°

cos 225° = cos (180° + 45°)

The acute angle is 45°.

Cos is negative in Quadrant III.

$\therefore$ cos 225° = –cos 45°

$= -\frac{1}{\sqrt{2}}$

Example 3: If $\tan\theta = -\sqrt{3}$ find both possible solutions for $0^o \le \theta \le 360^o$.

$\tan\theta$ is negative in QII and QIV.

Therefore $\theta = 120^o$ or 300°

Example 4: cosec 315°

cosec 315º = cosec (360º – 45º)

The acute angle is 45º

Cosec is negative in Quadrant IV.

$\therefore$ cosec 315° = –cosec 45°

$= -\sqrt{2}$

TRIGONOMETRIC GRAPHS

If we draw up a table of values for $0^{\circ} \leq \theta \leq 360^{\circ}$, then we will obtain the following 3 graphs.

$y = \sin \theta$

0°	30°	60°	90°	120°	150°	180°	210°	240°	270°	300°	330°	360°
0	0.5	0.87	1.0	0.87	0.5	0	−0.5	−0.87	−1.0	−0.87	−0.5	0

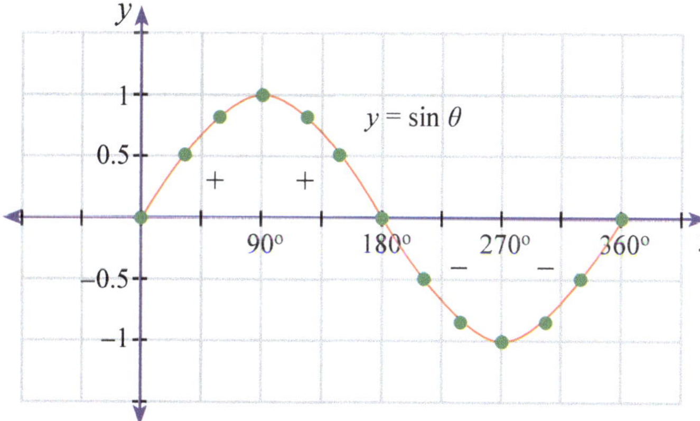

The curve will keep on repeating itself after $\theta = 360^{\circ}$. Note that the curve is positive in the first 2 quadrants upto 180°, and then it is negative in the next 2 quadrants. We would expect this from ALL STATIONS TO CENTRAL.

$y = \cos \theta$

0°	30°	60°	90°	120°	150°	180°	210°	240°	270°	300°	330°	360°
1.0	0.87	0.5	0	−0.5	−0.87	−1.0	−0.87	−0.5	0	0.5	0.87	1.0

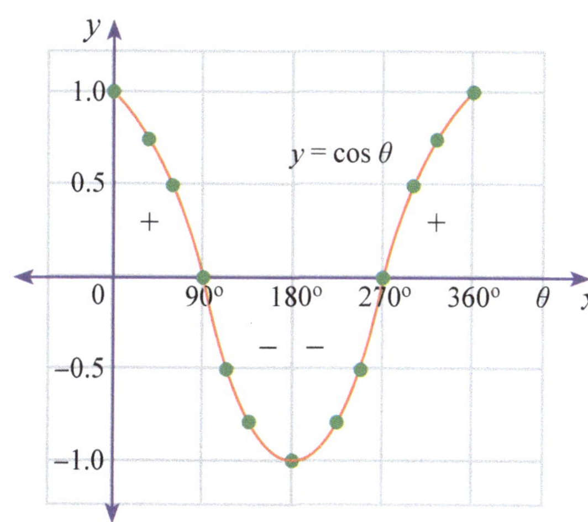

Once again, the graph shows 1 full cycle of the curve. After 360°, it will simply keep on repeating itself. Note that the curve is positive in quadrants 1 and 4, and negative in quadrants 2 and 3 from 90° to 270°.

$y = \tan$

0°	30°	60°	90°	120°	150°	180°	210°	240°	270°	300°	330°	360°
0	0.58	1.73	*u*	−1.73	−0.58	0	0.58	1.73	*u*	−1.73	−0.58	0

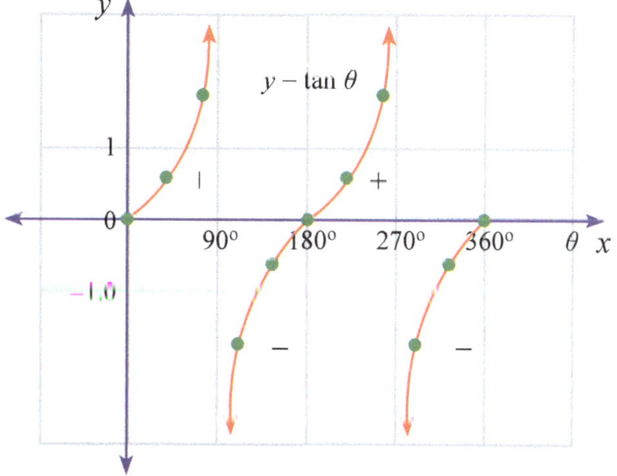

The value for tan 90° and tan 270° are undefined and you will get an error sign if you try to obtain a value on your calculator. You can however enter angles close to 90° such as 89° and 91°. Also you can enter angles close to 270° such as 269° and 271°.

THE SINE RULE

The trigonometry you have learned so far has been related to finding unknown sides and angles in RIGHT-ANGLED TRIANGLES.

The remainder of this chapter explains how to calculate unknown sides and angles in NON RIGHT-ANGLED TRIANGLES.

With non right-angled triangles, the sides are named in a different way to right-angled triangles. The vertices (corners) of the triangle are denoted by capital letters, while the sides directly opposite to them are denoted by the corresponding small letters.

$$\frac{a}{\sin A} = \frac{b}{\sin B} = \frac{c}{\sin C}$$

OR

$$\frac{\sin A}{a} = \frac{\sin B}{b} = \frac{\sin C}{c}$$

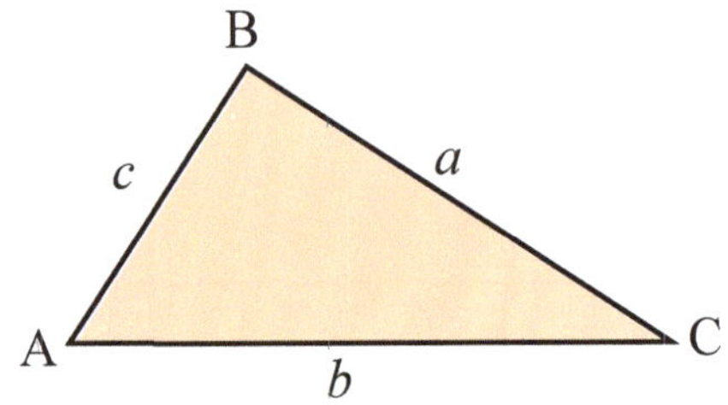

Note: Only two of these formulae are used at any one time.

Example: Find the length of AC correct to 2 decimal places.

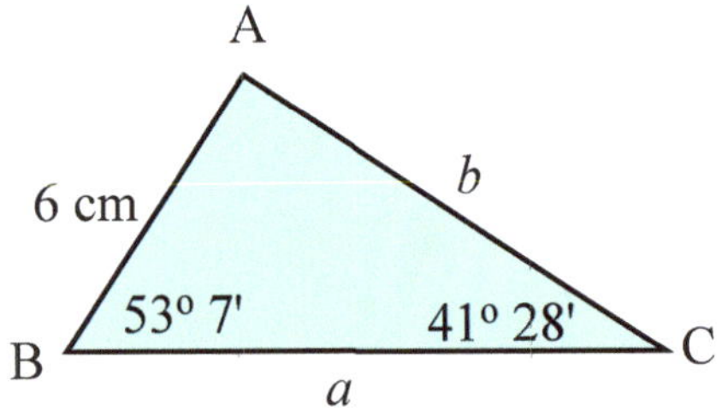

Solution: This problem obviously involves b, $\angle$B, c, $\angle$C.

The problem does not involve $\angle$A or side a.

i.e. $\dfrac{b}{\sin B} = \dfrac{c}{\sin C}$ Formula.

$\therefore \dfrac{b}{\sin 53^\circ 7'} = \dfrac{6}{\sin 41^\circ 28'}$ Substitute in known values.

$\therefore b = \dfrac{6 \times \sin 53^\circ 7'}{\sin 41^\circ 28'}$ Multiply both sides by sin 53° 7'.

$\therefore b = 7.25$ cm From calculator.

THE COSINE RULE

This is used when sufficient information is not available to use the SINE RULE. It is used specifically in these 2 types of questions:

(1) You are given 2 sides and the included angle of a triangle, and you are asked to calculate the third side.

(2) You are given the lengths of 3 sides of a triangle, and you are asked to find the size of any one of the angles.

Note: The included angle is the angle between the two given sides.

$$a^2 = b^2 + c^2 - 2bc \cos A$$

$$b^2 = a^2 + c^2 - 2ac \cos B$$

$$c^2 = a^2 + b^2 - 2ab \cos C$$

It is easy to learn the formulae above once you have seen the pattern to them.

Example: In $\triangle ABC$, $\angle C = 47° 12'$, $b = 7$ cm and $a = 9$ cm; find the length of c correct to 2 d.p.

Solution: The first step is to sketch a diagram from the facts given in the question. In this problem, you have been given 2 sides and the included angle, and have been asked to calculate the third side (c).

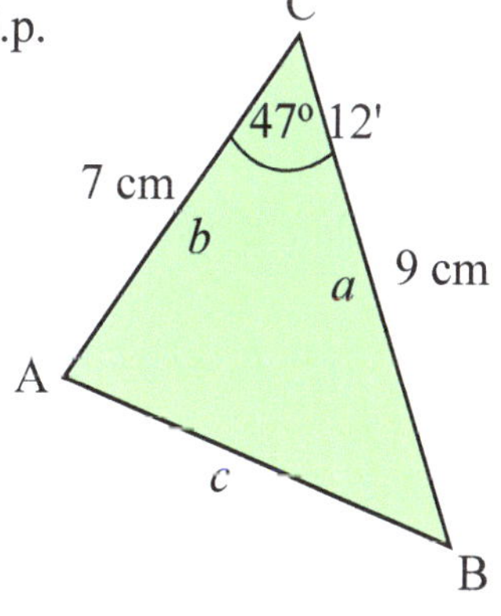

$c^2 = a^2 + b^2 - 2ab \cos C$	Formula to find 'c'.
$\therefore \; c^2 = 9^2 + 7^2 - 2 \times 9 \times 7 \times \cos 47° 12'$	Substitute in known data.
$\therefore \; c^2 = 44.390\,396$	From calculator
$\therefore \; c = 6.66$ cm	take $\sqrt{}$ of both sides.

THE COSINE RULE TRANSPOSED

The three formulae in the previous section, relating to the 'cosine rule', can easily be transposed to make each of the angles the subject of the formula.

It is a much more useful format when:

> You are given 3 sides of a triangle, and asked to calculate one of the angles.

For example: $a^2 = b^2 + c^2 - 2bc \cos A$ — 'a^2' is the subject.

$\therefore\ a^2 + 2bc \cos A = b^2 + c^2$ — Add $2bc \cos A$ to both sides.

$\therefore\ 2bc \cos A = b^2 + c^2 - a^2$ — Subtract a^2 from both sides.

$\therefore\ \cos A = \dfrac{b^2 + c^2 - a^2}{2bc}$ — Divide both sides by $2bc$ to make cos A the subject.

$$\cos A = \frac{b^2 + c^2 - a^2}{2bc}$$

$$\cos B = \frac{a^2 + c^2 - b^2}{2ac}$$

$$\cos C = \frac{a^2 + b^2 - c^2}{2ab}$$

The student should be able to make 'cos B' and 'cos C' the subject in the same way.

Example: Calculate the size of $\angle B$ to the nearest minute.

$a = 9$ cm, $b = 10$ cm, $c = 7$ cm

$$\cos B = \frac{a^2 + c^2 - b^2}{2bc}$$

$$\cos B = \frac{9^2 + 7^2 - 10^2}{2 \times 9 \times 7}$$

$\therefore\quad B = 76^\circ\ 14'$

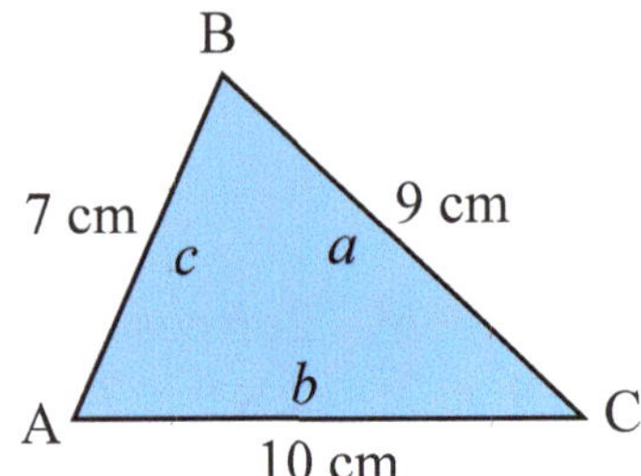

AREA OF A TRIANGLE

From earlier years, the student should recall that the area of a triangle, given the 'base' and 'vertical height', is determined by using:

$$\text{Area} = \frac{1}{2} \times \text{base} \times \text{height}$$

However, if two sides of a triangle and the included angle are given, then one of the following three formulae may be used to calculate the area:

$$\text{Area} = \frac{1}{2}ab \sin \text{C}$$
$$\text{Area} = \frac{1}{2}bc \sin \text{A}$$
$$\text{Area} = \frac{1}{2}ac \sin \text{B}$$

Once again, they are easy to learn because there is a simple pattern to them.

Example: In ΔPQR, $\angle$Q = 42°, r =7 cm and p = 9 cm. Find the area of ΔPQR correct to 2 d.p.

Solution: The first step is to draw a sketch.

$\therefore$ Area $= \frac{1}{2}pr \sin \text{Q}$

$= \frac{1}{2} \times 9 \times 7 \times \sin 42^{\circ}$

$= 21.08 \text{ cm}^2$.

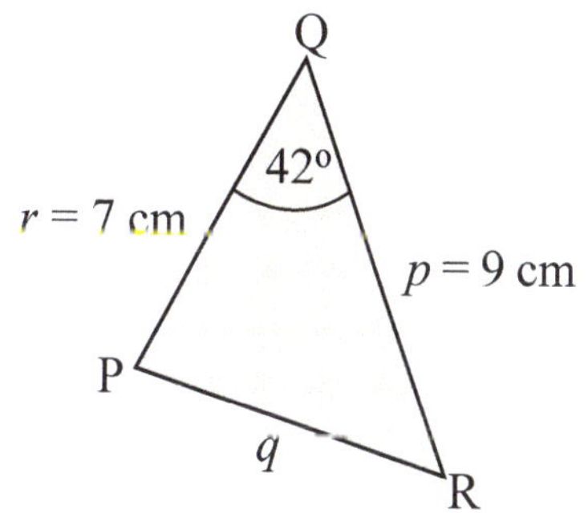

ALTERNATIVE: METHOD Many students prefer to change the letters of the vertices to A, B and C so they can use the formulae shown above.

$\therefore$ Area $= \frac{1}{2}ab \sin \text{C}$

$= \frac{1}{2} \times 9 \times 7 \times \sin 42^{\circ}$

$= 21.08 \text{ cm}^2$

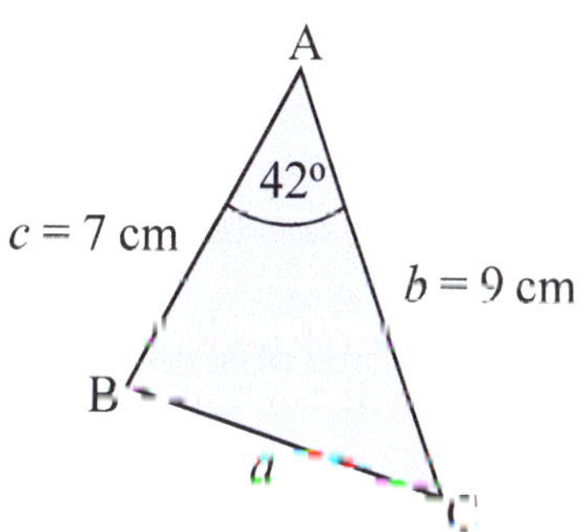

MORE DIFFICULT APPLICATIONS

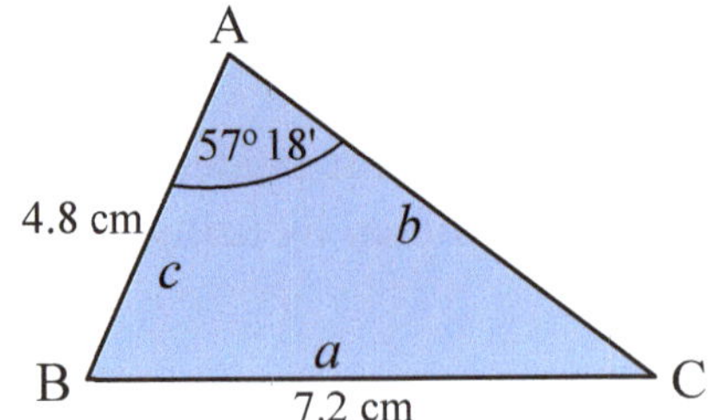

Example 1: Find the size of $\angle ABC$ to the nearest minute.

Solution: This problem involves c, $\angle C$, a, $\angle A$

$$\therefore \quad \frac{\sin C}{c} = \frac{\sin A}{a}$$ Sine rule.

$$\therefore \quad \frac{\sin C}{4.8} = \frac{\sin 57^\circ 18'}{7.2}$$ Substitute in data.

$$\therefore \quad \sin C = \frac{4.8 \times \sin 57^\circ 18'}{7.2}$$ Multiply both sides by 4.8.

$$\therefore \quad \sin C = 0.561\,0071$$ Decimal from calculator.

$$\therefore \quad C = 34^\circ 8'$$

$$\therefore \angle ABC = 180^\circ - 57^\circ 18' - 34^\circ 8'$$ Angle sum of $\triangle = 180^\circ$.

$$= 88^\circ 34'$$

Example 2: Given that A is the centre of the circle with dimensions shown, calculate correct to 1 d.p.

(i) the area of the circle
(ii) the area of the sector ABC
(iii) the area of the triangle ABC
(iv) the area of the grey shaded segment

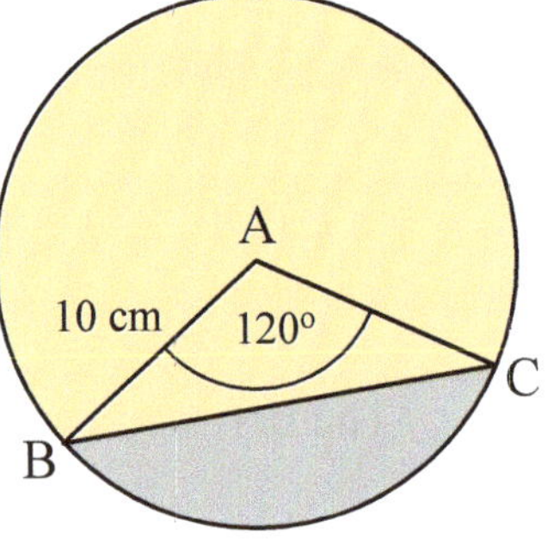

Solution:

(i) Area of circle $= \pi r^2$

$$= \pi \times 10 \times 10$$
$$= 314.2 \text{ cm}^2$$

(ii) Area of sector $= \dfrac{120}{360} \times$ area of circle

$$= 104.7 \text{ cm}^2$$

(iii) Area of $\triangle ABC = \dfrac{1}{2} bc \sin A$

$$= \frac{1}{2} \times 10 \times 10 \times \sin 120^\circ$$
$$= 43.3 \text{ cm}^2$$

(iv) Area of segment = (area of sector) – (area of $\triangle$)

$$= 104.7 - 43.3$$
$$= 61.4 \text{ cm}^2$$

Example 3: The three legs of a triangular sailing course are 700 m, 1000 m and 1400 m. Find the largest angle through which the boats have to turn when completing the course.

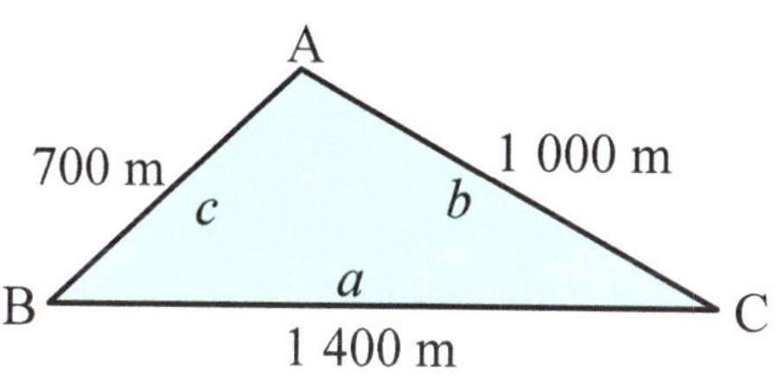
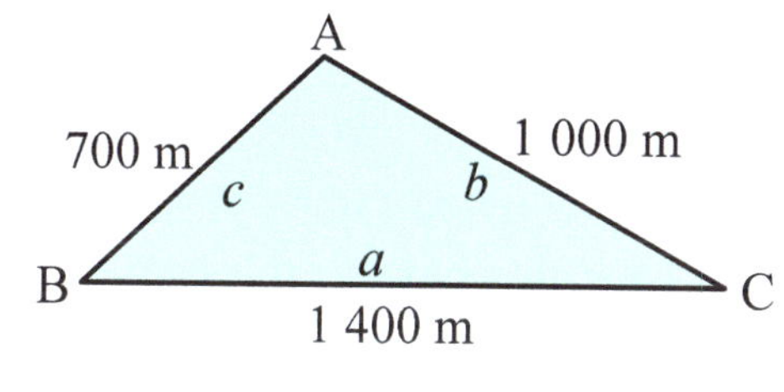

Solution: First draw a sketch as shown. Given 3 sides, and we have to find one of the angles ⇨ cosine rule.

$$\cos A = \frac{b^2 + c^2 - a^2}{2bc}$$

$$\therefore \quad \cos A = \frac{1\ 000^2 + 700^2 - 1\ 400^2}{2 \times 1\ 000 \times 700}$$

$$\therefore \quad \cos A = -0.335\ 7142$$

$$\therefore \quad \cos A = 109^o\ 36'$$

The largest angle is 109° 36'

Example 4: A man walking due East turns at A to avoid a lake. He walks 250 m on a bearing of 48° and then changes directions and follows a new bearing of 108° to C. Find the distance AC to 1 d.p. if C is due east of A.

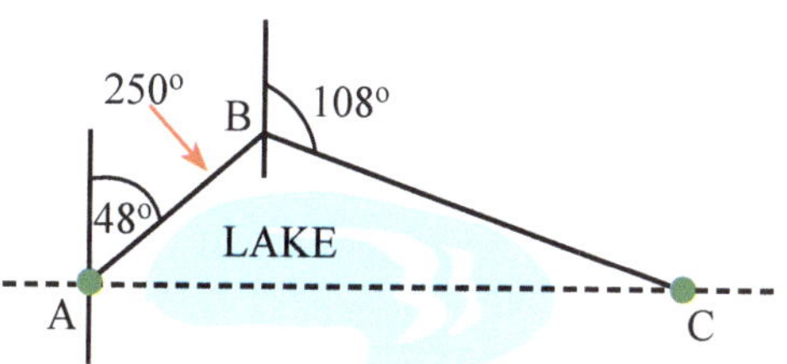

Solution: Firstly draw a sketch as shown above using your knowledge of bearings. By using basic geometric theorems, calculate all the angles in ΔABC. The problem involves b, ∠B, c, ∠C.

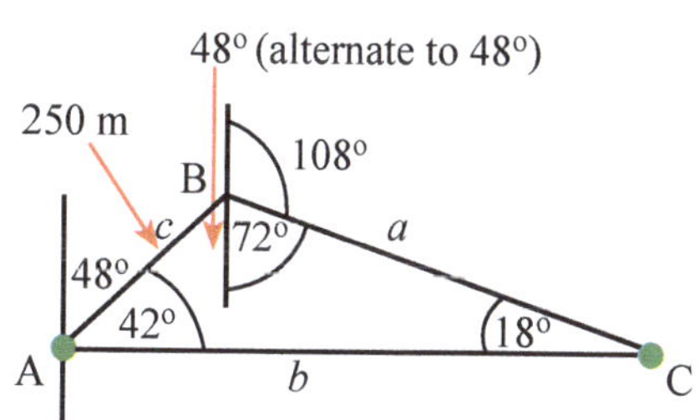

$$\therefore \quad \frac{b}{\sin B} = \frac{c}{\sin C}$$ Sine rule.

$$\therefore \quad \frac{b}{\sin 120^o} = \frac{250}{\sin 18^o}$$ Substitute in known values.

$$\therefore \quad b = \frac{250 \sin 120^o}{\sin 18^o}$$ Multiply both sides by sin 120°.

$$\therefore \quad b = 700.6$$ From calculator.

$$\therefore \quad AC = 700.6 \text{ m}$$ Answer the original question.

CHAPTER SUMMARY

INTRODUCTION

The trigonometric rules covered in this summary are primarily used to calculate 'lengths of sides' and 'angle sizes' in right-angled and non right-angled triangles. It is an important topic which is used extensively in architecture, surveying, engineering, science, astronomy, and navigation. Sometimes you will be given the diagram, and sometimes you will have to make a rough sketch of the diagram from the facts and bearings given in the question.

NAMING THE SIDES

The first step in any trig problem involving right-angled triangles is to name the 3 sides correctly. Consider a right-angled triangle with a given reference angle 'θ' as shown in the diagram in the next section.

HYPOTENUSE	—	is always the side directly facing the right angle (as in Pythagoras' Theorem).
OPPOSITE SIDE	—	as the name implies, it faces directly opposite to the given acute reference angle.
ADJACENT SIDE	—	by elimination, it is the only side remaining.

THREE ESSENTIAL FORMULAE FOR RIGHT-ANGLED TRIANGLES

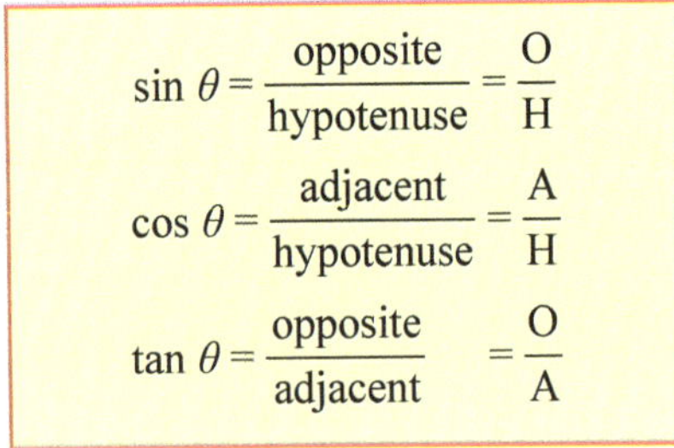

$$\sin\theta = \frac{\text{opposite}}{\text{hypotenuse}} = \frac{O}{H}$$

$$\cos\theta = \frac{\text{adjacent}}{\text{hypotenuse}} = \frac{A}{H}$$

$$\tan\theta = \frac{\text{opposite}}{\text{adjacent}} = \frac{O}{A}$$

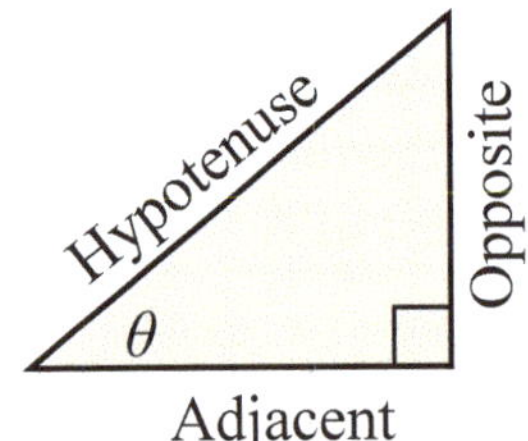

S O H — Some Old Houses
C A H — Can Always Hide
T O A — Their Old Age

TWO WAYS OF DESCRIBING ANGLES

ANGLE OF ELEVATION

From the horizontal,
the angle looking **UPWARDS**.

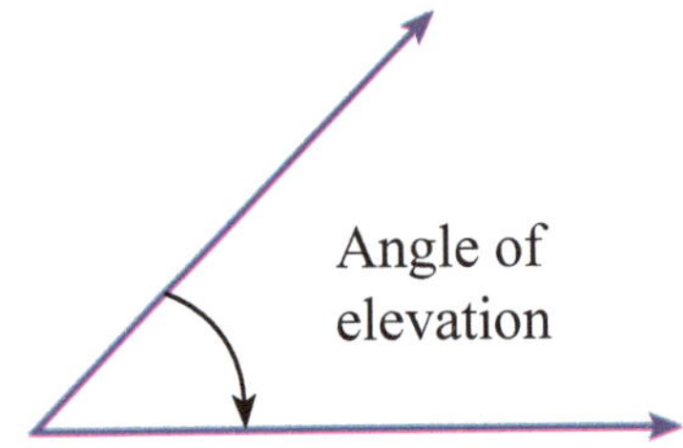

ANGLE OF DEPRESSION

From the horizontal, the
angle looking **DOWNWARDS**.

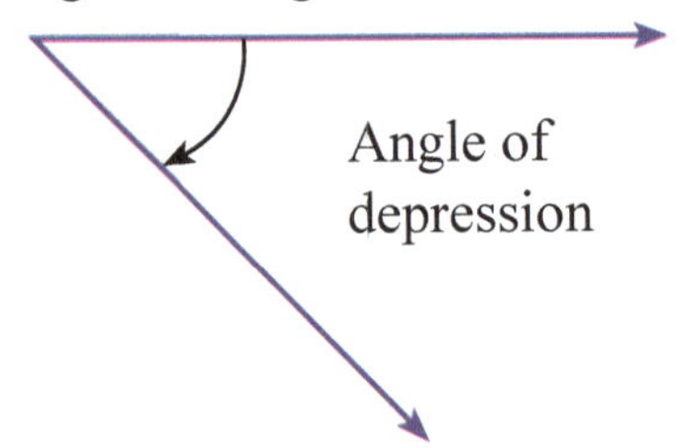

BEARINGS

The bearing of point B from point A gives the direction of point B from the original point A. There are two main methods of giving bearings:

1. A three figure angle measured in a clockwise direction from true North.
2. As an angle to the East or West of the main North/South axis.

By studying the 3 examples below, you should be able to understand both methods:

1.

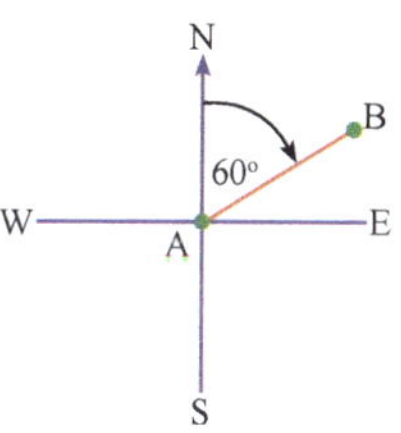

The bearing of B from A is 060°
OR: N60°E

2.

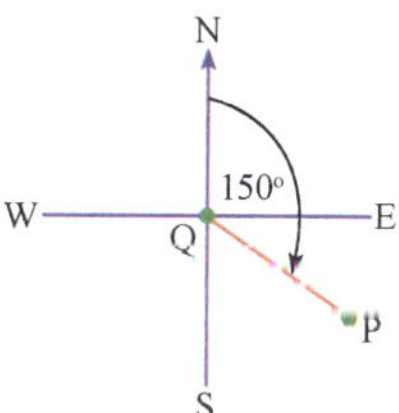

The bearing of P from Q is 150°
OR: S30°E

3.

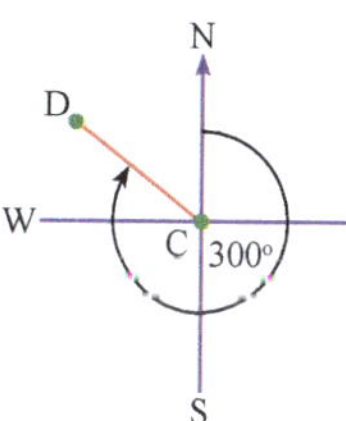

The bearing of D from C is 300°
OR: N60°W

It is particularly important to read the question carefully in order to determine at which point the bearing is being taken from.

SPECIAL ANGLES (30°, 45°, 60°)

Many angles in trigonometry larger than 90° can be reduced to one of these three special angles. You should remember how to draw the 2 triangles shown, and then it is easy to obtain any ratio simply by using **SOH CAH TOA**.

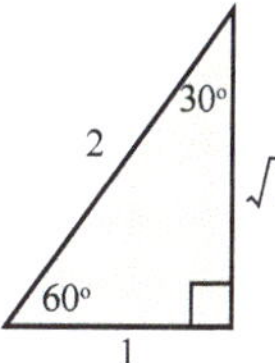

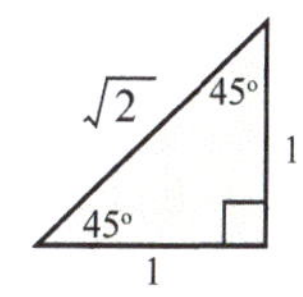

These triangles give the EXACT RATIOS, and calculators should not be used for these special angles.

Examples: $\tan 30^o = \dfrac{O}{A} = \dfrac{1}{\sqrt{3}}$, $\cos 45^o = \dfrac{1}{\sqrt{2}}$, $\sin 60^o = \dfrac{\sqrt{3}}{2}$

The reciprocal trigonometric functions:

$\text{cosec}\ \theta = \dfrac{1}{\sin\theta}$, $\quad \sec\theta = \dfrac{1}{\cos\theta}$, $\quad \cot\theta = \dfrac{1}{\tan\theta}$

ANGLES LARGER THAN 90°

For any angles greater than 90° the three trigonometric ratios of sin, cos, and tan (and similarly for cosec, sec, and cot) are either positive or negative, depending in which of four quadrants the angle lies.

ALL STATIONS TO CENTRAL

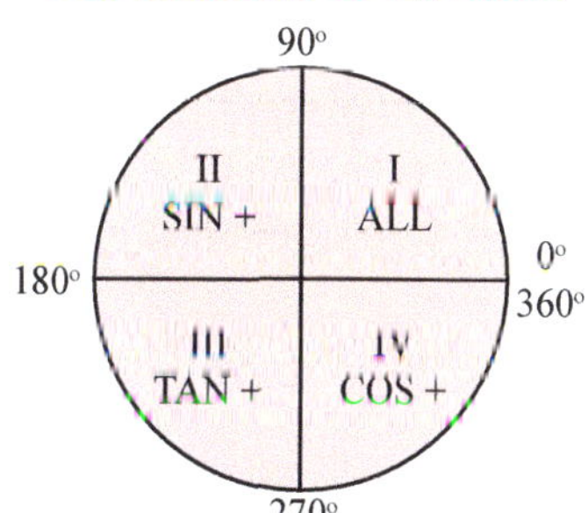

An easy way to remember the signs for each of the 4 quadrants.

NON-R1GHT ANGLED TRIANGLES

The sides are named in a completely different way to right-angled triangles. The vertices (or corners) are named with capital letters. The sides opposite are named with the corresponding small (or lower case) letters, as shown.

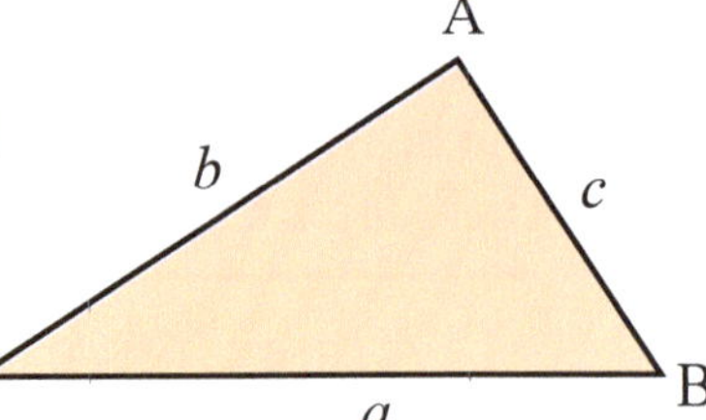

THE SINE RULE

$$\frac{a}{\sin A} = \frac{b}{\sin B} = \frac{c}{\sin C} \quad \text{OR} \quad \frac{\sin A}{a} = \frac{\sin B}{b} = \frac{\sin C}{c}$$

To solve a particular problem, 2 out of 3 equalities will be relevant.

THE COSINE RULE or COSINE RULE TRANSPOSED

$$a^2 = b^2 + c^2 - 2bc\cos A \quad \text{OR} \quad \cos A = \frac{b^2 + c^2 - a^2}{2bc}$$

$$b^2 = a^2 + c^2 - 2ac\cos B \quad \text{OR} \quad \cos B = \frac{a^2 + c^2 - b^2}{2ac}$$

$$c^2 = a^2 + b^2 - 2bc\cos C \quad \text{OR} \quad \cos C = \frac{a^2 + b^2 - c^2}{2ab}$$

There is a definite pattern to all these formulas.

WHICH RULES TO USE

The cosine rule is used for two types of questions:

(i) Given two sides and the included angle, find the third side.

(ii) Given three sides, find any one of the angles using one of the transposed formulae.

Note: All other trigonometric problems relating to non-right angled triangles will involve the use of the sine rule.

AREA OF A TRIANGLE

If we know the base and height of a triangle, then obviously we use the standard formula learned in earlier years: Area of triangle $= \frac{1}{2}(\text{base} \times \text{height}) = \frac{1}{2}bh$

However, in senior years you are more likely to be given two sides and the included angle, and one of the following three formulae will have to be used:

$$\text{Area} = \frac{1}{2}ab\sin C = \frac{1}{2}ac\sin B = \frac{1}{2}bc\sin C$$

It will be obvious which formula to use when the data is given.

Example: Find the size of $\angle ABC$ and hence find the area of the triangle.

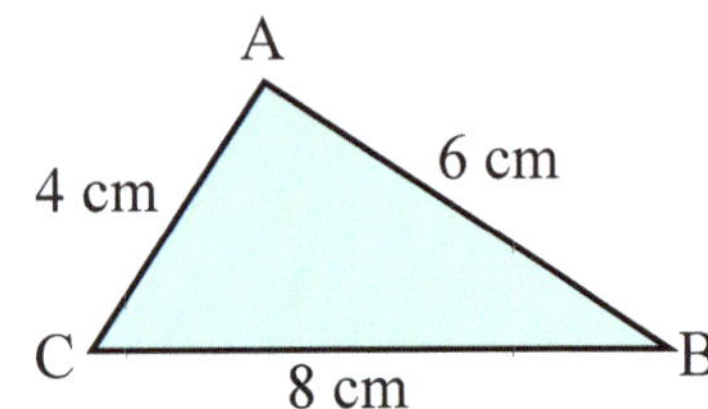

Solution : $\angle ABC = 28.96^\circ$, area $\Delta ABC = 11.62\text{cm}^2$

LEVEL 1 — FURTHER TRIGONOMETRY

EASIER QUESTIONS

Note: Only turn back to page number shown if you have difficulty.

	Page
Q1. Find correct to 3 d.p.: a) sin 50° 30' b) tan 7° 52' c) cos 83° 19' d) tan 68° 09' e) cos 21° 40' f) sin 36° 24'	225
Q2. Find θ correct to the nearest minute if $0° \le \theta \le 90°$: a) cos $\theta = 0.155$ b) sin $\theta = 0.689$ c) tan $\theta = 0.445$ d) sin $\theta = 0.927$ e) tan $\theta = 1.731$ f) cos $\theta = 0.802$	225
Q3. Find the value of x correct to 2 d.p.:	226
Q4. Find θ correct to the nearest minute:	227
Q5. Find the value of x correct to 2 d.p.:	236
Q6. Find the value of x correct to 2 d.p.:	237
Q7. Find the area of each triangle correct to 3 sig. figs.:	239
Q8. Write the bearing of B from A using 2 different methods:	229, 230

Q1. Find correct to 3 d.p.:

a) sin 50° 30'
b) tan 7° 52'
c) cos 83° 19'
d) tan 68° 09'
e) cos 21° 40'
f) sin 36° 24'

Q2. Find θ correct to the nearest minute if $0° \le \theta \le 90°$:

a) $\cos \theta = 0.155$
b) $\sin \theta = 0.689$
c) $\tan \theta = 0.445$
d) $\sin \theta = 0.927$
e) $\tan \theta = 1.731$
f) $\cos \theta = 0.802$

Q3. Find the value of x correct to 2 d.p.:

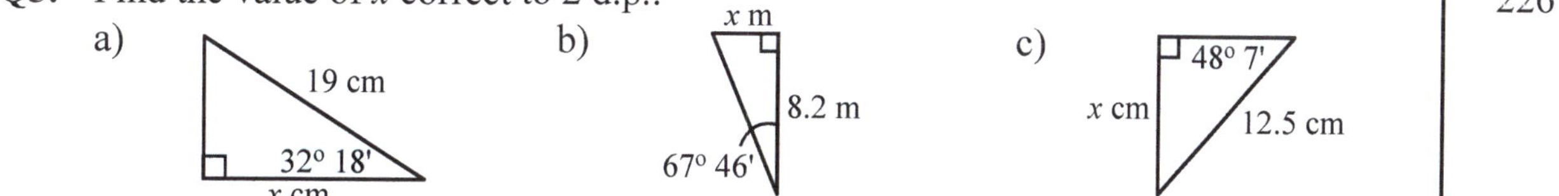

Q4. Find θ correct to the nearest minute:

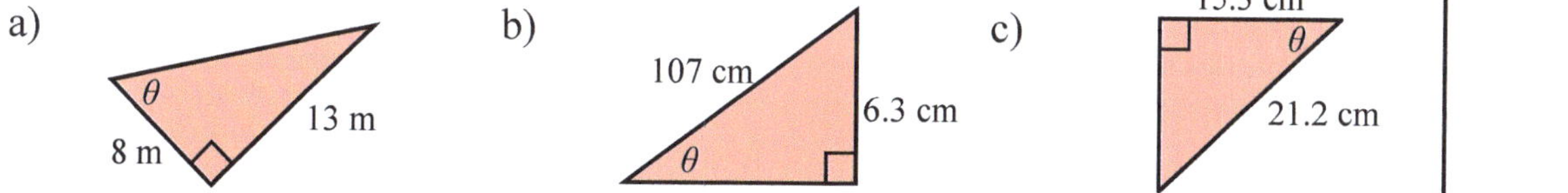

Q5. Find the value of x correct to 2 d.p.:

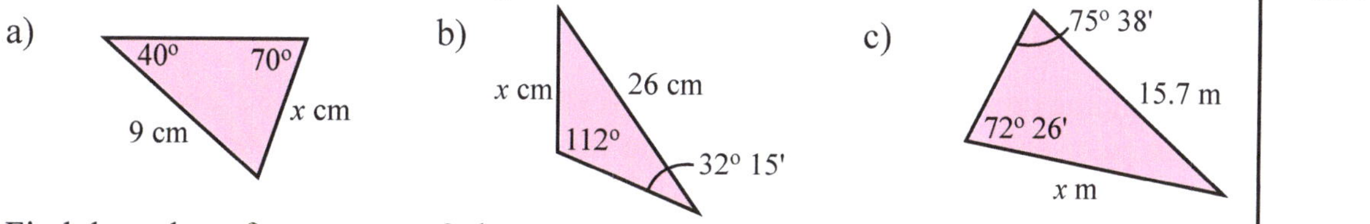

Q6. Find the value of x correct to 2 d.p.:

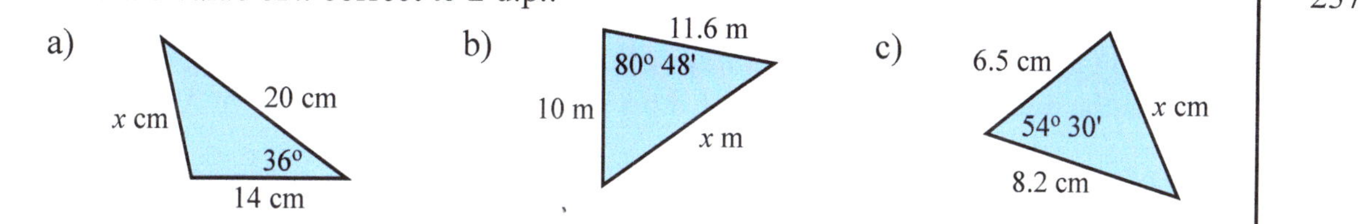

Q7. Find the area of each triangle correct to 3 sig. figs.:

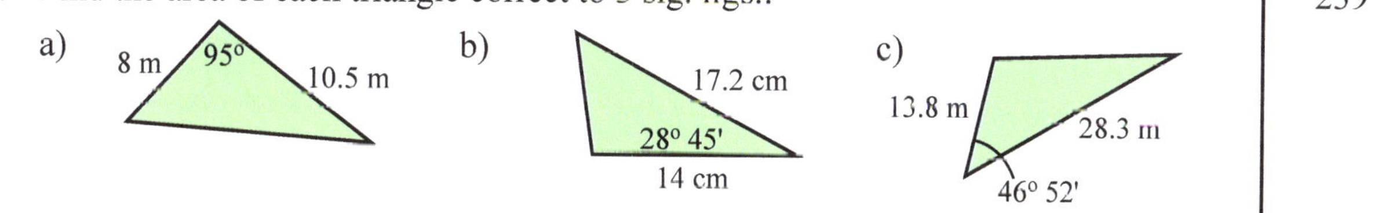

Q8. Write the bearing of B from A using 2 different methods:

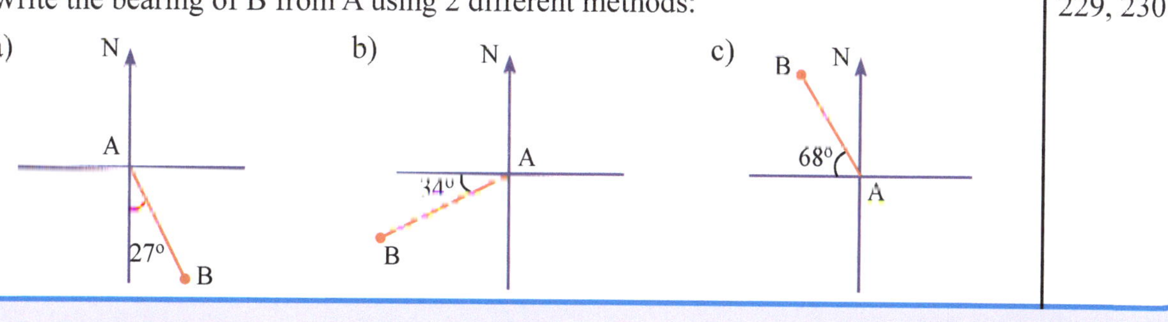

Note: Only turn back to page number shown if you have difficulty.

	Page
Q1. a) In ΔDEF, $\angle$D = 32° 09', d = 8.3 cm, f = 14.1 cm, find $\angle$F to the nearest minute. b) In ΔABC, $\angle$A = 62° 18', $\angle$C = 86° 14', c = 11.3 cm, find a to 2 d.p. c) In ΔXYZ, $\angle$Y = 123° 46', x = 7.3 cm, y = 12.5 cm, find $\angle$X to the nearest minute.	236
Q2. a) In ΔMNO, $\angle$M = 53° 48', n = 15.8 cm, o = 20.1 cm, find m to 1 d.p. b) In ΔABC, a = 10 cm, b = 15.4 cm, c = 12.6 cm, find $\angle$C to the nearest minute. c) In ΔEFG, $\angle$G = 15° 35', e = 16.3 cm, f = 14.2 cm, find g to 1 d.p.	237, 238
Q3. Find the area of each figure (correct to 2 d.p.): a) 73° 50', 13 cm, 15 cm, 42°, 12 cm b) 15.2 cm, 54°, 16.5 cm c) 57 cm, 86 cm, 63°, 41°	239
Q4. A building, B, is observed from two points P and Q which are 280 m apart. The angles PQB and QPB are found to be 67° and 73° respectively. What is the distance, to the nearest metre, of P from the building?	236
Q5. Point A is 52 km N29°W of point C, and point B is 65 km N46 E of C. What is the distance between A and B, correct to 2 d.p.	237, 238
Q6. A person standing on shore at point A observes a yacht on a bearing of 32°, while a second person on shore at B observes the same yacht on a bearing of 300°. If the distance between A and B is 2.2 km, find the distance of each observer from the yacht (answer to 1 d.p.).	229, 230
Q7. If the yacht in Question 6 moves so that it now lies 1.2 km from B on a bearing of 315°, what is the new distance (correct to 1 d.p.) between the yacht and point A?	229, 230
Q8. Two sides of a triangle have lengths 16.3 cm and 14.7 cm, and an included angle of 58° 42'. Find: a) the perimeter of the triangle correct to 3 significant figures. b) the area of the triangle correct to the nearest cm^2.	237, 239
Q9. Find the exact value of the following: a) tan 60° b) cos 30° c) sin 45° d) cot 30° e) cosec 60° f) sec 30°	232
Q10. In which quadrants do the following angles have a negative value? a) cos θ b) sin θ c) tan θ	233, 234
Q11. Find the exact value of the following: a) tan 150° b) cos 210° c) sin 135°	233, 234

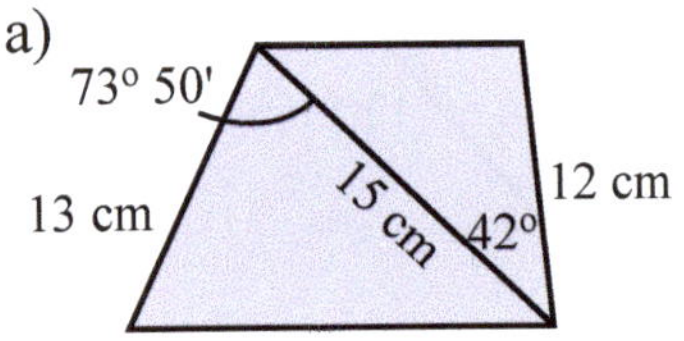

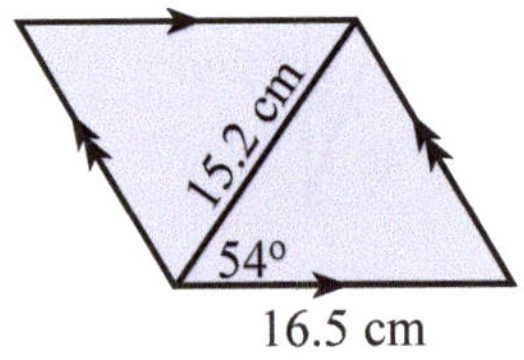

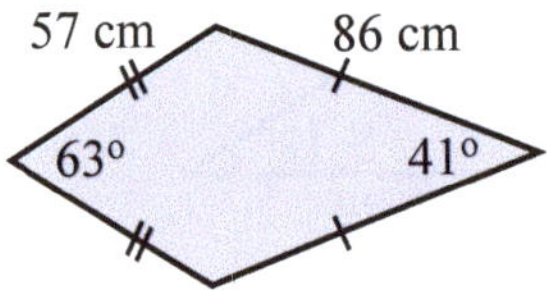

HARDER QUESTIONS

Q1. Find the exact value of the following angles:
a) sin 210° b) cos 300° c) tan 300° d) tan 150°

Q2. Find both angles ($0° \le \theta \le 360°$) which satisfy the following equations:

a) $\sin\theta = \dfrac{\sqrt{3}}{2}$ b) $\tan\theta = -\sqrt{3}$ c) $\cos\theta = \dfrac{-\sqrt{3}}{2}$ d) $\sin\theta = \dfrac{-1}{\sqrt{2}}$

Q3. Standing on top of a building 42 m high, an observer sights two objects at P and Q on opposite sides of the building. The angles of depression from the observer to P and Q are 23° 18' and 35° 33' respectively. What is the distance between P and Q to the nearest m?

Q4. A boat is observed from a lighthouse 55 m high at an angle of depression of 11° 42', and two minutes later sighted again at an angle of depression of 25° 07'. What distance has the boat travelled in two minutes? (Answer to the nearest m.) What is the speed of the boat to the nearest nautical mile per hour? (Note: 1 nautical mile = 1.852 km.)

Q5.

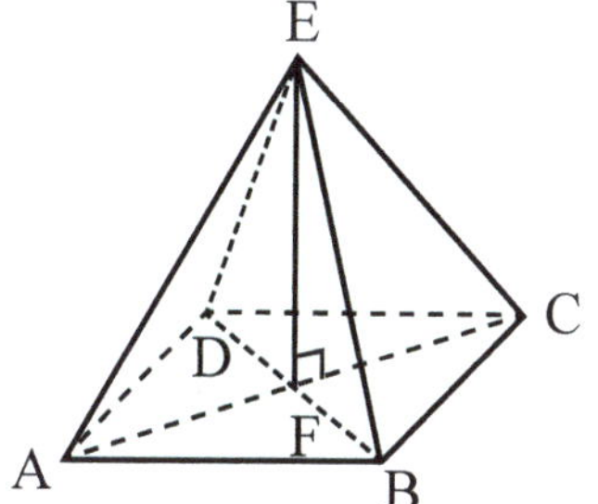

The figure shown to the left is a square pyramid with a base of side 6 cm and a vertical height of 9 cm. Find:
a) Length AC (correct to 2 d.p.)
b) Length EC (correct to 2 d.p.)
c) ∠ECF (correct to nearest minute)
d) ∠ECB (correct to nearest minute)

Q6. A, B and C are three towns, the bearing of B and C from A being 295 and 205°, and their distances from A are 280 km and 240 km respectively. Find the bearing of C from B to the nearest minute.

Q7. A bushwalking party leave P and walk on a bearing of 335° for 11.4 km until they reach Q. From Q they walk on a bearing of 65° for 18.7 km at which point they arrive at R.
a) What is the distance between R and P to 3 sig. figs.?
b) What is the bearing of R from P, correct to the nearest minute?

Q8. A ship sights the top of a cliff, height H, from a point A, due East at the angle of elevation of 8°. It then travels 12 kilometres on the bearing 150°T to the point B which is due South of the cliff. From point B the angle of elevation to the top of the cliff is θ°.

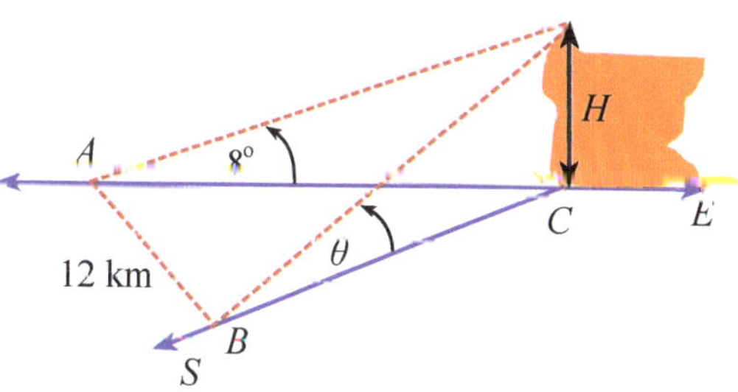

a) Find ∠ABC.
b) (i) Find the distance from the ship at point A to the base of the cliff, C.
(ii) Hence find the height of the cliff, H to the nearest centimetre.
c) (i) Find the distance from the ship at point B to the base of the cliff, C to the nearest metre.
(ii) Hence find the angle of elevation from point B to the top of the cliff, θ expressed to two decimal places.

 EXTENSION QUESTIONS

Q1. Find the unknown lengths correct to the nearest millimetre, or angles correct to one decimal place shown in the following triangles.

a)

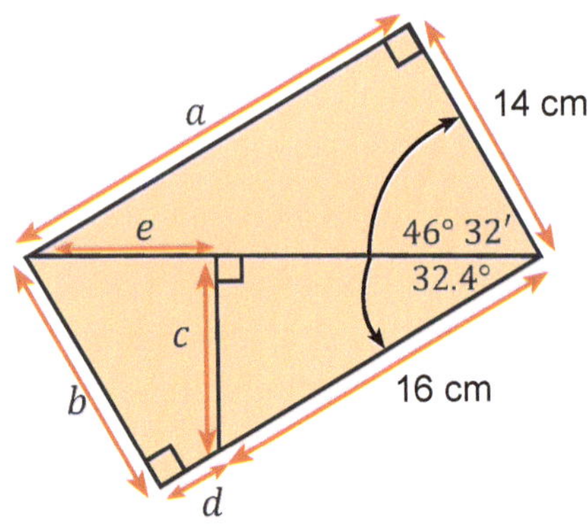

b)

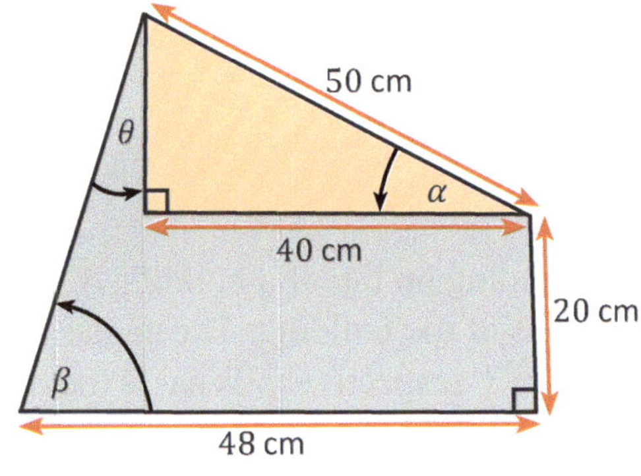

Q2. a) For triangle ABC, AC = 24m, BC = 16m and ∠BAC = 30°.

(i) Use the sine rule to show that $\theta = 48.6°$ and $\theta = 101.4°$

(ii) For each value of θ find the corresponding value of α.

(iii) Use the cosine rule for the two lengths AB to the nearest centimetre and draw both triangles showing all dimensions.

(iv) Find the area for both triangles.

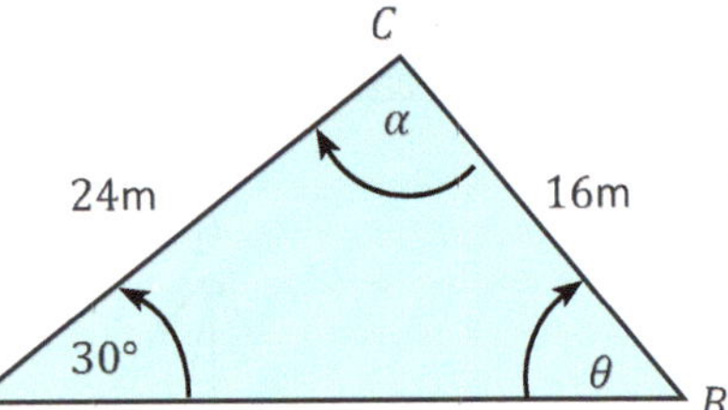

b) For triangle ABC, AC = 20 cm, BC = 18 cm and ∠BAC = 34°.

(i) Use the sine rule to show that $\theta = 38.4°$ and $\theta = 141.6°$

(ii) For each value of θ find the corresponding value of a.

(iii) Use the cosine rule for the two lengths AB and draw both triangle showing all dimensions.

(iv) Find the area for both triangles.

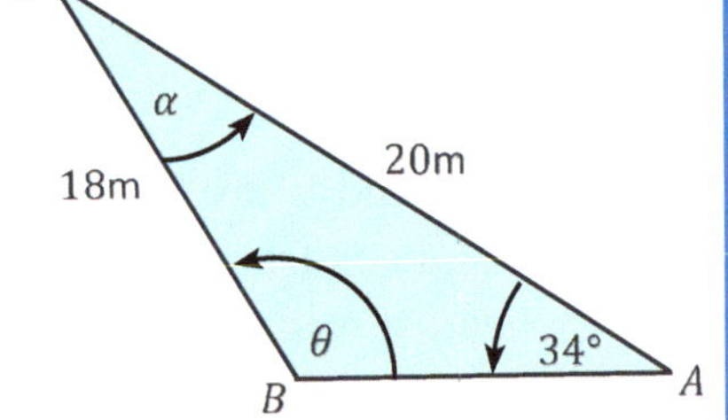

Q3. Two masses are suspended by ropes attached to the points A and B in a horizontal plane.
The masses are attached to the points C and D and both masses hang vertically.
Some angles and lengths are marked on the diagram.

a) Using the cosine rule find the length of BC to the nearest millimetre.

b) Using the sine rule find (i) ∠DCB (ii) ∠ACB.
(iii) Hence, find ∠ACD

c) Find ∠ABD.

d) Find length AC to the nearest millimetre, hence find the area of ABDC.

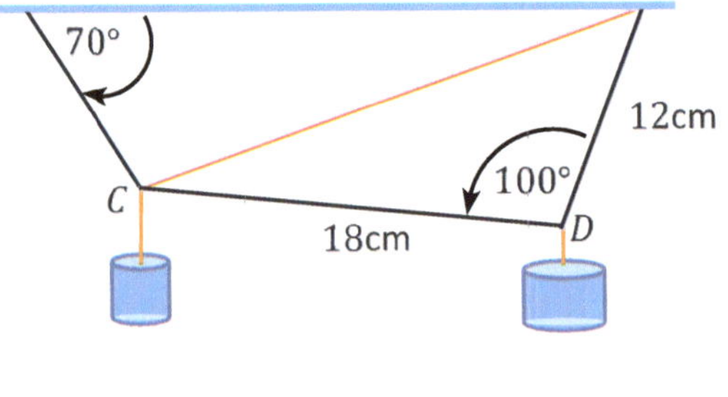

Q4. In the diagram AB = 36 m, ∠AED = 70°, ∠BAC = 24°, ∠ABD = 18°, ∠BCD = 95°.
Find the following lengths expressed to the nearest centimetre:

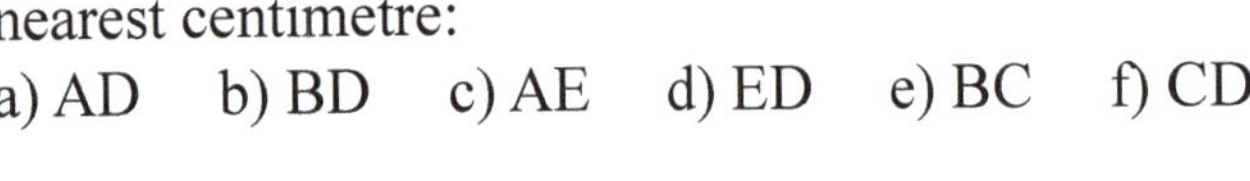
a) AD b) BD c) AE d) ED e) BC f) CD

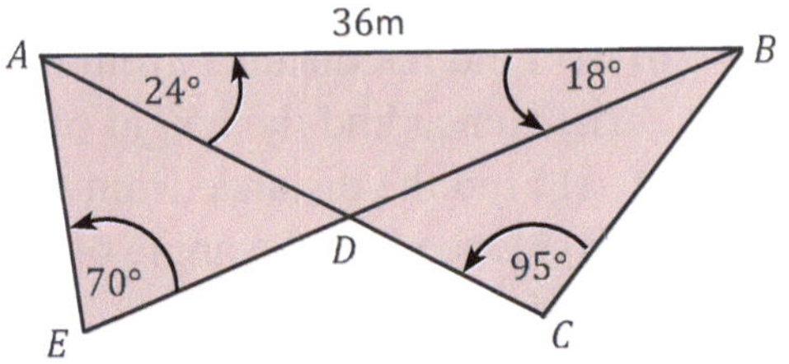

SOLUTIONS

Marie Curie (1867 – 1934)

She was a Polish and naturalized-French physicist and chemist who conducted pioneering research on radioactivity. She was the first woman to win a Nobel Prize, the first person to win the Nobel Prize twice, and the only person to win the Nobel Prize in two different scientific disciplines. She was also the first woman to become a professor at the University of Paris. She achieved her first award for Physics in 1903 for her pioneering work in developing the theory of radioactivity, and her second award in Chemistry in 2011 for her discovery of two new elements called radium and polonium (the latter named after her native country). Under her direction, the world's first studies were conducted into the treatment of neoplasms using radioactive isotopes. She was the founder of the Curie Institutes in Paris and Warsaw which still remain major centres of medical research. During World War 1 she developed mobile radiography units to provide X-Ray services to field hospitals. In 1995 she became the first woman to be entombed at the Pantheon in Paris, an extremely rare honour, and also the very highest that can be awarded in that country.

LEVEL 1 – Further Financial Mathematics

Q1. a) $I = P \times R \times N = \$12\ 000 \times 0.025 \times 1 = \$300$
b) $I = P \times R \times N = \$12\ 000 \times 0.025 \times 6 = \$1\ 800$
c) $I = P \times R \times N = \$12\ 000 \times 0.025 \times 4.75 = \$1\ 425$

Q2. a) $I = \$8\ 000 \times 0.035 \times 4 = \$1\ 120$
b) $I = \$12\ 000 \times 0.035 \times 3 = \$1\ 170$
c) $I = \$13\ 400 \times 0.02 \times 3.5 = \$938$
d) $I = \$18\ 000 \times 0.275 \times 1.5 = \$742.50$

Q3. $\$960 = \$8\ 000 \times R \times 4 \Rightarrow R = \frac{960}{4 \times 8\ 000} = 0.03 = 3\%$

Q4. $\$1\ 800 = \$12\ 000 \times 0.02 \times N \Rightarrow N = \frac{1\ 800}{12\ 000 \times 0.02} = 7.5 = 7\frac{1}{2}$ years

Q5. a) $40 \times \$22 + 3 \times \$22 \times 1.5 = \$979$
b) Net earnings = $\$979 \times 74\% = 724.46$
c) $(40 \times \$22 + 10 \times \$22 \times 1.5 + 4 \times \$22 \times 2) \times 74\% = \$1\ 025.64$
d) $\$1\ 342 - \$880 = \$462$ (1st 40 hours)
$\$462 - \$330 = \$132$ (Next 10 hours)
$\$132 \div \$22 \times 2 = 3$
She worked $40 + 10 + 3 = 53$ hours

Note: Subtracting 26% tax from \$979 is the same as finding 74% of \$979.

Q6. a) Deposit = $\$38\ 000 \times 0.175 = \$6\ 650$
b) Interest = $\$31\ 350 \times 0.0925 \times 4 = \$11\ 599.50$
c) Monthly instalment = $(\$31\ 350 + \$11\ 599.50) \div 48 = \$894.78$

Q7.

Date	Balance	Contribution	Interest for the Month
March 1st	\$60 000		
March 31st	\$64 160	\$4 000	\$160
April 1st	\$64 160		
April 30th	\$68 331.09	\$4 000	\$171.09
May 1st	\$68 331.09		
May 30th	\$72 513.31	\$4 000	\$182.22
June 1st	\$72 513.31		
June 30th	\$76 706.68	\$4 000	\$193.37

LEVEL 2 – Further Financial Mathematics

Q1. a) \$2 700.00 b) \$2 812.50 c) \$650.25 d) \$1 685.25
Q2. a) \$1 261.00 b) \$3 753.73 c) \$1 011.15 d) \$1 831.41
Q3. \$13 926.40 **Q4.** \$5 409.78
Q5. a) \$824.00 b) \$6 465.17 c) \$22 549.46 d) \$31 082.78
Q6. a) \$350.05 b) First option is the better choice by \$28.80
Q7. a) $40 \times \$21.40 + 7.5 \times \$21.40 \times 1.5 + 4 \times \$21.40 \times 2 = \$1\ 267.95$
b) $\$820 - \$148.90 - 0.065 \times \$820 - 0.08 \times \$820 = \$552.20$
Each week she will save $12\% \times \$552.20 = \$66.26$

Q8. a) Let x = total wages for 4 weeks.
$\therefore\ 0.175 \times x = \$602$
$\therefore\ x = \$602 \div 0.175 = \$3\ 440$
$\therefore$ Weekly wage = $\$3\ 440 \div 4 = \$860$
b) In 38 hours she gets paid = $38 \times \$18.4 = \$699.20$
$(\$795.80 - \$699.20) \div (18.4 \times 1.5) = 3.5$
She worked $3\frac{1}{2}$ hours overtime

Q9. a) Gross earnings = 48 × \$860 = \$41 280

Tax payable = \$3 572 + (\$41 280 – \$37 000) × 0.325

= \$4 963

b) Normal hourly wage = \$860 ÷ 40 = \$21.50

∴ Double-time wages = \$43 per hour.

\$90 000 – \$41 280 = \$48 720

\$48 720 ÷ 43 = 1 133 hours

She can work 1 133 hours overtime during the year and still remain in the same tax category.

LEVEL 3 – Further Financial Mathematics

Q1. a) \$338.85 b) \$4 136.93 c) \$140.89

Q2. 7.75% p.a.

Q3. a) \$1 993.22 b) \$3 068.57

Q4. a) \$17 726.01 b) \$3 870.67 c) \$6 742.92

Q5. Compound interest pays an extra \$366.71

Q6. 4.5% per 6 months or 9% p.a.

Q7. a) \$7 902.23 b) \$332.36 c) \$31.18

Q8. a) \$986.63

Q9. \$1 336.56/month

Q10. 6.3% p.a.

Q11. \$21 549.30 – \$10 344.50 = \$11 204.80

\$11 204.80 ÷ 0.47 = \$23 840

∴ Total gross wage = \$23 840 + \$34 000 = \$57 840

LEVEL 4 – Further Financial Mathematics

Q1. a) Total salary = $\$86\,560 + 0.175 \times \$86\,560 \times \frac{4}{52} + \$5\,000$

= \$92 725 (to the nearest \$)

Tax = \$20 797 + 0.37 (\$92 725 – \$90 000) = \$21 805

Note: The holiday loading of 17.5% is only applied to 4 weeks of the total yearly salary.

b) Wages = \$24.50 × 28 × 48 + \$24.50 × 4 × 48 × 1.5 = \$39 984

Holiday leave loading = \$24.50 × 28 × 4 × 0.175 = \$480.20

Total wages = \$39 984 + \$480.20 = \$40 464.20

Tax: \$3 572 + 0.325(\$40 462 – \$37 000) = \$4 697.15

Q2. a) Total salary = $\$135\,600 + 17.5\% \times \$135\,600 \times \frac{4}{52}$

= \$137 425 (to nearest \$)

Employers Superannuation amount = 9.5% × \$137 425 = \$13 055

Maximum salary sacrifice: \$25 000 – \$13 055 = \$11 945

Salary less salary sacrifice: \$137 425 – \$11 945 = \$125 480

Tax: \$20 797 + 0.37(\$125 480 – \$90 000) = \$33 924.60

b) Total wages = \$24 × 38 × 40 = \$36 480

Employers Superannuation amount = 9.5% × \$36 480 = \$3 465.60

Maximum salary sacrifice: \$25 000 – \$3 465.60 = \$21 534.40

Total wages less salary sacrifice = \$36 480 – \$21 534.40 = \$14 945.60

Tax = Zero (no tax payable)

Q3. a) (i) Initial price = \$308 859.

End of first year: \$308 859 × 0.8 = \$247 087.20

End of second year: \$247 087.20 × 0.9=\$222 378.48

End of third year: \$222 378.48 × 0.9 = \$200 140.63

End of fourth year: \$200 140.63 × 0.9 = \$180 126.57

End of fifth year: \$180 126.57 × 0.9 = \$162 113.91

Overall depreciation = $\frac{\$308\,859 - \$162\,113.91}{\$308\,859} \times 100 = 47.51\%$

(ii) Initial price = \$469 500.

End of first year: \$469 500 × 0.8 = \$375 600

End of second year: \$375 600 × 0.9 = \$338 040

End of third year: \$338 040 × 0.9 = \$304 236

End of fourth year: \$304 236 × 0.9 = \$273 812.40

End of fifth year: \$273 812.40 × 0.9 = \$246 431.16

Overall depreciation = $\frac{\$469\,500 - \$246\,431.16}{\$469\,500} \times 100 = 47.51\%$

Q4. a) Compounding annually: $20\,000 \times 1.06^5 - 20\,000 = \$6\,764.51$

Simple interest: 20 000 × 0.06 × 5 = \$6000

Compounding Interest investment gives \$764.51 more than the Simple Interest investment.

b) Compounding annually: $50\,000 \times 1.03^4 - 50\,000 = \$6\,275.44$

Compounding monthly: $50\,000 \times 1.0025^{48} - 50\,000 = \$6\,366.40$

Compounding monthly investment gives \$90.96 more than the Compounding annually investment

c) Compounding monthly: $120\,000 \times 1.0067^{120} - 120\,000 = \$146\,356.83$

Simple interest: 120 000 × 0.08 × 10 = \$96 000

Compounding Interest investment gives \$51 417.29 more than the Simple Interest investment.

LEVEL 1 – Surface Area and Volume

Q1. a) 1 400 m^2 b) 521.6 cm^2 c) 9.58 m^2 d) 1 596 cm^2
e) 8 516 m^2 f) 2 200 cm^2 g) 1 320 m^2 h) 254.6 cm^2
i) 113.71 cm^2 j) 2 261.95 cm^2 k) 30.91 m^2 l) 1 053.72 cm^2

Q2. a) 105 cm^2 b) 302 cm^2 c) 205 cm^2 d) 77 cm^2
e) 478 cm^2 f) 1 472 cm^2

Q3. a) 1 256.6 cm^2 b) 706.9 cm^2 c) 254.5 cm^2 d) 1 039.1 cm^2

LEVEL 2 – Surface Area and Volume

Q1. a) 1 485 cm^3 b) 1 188 cm^3 c) 880 cm^3 d) 1 580.8 cm^3
e) 140.4 cm^3 f) 3 255 cm^3

Q2. a) 3 402.786 m^3 b) 468.764 41 m^3 c) 2 014.8075 m^2

Q3. a) 152.88 cm^3 b) 1 216.42 cm^3 c) 381.70 cm^3 d) 3 324.63 cm^3
e) 54.52 cm^3 f) 2 424.52 cm^3

Q4. a) 106.20 m^3 b) 240.8 m^3 c) 5 150.02 m^3

LEVEL 3 – Surface Area and Volume

Q1. Area of end = $\pi \times 0.55^2 - \pi \times 0.45^2 = 0.1\pi$ m^2
Volume =$0.1\pi \times 3 = 0.94248 \approx 0.94$ m^3

Q2. a) (i) Volume = $10 \times \frac{1}{2}(2.2 + 1.2) \times 5 = 85$ m^3 (ii) 85 000 litres

b) With the height at the deep end being 1.3 m, the height of water at the shallow end = 0.3 m.

Volume = $10 \times \frac{1}{2}(1.3 + 0.3) \times 5 = 40$ m^3 = 40 000 litres

Q3. a) (i) $\frac{1}{3}\pi(2)^2 \times 3 + \pi(2)^2 \times 6 = 28\pi \approx 88$ m^3 (ii) 87 965 litres

b) Volume = $\frac{1}{2} \times \frac{4}{3}\pi(2)^3 + \pi(2)^2 \times 6 = \frac{16}{3}\pi + 24\pi = 29\frac{1}{3}\pi \approx 92$ m^3. The new silo with a hemispherical top will hold approximately 4m^3 more wheat than the silo with the conical top.

Q4. a) Radius for each ball = $\sqrt[3]{\frac{3}{4} \times 143.79 \div \pi} \approx 3.25$ cm so the diameter is $3.25 \times 2 = 6.5$ cm

b) Height of container = $4 \times 6.5 = 26$ cm, Radius of base of container = 3.25 cm.
Volume of four balls = $4 \times 143.79 = 575.16$ cm^3, Volume of container = $\pi(3.25)^2 \times 26 = 862.76$ cm^3.
Volume of air in the container = $862.76 - 575.16 = 287.6$ cm^3

Percentage of air in container = $\frac{287.6}{862.76} \times 100 = 33.3\%$

c) Letting h be the height of the cone, the base radius = h.

Volume = $\frac{1}{3}\pi h^2 \times h = \frac{1}{3}\pi h^3 = 143.79$

$h = \sqrt[3]{\frac{3 \times 143.79}{\pi}} = 5.16$ cm. Height of the trophy = 5.16 cm, Base diameter = $2 \times 5.16 = 10.32$ cm

LEVEL 1 – Statistics

Q1. a) 739 students b) (i) $\frac{178}{739} \times 100\% = 24.1\%$ (ii) $\frac{75}{739} \times 100\% = 10.1\%$

c) (i) $24.1\% \times 150 \approx 36$ students (rounded off to the nearest whole number).
(ii) $10.1\% \times 150 \approx 15$ students (rounded off to the nearest whole number).

Q2. a) Mean = 5
Median = 5.5
Mode = 6
Range = 6
$S_n = 1.94$

b) Mean = 14.6
Median = 14
Mode = 14
Range = 7
$S_n = 2.10$

Q3. a) $\bar{x} = 11.12$, $S_n = 1.21$ b) $\bar{x} = 17.26$, $S_n = 1.34$

Q4. Melbourne

x	f	fx
25	1	25
26	1	26
27	2	54
28	5	140
29	5	145
30	3	90
31	2	62
32	1	32

$\sum fx = 574$
Mean = $574 \div 20$
= 28.7

Sydney

x	f	fx
23	1	23
27	2	54
28	3	84
29	4	116
30	5	150
31	2	62
32	1	32
33	1	33
34	1	34

$\sum fx = 588$
Mean = $588 \div 20$
= 29.4

a) Melbourne:
Mean = 28.7
Median = 29
Mode = 28 and 29
Range = 32 – 25 = 7

b) Sydney:
Mean = 29.4
Median = 29.5 (Average of the 10th and 11th score.)
Mode = 30
Range = 34 – 23 = 11

b) Melbourne has more consistent temperatures. The range is only 7 whereas Sydney has a larger range of 11.

c) Melbourne has almost a normal distribution (bell shaped) while Sydney is negatively skewed.

d) The interquartile range = $Q_3 - Q_1$ = 16th score – 5th score

Melbourne: IQR = 30 – 28 = 2

Sydney: IQR = 31 – 28 = 3

e) Sydney has the warmer weather because its average temperature is 0.7°C higher than the average temperature of Melbourne. It would have been even higher of it wasn't for the unusual low 'temperature of 23°C'. This reduced the mean significantly.

LEVEL 2 – Statistics

Q1. a) 0.06, 0.14, 0.20, 0.31, 0.23, 0.06 b) 0.025, 0.100, 0.175, 0.300, 0.250, 0.150

Q2. a) $\bar{x} = 5.4$, range 7, $\sigma_n = 2.0$

b) $\bar{x} = 15.5$, range 9, $\sigma_n = 2.6$

c) $\bar{x} = 66$, range 60, $\sigma_n = 15.5$

d) $\bar{x} = 72.9$, range 42, $\sigma_n = 12.2$

Q3. a) $\bar{x} = 18.5$, median 19, $\sigma_n = 1.18$

b) $\bar{x} = 72.79$, median class 70 – 79, $\sigma_n = 11.83$

Q4. a) 95%

b) 34%

c) 0.15%

Q5. {20, 110, 120, 144, 150, 164, 165, 180, 182, 187, 190}

↑ Q_1 (120) ↑ Median (164) ↑ Q_3 (182)

a) Lowest value: 20, First quartile (Q_1): 120, Median: 164, Third quartile (Q_3): 182, Highest value: 190,
Interquartile range = 182 – 120 = 62.

b) As $20 < 120 - 1.5 \times 62$, then 20 is an outlier.

c)

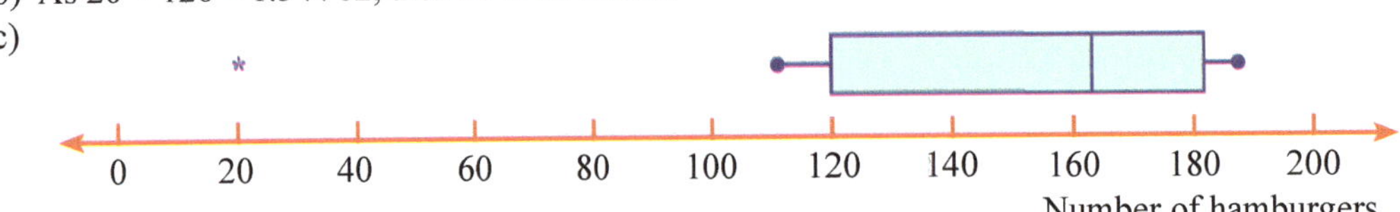

LEVEL 3 – Statistics

Q1. a) Restaurant A:
{3, 13, 14, 25, 25, 28, 30, 31, 36, 36, 37, 38, 44, 47, 47, 48, 55, 57, 59, 59, 62}.
Restaurant B:
{13, 23, 24, 35, 35, 38, 40, 41, 46, 46, 47, 48, 54, 57, 57, 58, 65, 67, 69, 69, 72}.

b) Restaurant A: (i) 37.8 (ii) 37 (iii) 25, 36, 47, 59
Restaurant B: (i) 47.8 (ii) 47 (iii) 35, 46, 57, 69

c) Restaurant A: (i) Range = 59 (ii) Standard Deviation (S_n) = 16.4
Restaurant B: (i) Range = 59 (ii) Standard Deviation (S_n) = 16.4

d) (i) From 25 to 57 (ii) The mean ± one standard deviation: 21.4 to 54.2. In whole values of data: 25 to 48.

Q2. a) 26

b)

Score	Frequency	Relative frequency
0	4	0.15
1	6	0.23
2	9	0.35
3	5	0.19
4	2	0.08

c) Mean: 1.8 Standard Deviation (S_n) =1.2

d) (i) $\frac{11}{26}$ (ii) $\frac{16}{26} = \frac{8}{13}$

Q3.

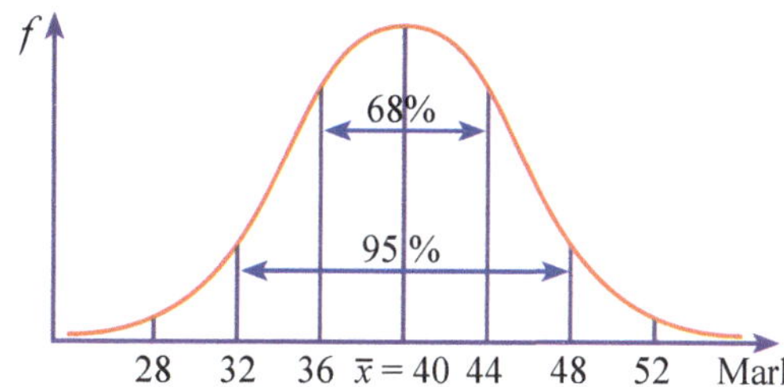

a) 4

b) (i) As (100% – 99.7%) = 0.03%, then 0.03% of the students is 48.

The total number of students is $\frac{48}{0.03\%} = 160\ 000$

(ii) Two standard deviations is 95% of the number of students. 95% of 160 000 = 152 000.

(iii) 40 – 4 = 36

Q4. a) Lower value: 21, First quartile (Q_1): 25, Median: 29, Third quartile (Q_3): 32, Highest value: 38

b)

0 10 20 30 40 Age

c) The trend of the data is negatively skewed as its median is closer to the upper quartile.

LEVEL 4 – Statistics

Q1. a) Class A: (i) 50 (ii) 10 (iii) 90 (iv) 20 (v) 80
Class B: (i) 50 (ii) 10 (iii) 80 (iv) 40 (v) 60
Class C: (i) 60 (ii) 40 (iii) 90 (iv) 50 (v) 80

b) Interquartile range: Class A = 60, Class B = 20, Class C = 30.
Least to most consistent: Class B, Class C, Class A

c) The results shown in the box plots can't be directly compared as the results of only five students' shown by Class C each student is shown as a position on the box plot whereas with twenty students in Class A, five students are shown in each of the four regions. The results for Class A are far more spread reflecting the larger number of students represented. It is not reasonable for the teacher of Class C to claim that his class outperformed the larger number of students in Class A.

Q2. a) (i) {43, 60, 61, 65, 65, 67, 71, 71, 72, 73, 78, 78, 78, 79, 80, 82, 83, 84, 84, 87}
Lowest: 43, Highest: 87, Median: 75.5, First quartile: 65, Third quartile: 82.

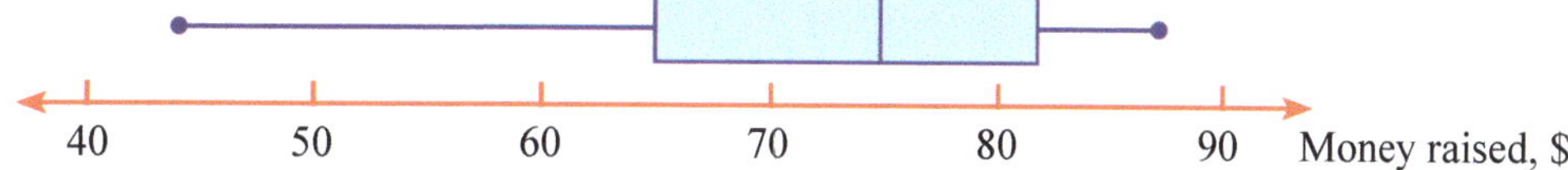

The box plot is not symmetrical.

b) Interquartile range (IQR) = 17.
As $43 > 65 - 1.5 \times 17$ then the value 43 is not considered to be an outlier. As $60 > 65 - 1.5 \times 17$ then the value 60 and the other values between it and the first quartile are considered not to be outliers.
As $65 + 1.5 \times 17 > 87$ then the value 87 and the other values between it and the third quartile are not considered to be outliers.

Q3. a) $\bar{x} = 14.75$, $\sigma_n = 3.18$.

b) (i) $\bar{x}_A = 13.4$, $S_A = 4.34$ and $\bar{x}_B = 17.8$, $S_B = 2.28$.

(ii) The mean from sample A is 4.4 smaller than the mean for sample B, while the sample standard deviation from sample A is 2.06 greater than that for sample B. The values shown in sample A are lower than those from sample B hence the difference in the means of the samples. The values shown in sample A are more spread out than those from sample B hence the difference in the standard deviations of the samples. As each sample is a different collection of values then the statistical measures will differ.

(iii) Sample A is more similar to the original population as the difference in the mean for sample A is only 1.35 different compared to a difference of 3.05 from the mean of Sample B and the population mean. The values of Sample A are more widely spread compared to those in the population almost to the same degree as the values in Sample B are more tightly grouped.

Q4. a) (i) $x = 6$ where median and mode = 6 (ii) $x = 5.5$ where median = 5.5 and mode = 4, 5.5
(iii) $x = 3$ where the data set is {3, 5, 6, 6}, Median = 5.5 and mode = 6

b) (i) For $x = 5$, Mean = 6 (ii) For $x = 4$, Mean = $5\frac{1}{2}$
(iii) For $x = 2\frac{1}{2}$, Mean = $4\frac{5}{8}$

LEVEL 1 – Probability

Q1. a) $\frac{1}{2}$ b) $\frac{1}{6}$ c) $\frac{8}{15}$ d) $\frac{4}{15}$ e) $\frac{1}{10}$ f) $\frac{3}{5}$

Q2. a) $\frac{1}{5}$ b) $\frac{4}{5}$ c) $\frac{7}{10}$

Q3. a) $\frac{1}{4}$ b) $\frac{1}{16}$ c) $\frac{15}{16}$

Q4. a) $\frac{3}{8}$ b) $\frac{3}{8}$ c) $\frac{1}{8}$ d) $\frac{7}{8}$

Q5. a) $\frac{1}{9}$ b) $\frac{5}{18}$ c) $\frac{5}{12}$ d) $\frac{1}{2}$ e) $\frac{11}{36}$ f) $\frac{1}{36}$

Q6. a) $\frac{1}{4}$ b) $\frac{1}{17}$ c) $\frac{25}{102}$ d) $\frac{4}{17}$ e) $\frac{1}{221}$

Q7. a) $\frac{1}{4}$ b) $\frac{3}{4}$ c) $\frac{9}{16}$

Q8. a) $\frac{4}{20} = 20\%$ b) $\frac{5}{20} = 25\%$ c) $\frac{10}{20} = 50\%$ d) $\frac{1}{20} = 5\%$ e) 10 blue balls

f) 5 red balls g) 30 blue balls

LEVEL 2 – Probability

Q1. a) 37% b) 28% c) 61% d) 21%

Q2. We cannot have fractions of a marble, and therefore we have to round off the percentages to the nearest 10.

Red marbles = 60% of 40 = 24 Blue = 30% of 40 = 12 Yellow = 10% of 40 = 4

Q3. a) $\frac{1}{2}$ b) $\frac{1}{4}$ c) $\frac{2}{3}$

Q4. a) 16 different seating arrangements b) 36 different seating arrangements

Q5. a)

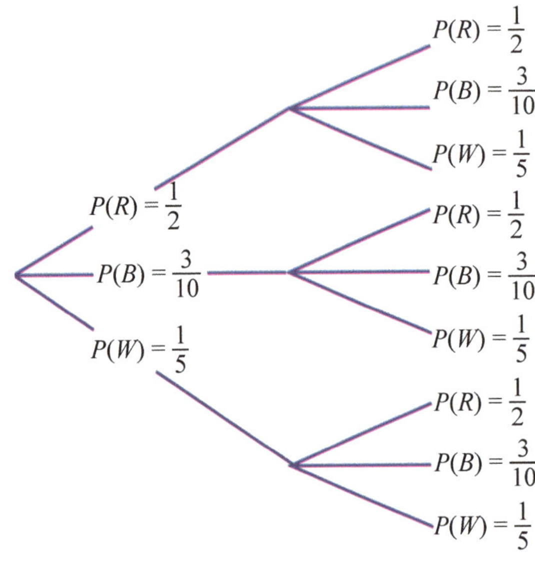

i) P(same colour)

$= \frac{1}{2} \times \frac{1}{2} + \frac{3}{10} \times \frac{3}{10} + \frac{1}{5} \times \frac{1}{5}$

$= \frac{1}{4} + \frac{9}{100} + \frac{1}{25}$

$= \frac{38}{100} = \frac{19}{50}$

ii) P(different colours)

= 1 – P(same colours)

$= 1 - \frac{19}{50} = \frac{31}{50}$

iii) P(at least 1 blue or red ball)

= 1 – P(2 white balls)

$= 1 - \frac{1}{5} \times \frac{1}{5}$

$= \frac{24}{25}$

b)

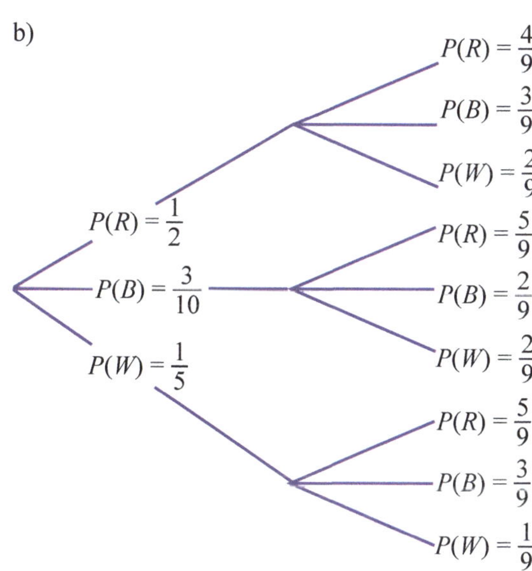

i) P(same colour)

$=\frac{1}{2}\times\frac{4}{9}+\frac{3}{10}\times\frac{2}{9}+\frac{1}{5}\times\frac{1}{9}$

$=\frac{28}{90}=\frac{14}{45}$

ii) P(different colours)

$=1-$ P(same colours)

$=1-\frac{14}{45}=\frac{31}{45}$

iii) P(at least 1 blue or red ball)

$=1-P$(2 white balls)

$=1-\frac{1}{5}\times\frac{1}{9}$

$=\frac{44}{45}$

iv) $P(R \text{ then } B)=\frac{5}{10}\times\frac{3}{9}=\frac{15}{90}=\frac{1}{6}$

v) $P(R \text{ then } W)=\frac{5}{10}\times\frac{2}{9}=\frac{1}{9}$

vi) $P(R \text{ then } R)=\frac{1}{2}\times\frac{4}{9}$

$=\frac{4}{18}=\frac{2}{9}$

LEVEL 3 – Probability

Q1. a) $\frac{1}{2}\times\frac{1}{2}\times\frac{1}{4}=\frac{1}{16}$ b) $\frac{1}{3}\times\frac{1}{2}\times\frac{4}{13}=\frac{2}{39}$ c) $\frac{1}{2}\times 1\times\frac{1}{2}=\frac{1}{4}$

Q2. The angle needs to be between 144° and 270°.

Q3. a) As $0.1+0.05+0.2+0.15+x+0.25=1$, then $x=0.25$

b) As the probability for tossing a 5 and a 6 are the same at 0.25 then it is expected that either of these numbers are more likely to occur more than the other values.

c) Outcomes in ascending order: 2, 1, 4, 3, (5, 6).

Number	1	2	3	4	5	6
Expected number in 1 000 rolls	100	50	200	150	250	250

d) (i) $120\times 0.05=6$ (ii) The probability of a 2 occurring being 0.05 is a long-term measure. A short term sample of 120 is prone to unusual results as the sample is so small. William needs to understand that a small sample of 120 rolls is not enough to suspect that something is wrong with the dice. If the probability values in the table are accurate then in the long term the number 2 will occur 5% of the time.

Q4. a)

Number	2	4	6	8	10
Probability	$\frac{2}{30}=\frac{1}{15}$	$\frac{4}{30}=\frac{2}{15}$	$\frac{6}{30}=\frac{1}{5}$	$\frac{8}{30}=\frac{4}{15}$	$\frac{10}{30}=\frac{1}{3}$

b) (i) $\frac{2}{15}\times\frac{2}{15}=\frac{4}{225}$ (ii) $\frac{1}{15}\times\frac{1}{15}+\frac{2}{15}\times\frac{2}{15}+\frac{1}{5}\times\frac{1}{5}+\frac{4}{15}\times\frac{4}{15}+\frac{1}{3}\times\frac{1}{3}=\frac{11}{45}$

(iii) $\frac{2}{15}\times\frac{2}{15}+\frac{4}{15}\times\frac{4}{15}=\frac{4}{45}$

(iv) $2\times\left(\frac{1}{15}\times\frac{1}{5}+\frac{1}{15}\times\frac{4}{15}+\frac{1}{15}\times\frac{1}{3}+\frac{2}{15}\times\frac{1}{5}+\frac{2}{15}\times\frac{4}{15}+\frac{2}{15}\times\frac{1}{3}\right)=\frac{8}{25}$

Q5. a) $\frac{3}{500}$ b) $\frac{3}{499} \times \frac{2}{498} = \frac{1}{41\,417}$ c) $\frac{3}{500} \times \frac{2}{498} = \frac{1}{41\,500}$

d) Not winning 1st prize $= \frac{497}{500}$

Not winning 2nd prize $= \frac{496}{499}$

Not winning 3rd prize $= \frac{495}{498}$

$\therefore$ Not winning any prize $= \frac{497}{500} \times \frac{496}{499} \times \frac{495}{498} = \frac{122\,023\,440}{124\,251\,000} = \frac{1\,016\,862}{1\,035\,425}$

e) $\frac{3}{500} \times \frac{2}{499} \times \frac{1}{498} = \frac{1}{20\,708\,500}$

Q6. Number of possible treifectas $= 24 \times 23 \times 22 = 12\,144$

P(choosing correct trifecta) $= \frac{1}{12\,144}$

LEVEL 1 – Graphing Lines and Curves

Q1.

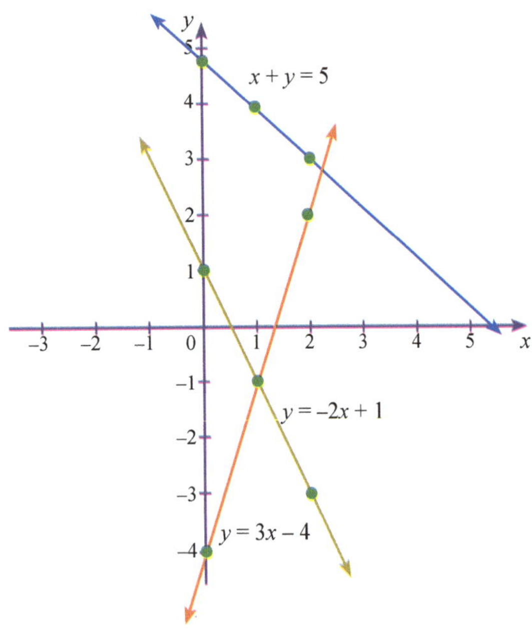

Q2.

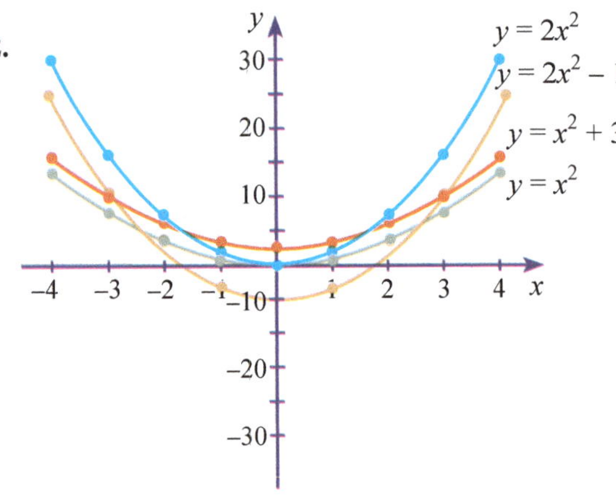

Q3. y values: –16, –2, –1.024, –0.432, –0.128, –0.16, 0, 0.16, 0.128, 0.432, 1.024, 2, 16

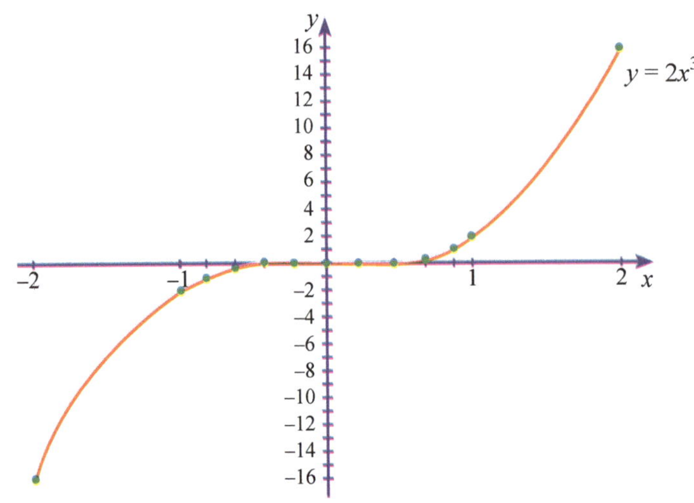

Q4. y values: –0.75, –1, –1.5, –3, –6, error, 6, 3, 1.5, 1, 0.75

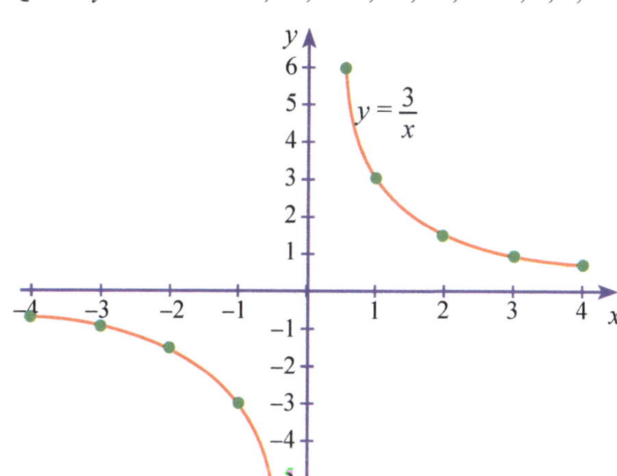

Q5. a) (1, 4) b) (3, 5) c) (1, –1)

Q6. a) $x^2 + y^2 = 25$ b) $x^2 + y^2 = 7$

c) $(x - 3)^2 + (y + 5)^2 = 16$ d) $(x + 2)^2 + (y + 1)^2 = 10$

e) $(x + 4)^2 + y^2 = 16$ f) $x^2 + (y - 6)^2 = 3$

LEVEL 2 – Graphing Lines and Curves

Q1. a)

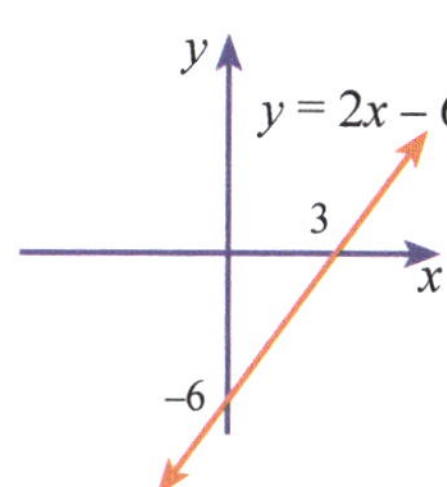

b)

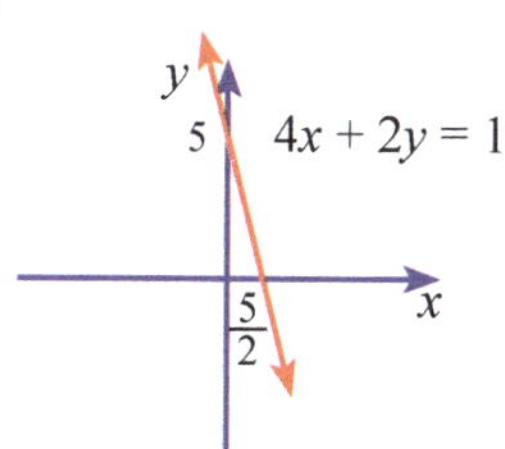

c)

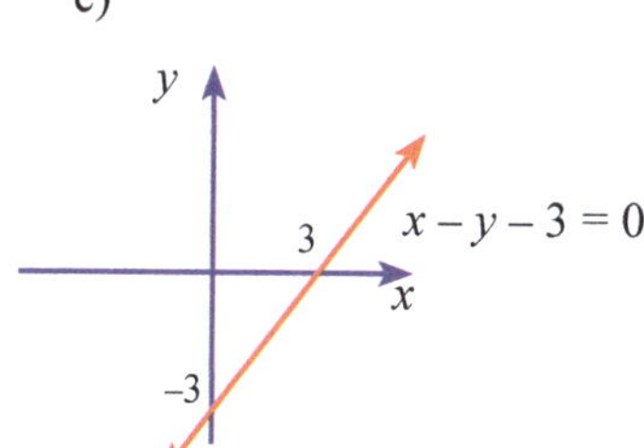

Q2. a)

x	–4	–3	–2	–1	0	1	2	3
y	4	1	0	1	4	9	16	25

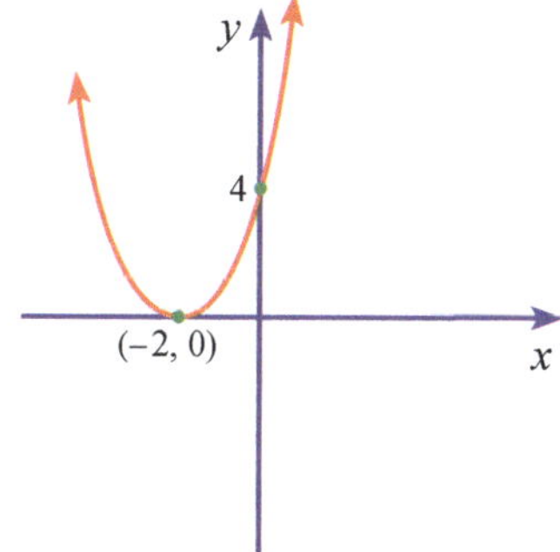

b)

x	–2	–1	0	1	2	3	4
y	5	0	–3	–4	–3	0	5

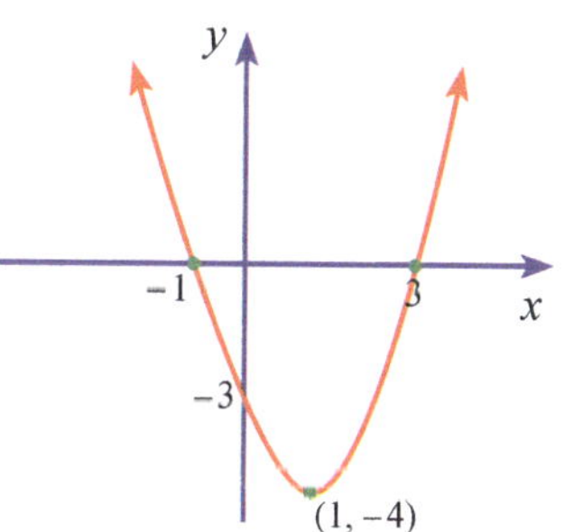

Q3. a) $x = -1$, (–1, 0) min. b) $x = 2$, (2, 1) min. c) $x = \frac{-3}{2}$, $(\frac{-3}{2}, \frac{-49}{2})$ min.

Q4. a)

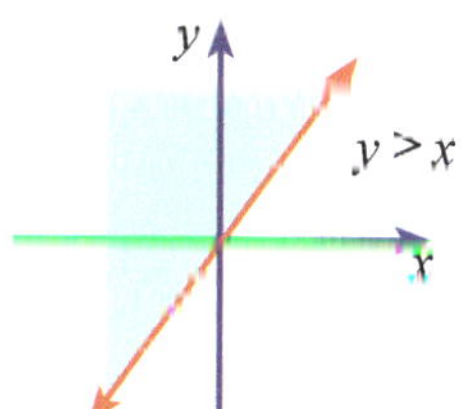

b)

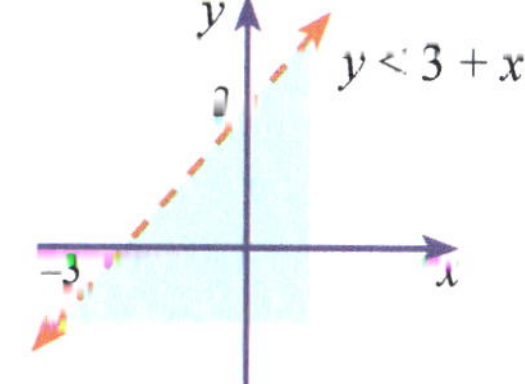

c)

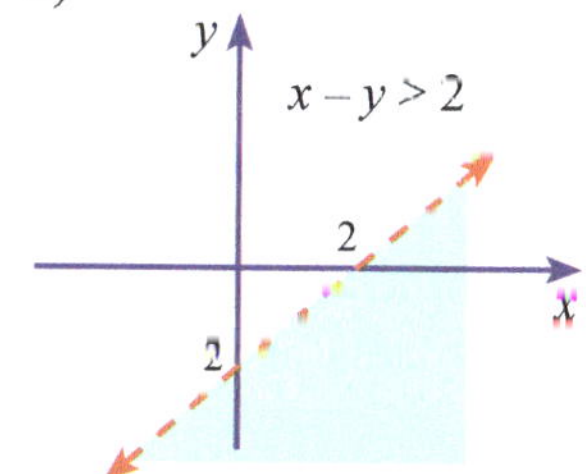

d)

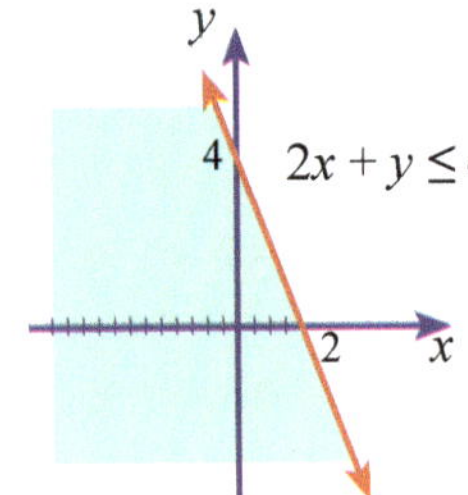

e)

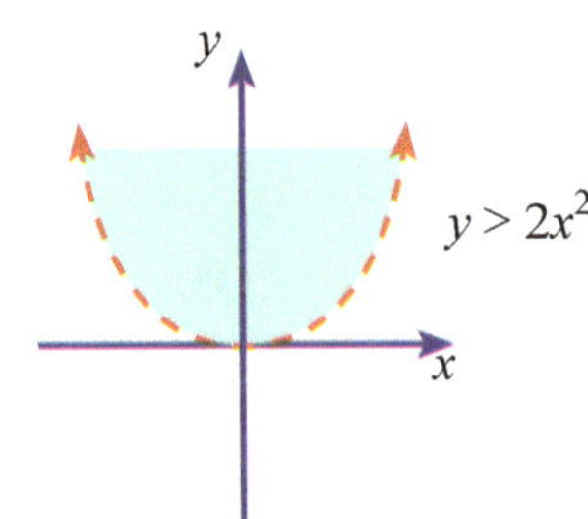

f)

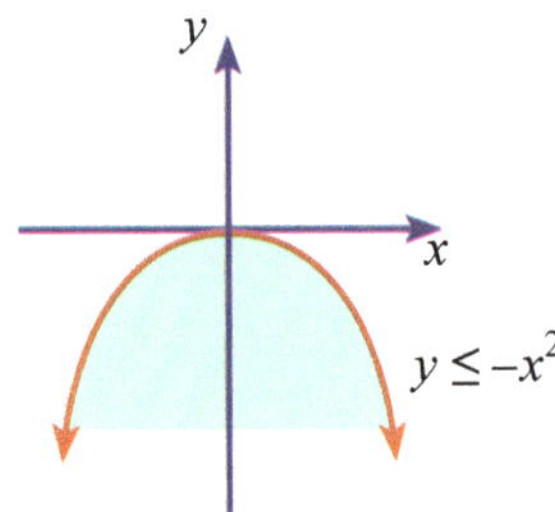

Q5. a)

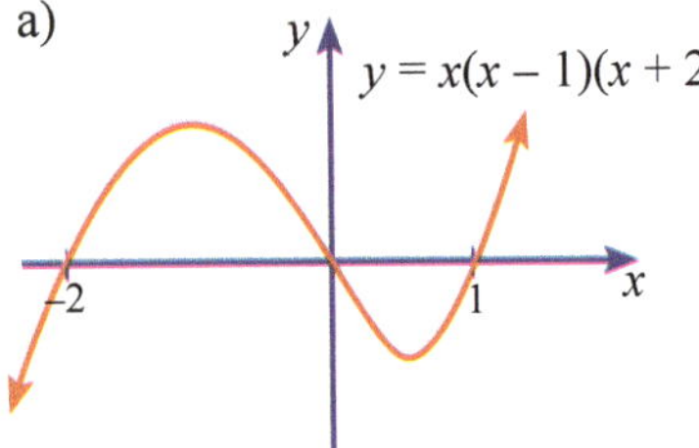

b)

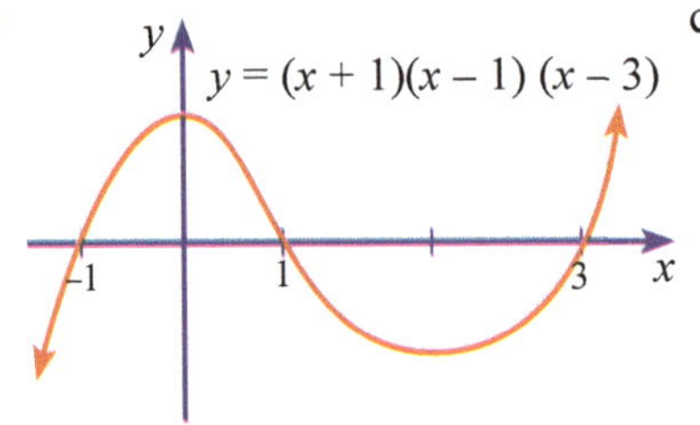

c)

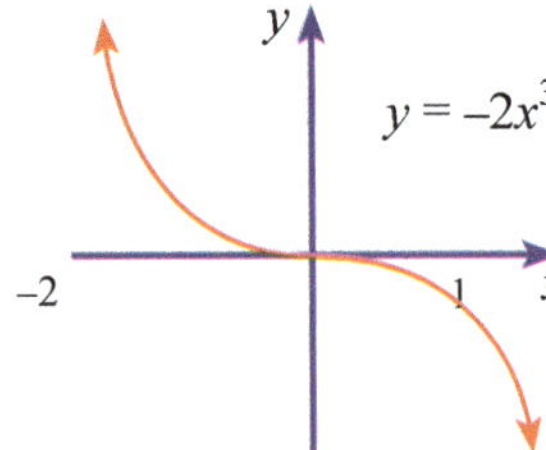

Q6.

a) (0, 0)

b) (2, 0.5) (–2, –0.5)

c) (0, 0)

d) (–1, 1) (1, –1)

e) (1, 1) (–1, –1)

f) (0, 0) (–0.5, –0.125) (0.5, 0.125)

g) (1, 1) (–1, –1)

h) (–1, –1) (0, 0) (1, 1)

LEVEL 3 – Graphing Lines and Curves

Q1. a) (i) $a = 4, b = 2, c = 0$ (ii) $y = 2x + 4$

(iii) Algebraically:

For $x = 1\frac{1}{2}$, $y = 2 \times 1\frac{1}{2} + 4 = 7$ so the point on the line with $x = 1\frac{1}{2}$ is $(1\frac{1}{2}, 7)$ not $(1\frac{1}{2}, 7\frac{1}{2})$.

Graphically:

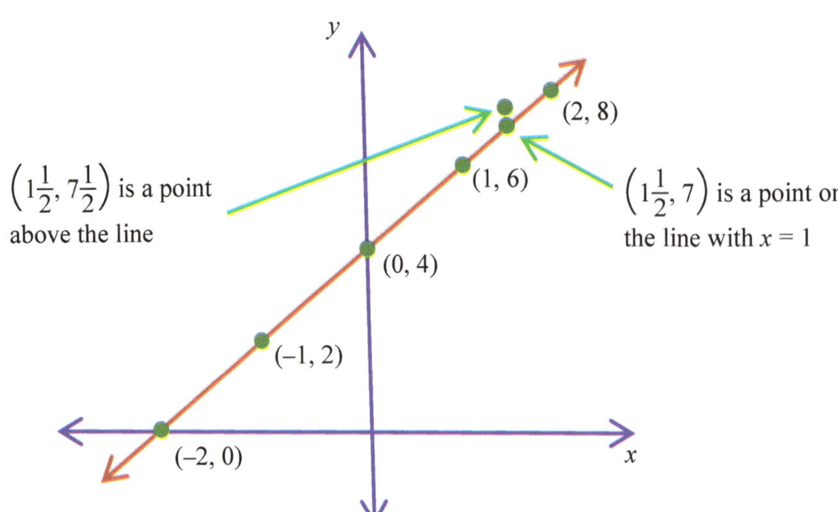

b) (i) $a = 4\frac{1}{2}, b = 4, c = 3\frac{1}{2}$

(ii) $y = -\frac{1}{2}x + 4$

(iii) Algebraically:

For $x = -1\frac{1}{2}, y = -\frac{1}{2} \times -1\frac{1}{2} + 4 = 4\frac{3}{4}$ so the point on the line with $x = -1\frac{1}{2}$ is $(-1\frac{1}{2}, 4\frac{3}{4})$ not $(-1\frac{1}{2}, 4)$.

Graphically:

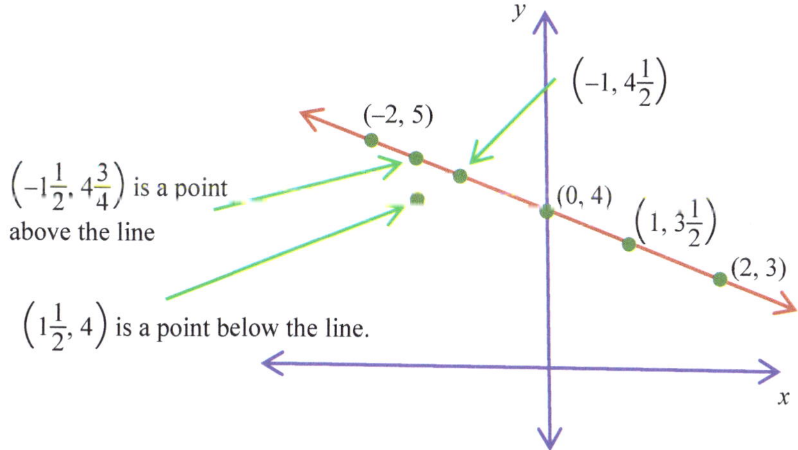

Q2. a) (i) Line A is $y = -\frac{1}{2}x + 5$ as the gradient is $-\frac{1}{2}$ and the y-intercept is 5.

Line B is $y = \frac{1}{2}x + 3$ as the gradient is $\frac{1}{2}$ and the y-intercept is 3.

(ii) For $x = 2$, the y-value for Line A is $y = -\frac{1}{2} \times 2 + 5 = 4$

For $x = 2$, the y-value for Line B is $y = \frac{1}{2} \times 2 + 3 = 4$

b) (i) For $x = 2, y = 4$ so for $y = ax + 1$, $4 = a \times 2 + 1 = 2a + 1$ where $a = 1\frac{1}{2}$.

The equation is: $y = 1\frac{1}{2}x + 1$.

For $x = 2, y = 4$ so for $y = -ax + 7$, $4 = -a \times 2 + 7 = -2a + 7$ where $a = 1\frac{1}{2}$.

The equation is: $y = -1\frac{1}{2}x + 7$.

(ii) For $y = 1\frac{1}{2}x + 1$

x	-2	-1	0	1	2
y	-2	$-\frac{1}{2}$	1	$2\frac{1}{2}$	4

For $y = -1\frac{1}{2}x + 7$

x	-2	-1	0	1	2
y	10	$8\frac{1}{2}$	7	$5\frac{1}{2}$	4

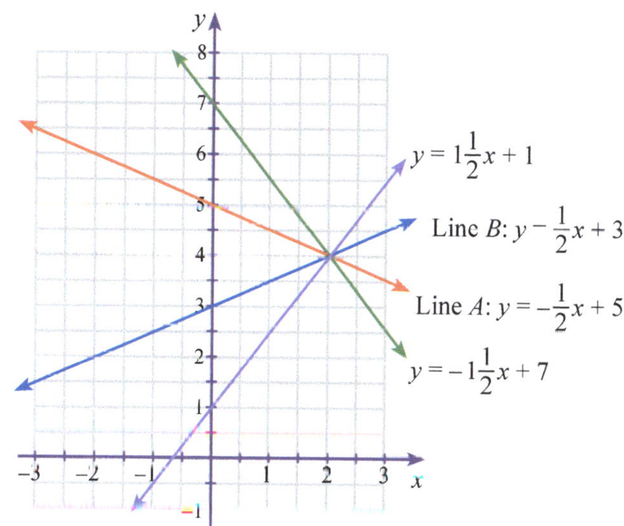

Q3. a) $2(x - 2)(x + 4) = 2(x^2 + 2x - 8) = 2x^2 + 4x - 16$

b)

x	–5	–4	–3	–2	–1	0	1	2	3
y	14	0	–10	–16	–18	–16	–10	0	14

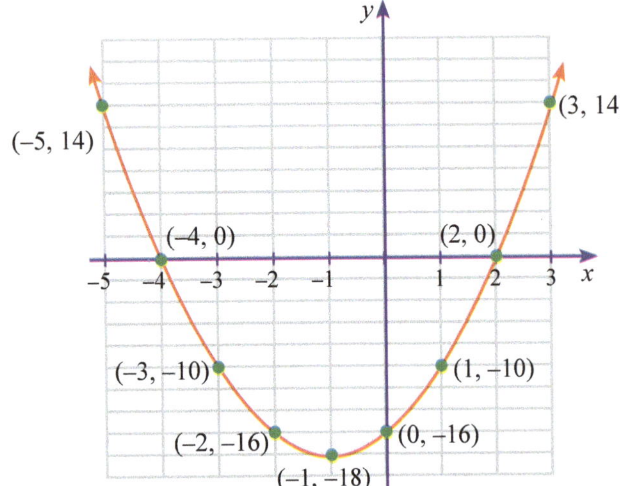

c) The x-intercepts for a curve is/are the points on the curve where it crosses the x-axis. The y value for all x-intercepts is zero.

The table of values shows that the values of x for which the value of y is zero are –4 and 2.

d) (i) The lowest point on the graph occurs at the point (–1, –18). This is the point at which the y value is a minimum, where $y = -18$ for $x = -1$.

(ii) The x-value for the minimum point $= \dfrac{-4 + 2}{2} = -1$

e) (i)

x	–1	0	1	2	3
y	–18	–12	–6	0	6

(ii)

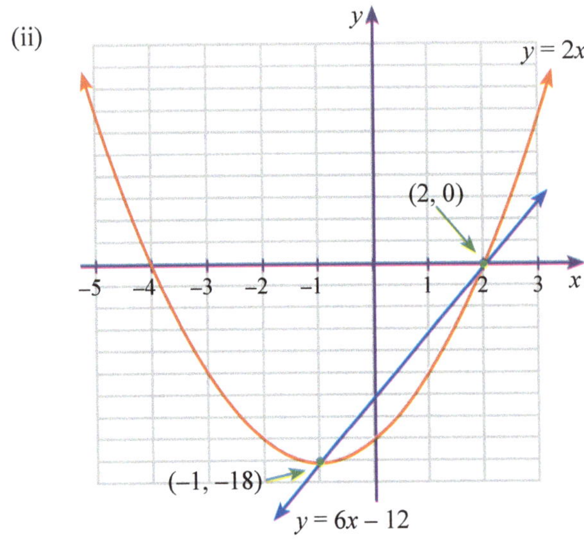

The 2 points of intersection are (–1, –18) and (2, 0).

LEVEL 4 – Graphing Lines and Curves

Q1. a) $x(x + 2)(x - 4) = x(x^2 - 2x - 8) = x^3 - 2x^2 - 8x$

b)

x	–3	–2	–1	0	1	2	3	4	5
y	–21	0	5	0	–9	–16	–15	0	35

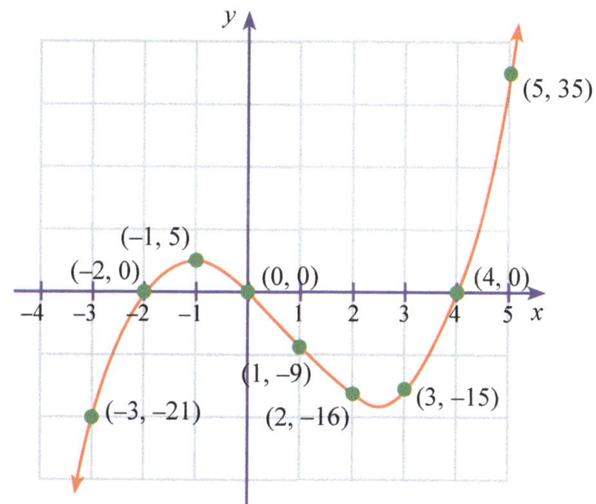

c) The x-intercepts occur for $y = 0$.

For $x = 0$, $y = 0 \times (0 + 2)(0 - 4) = 0 \times 2 \times -4 = 0$.

For $x = -2$, $y = -2 \times (-2 + 2)(-2 - 4) = -2 \times 0 \times -6 = 0$.

For $x = 4$, $y = 4 \times (4 + 2)(4 - 4) = 4 \times 6 \times 0 = 0$.

d) (i)

x	–3	–2	–1	0	1	2	3	4	5
y	–21	–18	–15	–12	–9	–6	–3	0	3

(ii)

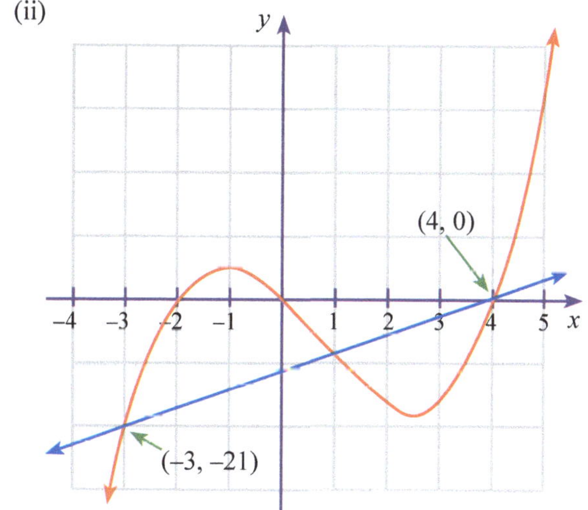

Q2. a)

x	–2	–1	0	1	2	3	4	5	6
y	3.75	3.67	3.5	3	undefined	5	4.5	4.33	4.25

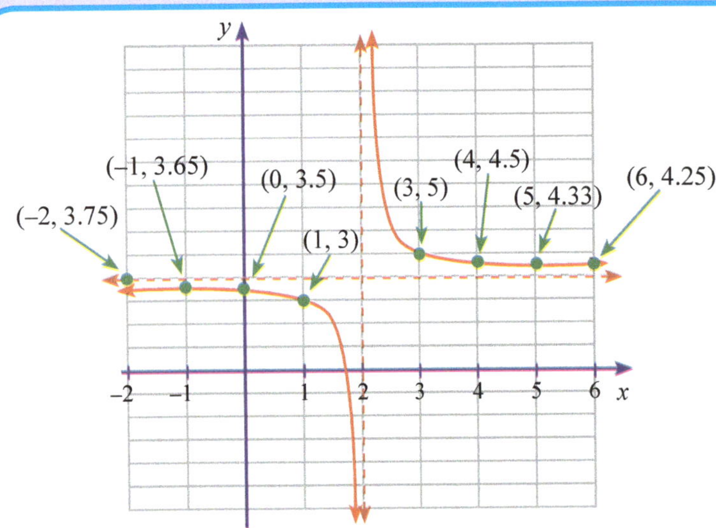

b) (i) For $y = \frac{1}{x-2} + 4$, when $x = 2$, $y = \frac{1}{2-2} + 4 = \frac{1}{0} + 4$. As the value of $\frac{1}{0}$ is not defined then the point for $x = 2$ doesn't exist.

(ii) For $y = \frac{1}{x-2} + 4$, when $y = 4$, $4 = \frac{1}{x-2} + 4$, $\frac{1}{x-2} = 0$, $x - 2 = \frac{1}{0}$

As the value of $\frac{1}{0}$ is not defined then the point for $y = 4$ doesn't exist.

c) An asymptote is a line that a graph approaches but cannot touch.
In this graph there are 2 lines which are asymptotes.
The lines $x = 2$ and $y = 4$ are asymptotic lines.

Q3. a) (i) $y = \frac{12}{x}$

x	−4	−3	−2	−1	0	1	2	3	4
y	−3	−4	−6	−12	undefined	12	6	4	3

(ii) $y = 6x + 6$

x	−4	−3	−2	−1	0	1	2	3	4
y	−18	−12	−6	0	6	12	18	24	30

(iii) $y = x^2 + 7x + 4$

x	−4	−3	−2	−1	0	1	2	3	4
y	−8	−8	−6	−2	4	12	22	34	48

b) The points of intersection for the three curves are (−2, −6) and (1, 12).

c)

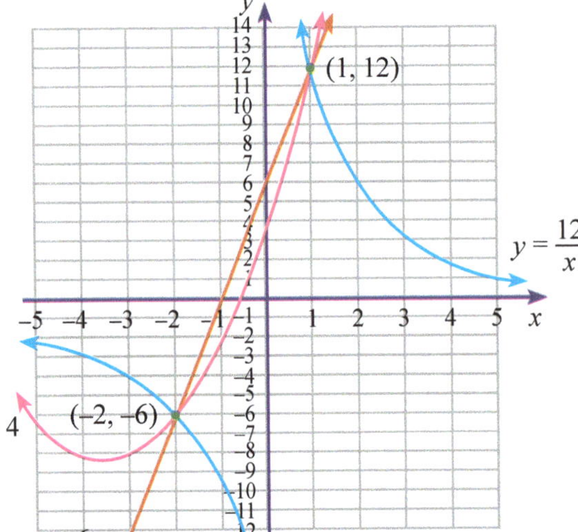

LEVEL 1 – Circle Geometry

Q1. a) $x = 74^\circ, y = 68^\circ$ b) $x = 39^\circ$ c) $a = 82^\circ$ d) $m = 60^\circ$ e) $p = 65^\circ, q = 98^\circ$
f) $x = 95^\circ, y = 85^\circ$ g) $n = 65^\circ$ h) $x = 44^\circ$ i) $a = 90^\circ, b = 90^\circ$

Q2. a) RS = 15 cm b) OE = 3.2 cm c) OA = 13 cm (using Pythagoras' theorem)

Q3. a) $a = 47^\circ, b = 47^\circ$ b) $a = 18^\circ$ c) $x = 38^\circ, y = 52^\circ$

LEVEL 2 – Circle Geometry

Q1.

a) $x = 220^\circ$ ($\angle$ subtended at centre equals twice $\angle$ subtended at circumference)

b) $x = 180^\circ - 125^\circ$ (opp. $\angle$'s of a cyclic quad. are supplementary)
$= 55^\circ$
$y = 2 \times 55^\circ$ ($\angle$ subtended at centre equals twice $\angle$ subtended at circumference)
$= 110^\circ$

c) $x = 28^\circ$ (base $\angle$ of isos. ΔABO)
$x + y = 90^\circ$ ($\angle$ in semicircle is right $\angle$)
$\therefore\ y = 62^\circ$

d) X is midpoint of AB (perp. from O bisects AB)
$\therefore$ AB = 2×7 = 14 cm
$\therefore$ AB = CD
$\therefore$ OX = OY (eq. chords arc same distance from centre)
= 3 cm

e) AB = DC (equal chords subtend equal $\angle$'s at centre, i.e. $\angle$AOB = $\angle$DOC)
$\therefore$ DC = 2.3 cm

f) $x + y + \angle$COB $= 180^\circ$ ($\angle$'s on straight line)
$x = y = \angle$COB (eq. chords subtend eq. $\angle$'s at centre)
$x = y = 60^\circ$

g) $m = 90^\circ$ (tang. is perp. to radius at point of contact)
$\angle$OBC $= 90^\circ$ (tang. is perp. to radius at point of contact)
$\therefore\ n = 360^\circ - 2 \times 90^\circ - 140^\circ$
$= 40^\circ$ ($\angle$ sum of quad.)

h) $x = 180^\circ - 110^\circ - 40^\circ$ ($\angle$ sum of $\Delta = 180^\circ$)
$= 30^\circ$

i) $x + 63^\circ = 90^\circ$ (tang. perp. to radius at point of contact)
$\therefore\ x = 27^\circ$
$y = x$ (base $\angle$'s of isos. ΔOAB)
$\therefore\ y = 27^\circ$

j) $x + 32^\circ = 90^\circ$ (tang. pcip. to radius at point of contact)
$\therefore\ x = 58^\circ$
$y = x = 58^\circ$ (base $\angle$'s of isos. ΔOCD)
$\angle$CDB $= 180^\circ - y = 122^\circ$ ($\angle$'s on st. line supp.)
$z = 180^\circ - 122^\circ - 32^\circ$ ($\angle$ sum of $\Delta - 180^\circ$)
$= 26^\circ$

k) XY = YZ (tangents drawn from ext. pt. to circle are equal)
$\therefore\ a = 8.4$ cm
$x = \angle$ZXY (base $\angle$'s of isos. ΔXYZ)
$= 62^\circ$

l) $x + 54^\circ = 90^\circ$ (tang. perp. to radius at point of contact)
$\therefore\ x = 36^\circ$
$y = 180^\circ - 2x$ (ΔCOB is isosceles)
$\therefore\ y = 108^\circ$
$z = \frac{1}{2}y^\circ$ ($\angle$ subtended at centre is twice $\angle$ subtended at circumference)
$= 54^\circ$

m) $x + 48^\circ = 90^\circ$ (tang. perp. to radius at point of contact)
$\therefore\ x = 42^\circ$
$x + y = 90^\circ$ ($\angle$ in a semicircle is a right $\angle$)
$y = 48^\circ$

n) OB $= \frac{1}{2}$AB (radius $= \frac{1}{2}$ diameter)
= 5 cm
OX = OB + BX
= 10 cm
$\angle$OYX $= 90^\circ$ (tang. perp, to radius at point of contact)
Using Pythagoras' theorem:
$XY^2 = OX^2 - OY^2$
$= 75$
$\therefore$ XY = 8.66 cm (correct to 2 d. p.)

o) $\angle$XAO = $\angle$AXO = 31° (base $\angle$'s of isos. ΔAXO)
$\angle$OXY $= 90^\circ$ (tang. perp. to radius at point of contact)
$\therefore\ x = 180^\circ - 31^\circ - 31^\circ - 90^\circ$
$= 28^\circ$ ($\angle$ sum of $\Delta = 180^\circ$)

LEVEL 3 – Circle Geometry

Q1. a) (i) $\angle OAB = \angle OBA$ as they are base angles of the isosceles triangle OAB.

(ii) $2 \times \angle OBA + \alpha = 180°, \angle OBA = \frac{1}{2}(180° - \alpha)$

b) (i) $\angle OCB = \angle OBC$ as they are base angles of the isosceles triangle OBC.

(ii) $2 \times \angle OBC + \beta = 180°, \angle OBC = \frac{1}{2}(180° - \beta)$

c) (i) $\angle ABC = \angle OBA + \angle OBC = 90° - \frac{1}{2}\alpha + 90° - \frac{1}{2}\beta = 180° - \frac{1}{2}(\alpha + \beta)$

(ii) $\angle AOC = \alpha + \beta = 180°$

(iii) $\angle ABC = 180° - \frac{1}{2}(180°) = 90°$

(iv) The angle subtended by a semi-circle is a right-angle.

Q2. a) (i) $\angle OAB = \angle OBA$ as they are base angles of the isosceles triangle OAB.

(ii) $2 \times \angle OBA + \alpha = 180°, \angle OBA = \frac{1}{2}(180° - \alpha)$

b) (i) $\angle OCB = \angle OBC$ as they are base angles of the isosceles triangle OBC.

(ii) $2 \times \angle OBC + \beta = 180°, \angle OBC = \frac{1}{2}(180° - \beta)$

c) (i) $\angle ABC = \angle OBA + \angle OBC = 90° - \frac{1}{2}\alpha + 90° - \frac{1}{2}\beta = 180° - \frac{1}{2}\alpha - \frac{1}{2}\beta = 180° - \frac{1}{2}(\alpha + \beta)$

(ii) $\angle AOC = 360° - (\alpha + \beta)$

(iii) $\angle AOC = 2\left(180° - \frac{1}{2}(\alpha + \beta)\right) = 2 \times \angle ABC$

(iv) The angle at the centre of a circle is double the angle at the circumference standing on the same arc.

Q3. a) (i) $\angle ABD = \alpha$. The angle at the centre of a circle is double the angle at the circumference standing on the same arc.

(ii) $\angle ACD = \alpha$. The angle at the centre of a circle is double the angle at the circumference standing on the same arc.

b) $\angle ABD = \angle ACD$.

c) The angles at the same segment are equal.

Q4. a) $\angle ABC = \frac{1}{2}\alpha$

b) $\angle ADC = \frac{1}{2}\beta$

c) $\angle ABC + \angle ADC = \frac{1}{2}(\alpha + \beta)$

d) $\angle ABC + \angle ADC = \frac{1}{2}(\alpha + \beta) = \frac{1}{2} \times 360°, \angle ABC + \angle ADC = 180°$.

e) The sum of the opposite angles in a quadrilateral is 180°.

LEVEL 1 – Further Algebra and Factorisation

Q1.
a) $2a + 3b$
b) $y - 4x$
c) $m + 3n - 3$
d) $3x^2 + 3x + 6$
e) $5ab - a + 2b$
f) $2m^2 + 5m - 3$
g) $2a^2 - 5a$
h) $-x^2 - 10x$
i) $4m^2 - 4m + 3$

Q2.
a) $5a + 15 - 2a - 2$
$= 3a + 13$
b) $4m - 28 - 3m + 6$
$= m - 22$
c) $8x - 40 - 4 + 3x$
$= 11x - 44$
d) $m^2 + 5m - 14$
e) $x^2 - 12x + 35$
f) $x^2 - 5x - 24$
g) $4a^2 - 28a + 3a - 21$
$= 4a^2 - 25a - 21$
h) $12k^2 - 8k - 21k + 14$
$= 12k^2 - 29k + 14$
i) $12 + 6y - 20y - 10y^2$
$= 12 - 14y - 10y^2$

Q3.
a) $m^2 + 14m + 49$
b) $n^2 - 16n + 64$
c) $x^2 - 100$
d) $4x^2 - 12x + 9$
e) $25p^2 + 40p + 16$
f) $9x^2 - 49$
g) $12x^2 + 60x + 75$
h) $36m^3 - 16m$
i) $25 - 30x + 9x^2$

Q4.
a) $(m + 9)(m - 9)$
b) $(7 + k)(7 - k)$
c) $(3n + 5)(3n - 5)$
d) $(x + 1)(x - 1)$
e) $(4 - 5x)(4 + 5x)$
f) $2(m^2 - 9)$
$= 2(m + 3)(m - 3)$

Q5.
a) $\alpha\beta = +20$
$\alpha + \beta = +9$
$\therefore\ \alpha = +5, \beta = +4$
$\therefore\ (x + 5)(x + 4)$

b) $\alpha\beta = +12$
$\alpha + \beta = +7$
$\therefore\ \alpha = +4, \beta = +3$
$\therefore\ (x + 4)(x + 3)$

c) $\alpha\beta = +2$
$\alpha + \beta = -3$
$\therefore\ \alpha = -1, \beta = -2$
$\therefore\ (x - 1)(x - 2)$

d) $\alpha\beta = -10$
$\alpha + \beta = -3$
$\therefore\ \alpha = -5, \beta = +2$
$\therefore\ (x - 5)(x + 2)$

e) $\alpha\beta = -24$
$\alpha + \beta = +5$
$\therefore\ \alpha = +8, \beta = -3$
$\therefore\ (b + 8)(b - 3)$

f) $\alpha\beta = +15$
$\alpha + \beta = -8$
$\therefore\ \alpha = -5, \beta = -3$
$\therefore\ (a - 5)(a - 3)$

Q6.
a) $\dfrac{3(2a + 9)}{3}$
$= 2a + 9$

b) $\dfrac{{}^{2}10(m + 3)}{5}$
$= 2(m + 3)$

c) $\dfrac{x(3x + 1)}{x}$
$= 3x + 1$

d) $\dfrac{m(4m - 1)}{6m}$
$= \dfrac{4m - 1}{6}$

e) $\dfrac{(x + 2)(x - 2)}{(x + 2)}$
$= x - 2$

f) $\dfrac{(x + 3)(x + 2)}{(x + 3)}$
$= x + 2$

g) $\dfrac{(x + 3)(x - 2)}{(x - 2)}$
$= x + 3$

h) $\dfrac{(x - 7)(x + 5)}{(x - 7)}$
$= x + 5$

i) $\dfrac{(x - 5)(x - 3)}{(x - 5)}$
$= x - 3$

Q7.
a) $\dfrac{3(m - 5)}{12} + \dfrac{4(4 + m)}{12}$
$= \dfrac{3m - 15 + 16 + 4m}{12}$
$= \dfrac{7m + 1}{12}$

b) $\dfrac{5(x + 4)}{10} - \dfrac{2(8 + x)}{10}$
$= \dfrac{5x + 20 - 16 - 2x}{10}$
$= \dfrac{3x + 4}{10}$

c) $\dfrac{5(n + 1)}{15} - \dfrac{3(n - 3)}{15}$
$= \dfrac{5n + 5 - 3n + 9}{15}$
$= \dfrac{2n + 14}{15}$

d) $\dfrac{3(2a + 1) + 2(3a - 5)}{12}$
$= \dfrac{6a + 3 + 6a - 10}{12}$
$= \dfrac{12a - 7}{12}$

e) $\dfrac{8(2 - 5x)}{24} - \dfrac{3(5 + 4x)}{24}$
$= \dfrac{16 - 40x - 15 - 12x}{24}$
$= \dfrac{1 - 52x}{24}$

f) $\dfrac{6(6x + 5)}{42} - \dfrac{7(4x + 8)}{42}$
$= \dfrac{36x + 30 - 28x - 56}{42}$
$= \dfrac{8x - 26}{42}$
$= \dfrac{4x - 13}{21}$

LEVEL 2 – Further Algebra and Factorisation

Q1. a) $8(2m + 1)$ b) $5x(x - 3)$ c) $(x + 2)(x - 2)$
d) $4a(3 - 2a)$ e) $(3m + 4)(3m - 4)$ f) $2(4b^2 - 9)$
$= 2(2b + 3)(2b - 3)$

Q2. a) $(x + 3)(x + 2)$ b) $(m - 4)(m + 3)$ c) $(x + 5)(x - 1)$ d) $(y - 2)(y - 4)$
e) $(p - 8)(p + 1)$ f) $3(a + 6)(a - 1)$

Q3. a) $(b + c)(a + b)$ b) $(y - z)(x + y)$ c) $(r - 1)(p - q)$ d) $(x + y)(x - 3)$
e) $(3b + c)(2a + b)$ f) $(y - 5)(5x - y)$

Q4. a) $(2m + 1)(m - 5)$ b) $(3x - 1)(x - 3)$ c) $(5a + 2)(a + 4)$ d) $(3x - 2)(2x + 3)$
e) $(4p - 1)(3p - 4)$ f) $(4b - 1)(4b + 5)$

Q5. a) $\dfrac{4\cancel{(2m-1)}}{1\cancel{(2m-1)}}$ $= 4$ b) $\dfrac{\cancel{(m+2)}(m-2)}{\cancel{(m+2)}}$ $= m - 2$ c) $\dfrac{\cancel{3x}(3x+4)}{\cancel{3x}}$ $= 3x + 4$

d) $\dfrac{\cancel{(x-1)}(x+4)}{\cancel{(x-1)}}$ $= x + 4$ e) $\dfrac{3\cancel{(2x-1)}}{5x\cancel{(2x-1)}}$ $= \dfrac{3}{5x}$ f) $\dfrac{(x+4)\cancel{(x-4)}}{(x-4)\cancel{(x-4)}}$ $= \dfrac{(x+4)}{(x-4)}$

Q6. Area $= \dfrac{h(a+b)}{2} = \dfrac{(4x-1)(8x+8)}{2} = (16x^2 + 12x - 4)\text{ cm}^2$

Q7. a) Area $= (7x - 3)^2$
$= (49x^2 - 42x + 9)\text{ cm}^2$
b) Area $= \dfrac{(x-2)(6x+8)}{2}$
$= (3x^2 - 2x - 8)\text{ cm}^2$
c) Area $= \dfrac{1}{2} \times (4x + 2)(2x - 5)$
$= (4x^2 - 8x - 5)\text{ cm}^2$

Q8. Time $= \left(\dfrac{12}{x} + \dfrac{8}{x+10}\right)$ hours **Q9.** $\dfrac{12y + 65}{13}$

Q10. $Y = 150A + 250B$ ➡ $Y = 600B + 250B$ ➡ $Y = 850B$ $\therefore B = \dfrac{Y}{850}$

LEVEL 3 – Further Algebra and Factorisation

Q1. a) $6(x + 4)$ b) $5(1 - 3x)$ c) $6(3x + 5y)$
d) $2x(2 + 3x)$ e) $3p(3 - 5q)$ f) $2xy(7 - 4y)$

Q2. a) $(x - 4)(x + 4)$ b) $(2x - y)(2x + y)$ c) $(7 - 5m)(7 + 5m)$
d) $(3a - 2b)(3a + 2b)$ e) $(mn - 6)(mn + 6)$ f) $2(x - 1)(x + 1)$

Q3. a) $(x - 3)(x - 5)$ b) $(m - 9)(m + 8)$ c) $4(y + 5)(y - 4)$
d) $(m + 8)(m - 2)$ e) $x(x + 1)(x - 1)$ f) $(p^2 + q^2)(p + q)(p - q)$

Q4. a) $(a + b)(b - c)$ b) $(x + 2y)(3z + 1)$ c) $(6 - m)(2n + m)$
d) $(x - 3)(x + y)$ e) $(2a + b)(3b + c)$ f) $(5x - y)(y - 5)$

Q5. a) $(2x + 1)(x - 5)$ b) $(x - 3)(3x - 1)$ c) $(5x + 2)(x + 4)$
d) $(2x + 3)(3x - 2)$ e) $(3x - 4)(4x - 1)$ f) $(4x - 1)(4x + 5)$

Q6. a) $\dfrac{x\cancel{(x-4)}}{(x+4)\cancel{(x-4)}}$ $= \dfrac{x}{x+4}$ b) $\dfrac{3\cancel{(2x-1)}}{5\cancel{(2x-1)}}$ $= \dfrac{3}{5}$ c) $\dfrac{x(y+2) - y(y+2)}{x(x+2) - y(x+2)}$ $= \dfrac{\cancel{(x-y)}(y+2)}{\cancel{(x-y)}(x+2)}$ $= \dfrac{y+2}{x+2}$

d) $\dfrac{(x+3)\cancel{(x-2)}}{(x-2)\cancel{(x-2)}}$ $= \dfrac{x+3}{x-2}$ e) $\dfrac{\cancel{(3x+1)}(x-3)}{x^2\cancel{(3x+1)}}$ $= \dfrac{x-3}{x^2}$ f) $\dfrac{4\cancel{(x+1)}(x-1)}{3\cancel{(x+1)}(x-4)}$ $= \dfrac{4(x-1)}{3(x-4)}$

Q7. a) $\dfrac{\cancel{x+5}}{(x+1)\cancel{(x-5)}} \times \dfrac{x\cancel{(x-5)}}{(x+5)\cancel{(x+5)}}$ $= \dfrac{x}{(x+1)(x+5)}$

b) $\dfrac{\cancel{2x+1}}{(2x+3)\cancel{(x-3)}} \times \dfrac{2(x-3)\cancel{(x-3)}}{5\cancel{(2x+1)}}$ $= \dfrac{2(x-3)}{5(2x+3)}$

c) $\dfrac{\cancel{(a+6)}\,(a-6)}{(a+2)\cancel{(a-6)}} \times \dfrac{2\cancel{(a-6)}}{3\cancel{(a+6)}}$ $= \dfrac{2(a-6)}{3(a+2)}$

d) $\dfrac{\cancel{(x-5)}\,\cancel{(x+4)}}{\cancel{(x-5)}\cancel{(x+5)}} \times \dfrac{x\cancel{(x+5)}}{\cancel{(x+4)}}$ $= x$

Q8. a) $\dfrac{1(2x+1)-1(2x-1)}{(2x-1)(2x+1)}$

$= \dfrac{2}{(2x-1)(2x+1)}$

b) $\dfrac{x}{x+2} - \dfrac{8}{(x+2)(x-2)}$

$= \dfrac{x(x-2)-8}{(x+2)(x-2)}$

$= \dfrac{x^2-2x-8}{(x+2)(x-2)}$

$= \dfrac{\cancel{(x+2)}(x-4)}{\cancel{(x+2)}(x-2)}$

$= \dfrac{x-4}{x-2}$

c) $\dfrac{6x}{(x+3)(x-2)} + \dfrac{3x}{x-2}$

$= \dfrac{6x+3x(x+3)}{(x+3)(x-2)}$

$= \dfrac{3x^2+15x}{(x+3)(x-2)}$

$= \dfrac{3x(x+5)}{(x+3)(x-2)}$

LEVEL 4 – Further Algebra and Factorisation

Q1. a) $2x+18$ b) $-3a^2+11a+2$ c) $-m-2m^2$ d) $b^2-3b+32$
e) $6-4m$ f) $2xy-6x+2y$ g) $24a^2-10a-16$ h) $x-xy^2$ i) $6\text{m}-18$

Q2. a) $2x^2-9x-5$ b) $16a^2+24a+9$ c) $9b^2-4$ d) $-5n^2+49n-36$
e) $21-19x-12x^2$ f) $81p^2-18p+1$ g) $36a^2-25$ h) $3x^2+xy-10y^2$
i) $16m^2-9n^2$ j) $25x^2-70x+49$ k) $9m^2+30m+25$ l) $20x^3-140x^2+245x$

Q3. a) $6m(4m+3)$ b) $a(a^2-ab-1)$ c) $(9x-7y)(9x+7y)$ d) $2(4n^2+6n-3)$
e) $-16x(3+x)$ f) $pq(p-q+1)$ g) $3b(2ab-5a+4)$ h) $2(4x-1)(4x+1)$
i) $-2(m-5)(m+5)$

Q4. a) $2(x-4)(x+9)$ b) $5(a-2)(a-4)$ c) $(4-x)(x+3)$ d) $(2b+3)(3b+2)$
e) $(3+4m)(5-3m)$ f) $(a-b)^2$ g) $4(2x-5)(x-4)$ h) $(2m+n)(m+2n)$
i) $(4m-1)(5m+1)$

Q5. a) $(x-y)(x+z)$ b) $(2-b)(4a-1)$ c) $(n^2+1)(n-1)$ d) $(m+n)(1-m)$
e) $(a+3c)(b-5)$ f) $(2y+z)(2x-3y)$

Q6. a) $\dfrac{y+2}{x+2}$ b) $\dfrac{x+3}{x-2}$ c) $\dfrac{x-3}{x^2}$ d) $\dfrac{4(x-1)}{3(x-4)}$
e) $\dfrac{3(2x-1)}{3x+1}$ f) $\dfrac{x+2}{x-2}$ g) $\dfrac{a-2b}{b+2}$ h) $\dfrac{a-b}{a(1+a)}$
i) $\dfrac{x-1}{1-x}$ or -1

Q7. a) $\dfrac{3}{(x+1)(x-1)} - \dfrac{1}{x+1}$

$= \dfrac{3-1(x-1)}{(x+1)(x-1)} = \dfrac{4-x}{(x+1)(x-1)}$

b) $\dfrac{x(x+6)+5x}{x+6}$

$= \dfrac{x^2+11x}{x+6}$

c) $\dfrac{5}{3(9-x)} \times \dfrac{2(9-x)}{9}$

$= \dfrac{10}{27}$

d) $\dfrac{2}{x(x-1)} + \dfrac{3}{x(1-x)}$

$= \dfrac{2(1-x)+3(x-1)}{x(x-1)(1-x)}$

$= \dfrac{(x-1)}{x(x-1)(1-x)}$

$= \dfrac{1}{x(1-x)}$

e) $\dfrac{3(x-2)}{4x^3} \times \dfrac{8}{2(2-x)}$

$= \dfrac{-3(2-x)}{4x^3} \times \dfrac{8}{2(2-x)}$

$= \dfrac{-3}{x^3}$

f) $\dfrac{x(x-1)}{(x-4)(x+2)} \times \dfrac{x(x-4)}{(x-1)}$

$= \dfrac{x^2}{x+2}$

LEVEL 5 – Further Algebra and Factorisation

Q1. a) $a = -1$ b) $a = 9$ c) $a = 4$ d) $a = b = 2$

Q2. a) $a = -3$ b) $a = 3\frac{3}{4}$ c) $a = \pm 4$ d) $a = 5$ e) $a = 2$ f) a = 6

Q3. a) $a = 3, b = 6$ b) $a = 3, b = 4$ c) $a = -6, b = -2$ d) $a = 2, b = -12$ e) $a = -5, b = 3$

f) $a = -2, b = 3$

Q4. a) For $a = 0, y = x^2$. For $a = 1, y = x^2 - x - 1$. For $a = 2, y = x^2 - 2x - 2$

b) For $a = 0, y = x^3 - x^2$. For $a = 1, y = x^3 + 2x^2 + x$. For $a = 2, y = x^3 + 5x^2 + 8x + 4$

c) For $a = 0$, y is undefined. For $a = 1, y = x^2 + x + \frac{1}{2}$. For $a = 2, y = \frac{1}{2}x^2 + \frac{1}{2}x + 1$

Q5. a) $a = 2, b = 3$ b) $a = 3, b = 2$ c) $a = 2, b = 1$ d) $a = 3, b = 4$ e) $a = 3, b = 1$

f) $a = \frac{1}{2}, b = 5$

Q6. a) $ax^2 - a$ b) $ax^2 - ab^2$ c) $ax^2 - 4ab^2$ d) $ax^2 - 2abx + ab^2$

e) $(x + b)^2 - 1^2 = x^2 + 2bx + b^2 - 1$

f) $\frac{1}{4}[(4x)^2 - (2a)^2] = \frac{1}{4}[16x^2 - 4a^2] = 4x^2 - a^2$

Q7. a) $(a + b)^2 = a^2 + ab + ab + b^2 = a^2 + 2ab + b^2$

b) (i) $(x + 3)^2 = x^2 + 2 \times 3 \times x + 3^2 = x^2 + 6x + 9$

$(x + 2)^2 + 2(x + 2) + 1 = x^2 + 4x + 4 + 2x + 4 + 1 = x^2 + 6x + 9$

(ii) $(x + 3)^2 = ((x + 2) + 1)^2 = (x + 2)^2 + 2 \times 1(x + 2) + 1^2 = (x + 2)^2 + 2(x + 2) + 1$

c) (i) $(x - 1)^2 = x^2 + 2 \times -1 \times x + (-1)^2 = x^2 - 2x + 1$

$(x - 2)^2 + 2(x - 2) + 1 = x^2 - 4x + 4 + 2x - 4 + 1 = x^2 - 2x + 1$

(ii) $(x - 1)^2 = ((x - 2) + 1)^2 = (x - 2)^2 + 2 \times 1(x - 2) + 1^2 = (x - 2)^2 + 2(x - 2) + 1$

Q8. $\dfrac{(x^2 + (a + 1)x + a)}{(x^2 + (a + 2)x + 2a)} = \dfrac{(x + a)(x + 1)}{(x + a)(x + 2)}$

By cancelling the denominator and numerator by $(x + a)$: $\dfrac{(x + a)(x + 1)}{(x + a)(x + 2)} = \dfrac{(x + 1)}{(x + 2)}$

Q9. $\dfrac{x^2 + dx + b}{x^2 + cx + a} = \dfrac{x^2 + 5x + 6}{x^2 + 3x + 2} = \dfrac{(x + 2)(x + 3)}{(x + 2)(x + 1)} = \dfrac{(x + 3)}{(x + 1)} = \dfrac{x + c}{x + c - a}$

Q10. For $a = 2$, $x^2 + x - 2 = (x - 1)(x + 2)$ and $x^2 - x - 2 = (x + 1)(x - 2)$

For $a = 6$, $x^2 + x - 6 = (x - 2)(x + 3)$ and $x^2 - x - 6 = (x + 2)(x - 3)$

For $a = 12$, $x^2 + x - 12 = (x - 3)(x + 4)$ and $x^2 - x - 12 = (x + 3)(x - 4)$

For $a = 20$, $x^2 + x - 20 = (x - 4)(x + 5)$ and $x^2 - x - 20 = (x + 4)(x - 5)$

Q11. a) $(x + a)(x + b)$ b) $(x + a)(x - b)$ c) $(x - a)(x + b)$ d) $(x + a)(x + 2a)$

e) $-(x^2 + 3ax + 2a^2) = -(x + a)(x + 2a)$ f) $(x - a)(x + 2a)$

g) $-(x^2 + ax - 2a^2) = (x - a)(x + 2a)$ h) $(x + a)(x - 2a)$ i) $(x - a)(x - 2a)$

LEVEL 1 – Surds and Indices

Q1. a) 13 cm b) 31 m c) 26 cm d) 14 cm

Q2. a) 9 cm b) 14 cm c) 11 m d) 16 m

Q3. a) 36.0 cm b) 21.5 m c) 11.3 cm d) $a = 11.2$ m, $b = 10.0$ m

Q4. a) Yes b) No c) Yes d) No

Q5. 18.2 nautical miles

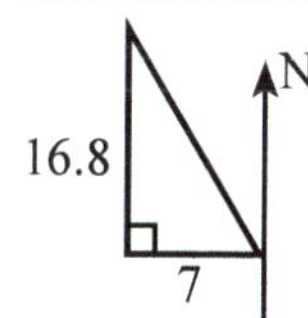

Q6. 52 m

48
52 m
12 20 12

Q7. Height is 11 m.

Q8. 30 km

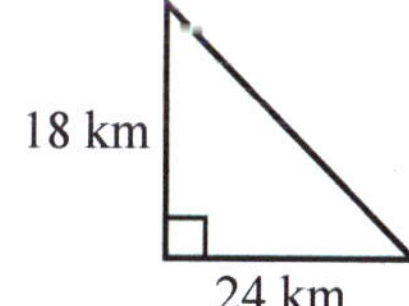

Q9. 933 m

LEVEL 2 – Surds and Indices

Q1. a) $10a^4b^3$ b) $9p^6q^5$ c) $28m^4n^5$ d) $8x^7y^4$ e) $6x^7y$ f) $15a^3b^4$

Q2. a) $\frac{m^2n}{4}$ b) $\frac{4y^3}{3}$ c) $3a^3b$ d) $6p^2q^2$ e) $\frac{m^3n^3}{2}$ f) $4x^7y$

Q3. a) $2x^{17}$ b) a^5b^4 c) $9m^6$ d) $8y^7$ e) $15x^5y^8$ f) $144a^{10}$

Q4. a) $8m^2$ b) $2a^7$ c) $9x^8$ d) $2y^{10}$ e) $6b$ f) $5m^3n$

Q5. a) $3x^3$ b) $\frac{12m^3}{5}$ c) $\frac{a^{15}}{2}$ d) $2b^4$ e) $2xy^2$ f) $\frac{3m^5n^3}{2}$

Q6. a) $\frac{5}{m^2}$ b) $\frac{1}{3a}$ c) $\frac{9}{100}$ or 0.09 d) $\frac{1}{(4x)^3} = \frac{1}{64x^3}$

Q7. a) $\sqrt{25} = 5$ b) $\sqrt[3]{8} = 2$ c) $\frac{1}{9^{\frac{1}{2}}} = \frac{1}{\sqrt{9}} = \frac{1}{3}$ d) $\frac{1}{1^2} = 1$ e) $\frac{1}{27^{\frac{1}{3}}} = \frac{1}{\sqrt[3]{27}} = \frac{1}{3}$

f) $4^{-3} = \frac{1}{4^3} = \frac{1}{64}$ g) $\frac{1}{16^{\frac{3}{2}}} = \frac{1}{\sqrt{16}^3} = \frac{1}{64}$ h) $\frac{1}{81^{\frac{3}{4}}} = \frac{1}{\sqrt[4]{81}^3} = \frac{1}{27}$ i) $3 \times 1 = 3$

j) $5^0 \times p^0 = 1 \times 1 = 1$ k) $10^{-6} = \frac{1}{10^6} = 0.000001$ or $\frac{1}{1\,000\,000}$

l) $\frac{1}{64^{\frac{2}{3}}} = \frac{1}{\sqrt[3]{64^2}} = \frac{1}{4^2} = \frac{1}{16}$

Q8. a) 9.2×10^3 b) 5.7×10^{-4} c) 1.5×10^{-4} d) 1.2×10^4

Q9.

a) $(2^3)^x \times 2^{4x}$
$= 2^{3x} \times 2^{4x}$
$= 2^{3x+4x}$
$= 2^{7x}$

b) $(3^2)^{\frac{x}{2}} \times (3^3)^{\frac{x}{3}}$
$= 3^x \times 3^x$
$= 3^{x+x}$
$= 3^{2x}$

c) $(2^5)^{x+2} \div (2^3)^{2x-1}$
$= 2^{5x+10} \div 2^{6x-3}$
$= 2^{(5x+10)-(6x-3)}$
$= 2^{13-x}$

LEVEL 3 – Surds and Indices

Q1. a) $\sqrt{18} = 3\sqrt{2}$ b) $\sqrt{5}$ c) $2\sqrt{56} = 4\sqrt{14}$ d) 4
e) $12\sqrt{6}$ f) 6

Q2. a) $5\sqrt{2}$ b) $3\sqrt{2}$ c) $3\sqrt{5}$ d) $6\sqrt{2}$
e) $2\sqrt{3}$ f) $2\sqrt{5}$ g) $8\sqrt{6}$ h) $9\sqrt{3}$

Q3. a) $2\sqrt{2}$ b) $\sqrt{5} + 3\sqrt{3}$ c) $3\sqrt{10} - 4\sqrt{5}$ d) $8\sqrt{3}$
e) $8\sqrt{3}$ f) $45\sqrt{5}$

Q4. a) $5\sqrt{6} + 15$ b) $\sqrt{6} - \sqrt{2}$ c) $4\sqrt{10} - 10\sqrt{2}$ d) $2\sqrt{10} + 10\sqrt{2}$
e) $14 + 3\sqrt{7}$ f) $14\sqrt{6} - 36$

Q5. a) $6 - 2\sqrt{5} + 3\sqrt{7} - \sqrt{35}$ b) -1 c) $2\sqrt{6} - 2\sqrt{2} + 3\sqrt{3} - 3$
d) $22 - 8\sqrt{6}$ e) $8 + 4\sqrt{3} - 4\sqrt{5} - 2\sqrt{15}$ f) 14

Q6. a) $\dfrac{\sqrt{2}}{2}$ b) $\sqrt{5}$ c) $\dfrac{5\sqrt{3}}{3}$ d) $\dfrac{\sqrt{3}}{3}$
e) $\dfrac{\sqrt{7}}{2}$ f) $\dfrac{\sqrt{30}}{15}$ g) $\dfrac{3\sqrt{2} - 2}{2}$ h) $\dfrac{\sqrt{15} + 4\sqrt{3}}{6}$

Q7. a) $\sqrt{2} - 1$ b) $\dfrac{30 + 5\sqrt{3}}{33}$ c) $\dfrac{3\sqrt{5} - \sqrt{15}}{6}$ d) $\dfrac{21\sqrt{6} - 14}{50}$
e) $\dfrac{-9 - 5\sqrt{3}}{2}$ f) $\sqrt{5} + \sqrt{3}$ g) $5\sqrt{3} + 5\sqrt{2}$ h) $4 + \sqrt{15}$

LEVEL 4 – Surds and Indices

Q1.

a) $3\sqrt{16 \times 3}$
$= 3\sqrt{16} \times \sqrt{3}$
$= 12\sqrt{3}$

b) $2\sqrt{9 \times 6}$
$= 2\sqrt{9} \times \sqrt{6}$
$= 6\sqrt{6}$

c) $4\sqrt{16 \times 6}$
$= 4\sqrt{16} \times \sqrt{6}$
$= 16\sqrt{6}$

d) $2\sqrt{36 \times 3}$
$= 2\sqrt{36} \times \sqrt{3}$
$= 12\sqrt{3}$

e) $3\sqrt{81 \times 2}$
$= 3\sqrt{81} \times \sqrt{2}$
$= 27\sqrt{2}$

f) $\dfrac{\sqrt{3}}{\sqrt{4}}$
$= \dfrac{\sqrt{3}}{2}$

g) $\dfrac{\sqrt{4} \times \sqrt{6}}{\sqrt{25}}$
$= \dfrac{2\sqrt{6}}{5}$

h) $\sqrt{x^2} \times \sqrt{x}$
$= x\sqrt{x}$

Q2.

a) $10\sqrt{5} - 6\sqrt{2} - 3\sqrt{5}$
$= 7\sqrt{5} - 6\sqrt{2}$

b) $7\sqrt{2} + 4\sqrt{2} - 3\sqrt{7}$
$= 11\sqrt{2} - 3\sqrt{7}$

c) $6\sqrt{3} - 2\sqrt{6} + 3\sqrt{3}$
$= 9\sqrt{3} - 2\sqrt{6}$

d) $10\sqrt{5} + 6\sqrt{5} - 4\sqrt{5}$
$= 12\sqrt{5}$

e) $4\sqrt{x} + 15\sqrt{x} - \sqrt{x}$
$= 18\sqrt{x}$

f) $4x\sqrt{y} - x\sqrt{y} + 9x\sqrt{y}$
$= 12x\sqrt{y}$

Q3.

a) $(2\sqrt{5})^2 + 2 \times 6\sqrt{15} + (3\sqrt{3})^2$
$= 20 + 12\sqrt{15} + 27$
$= 47 + 12\sqrt{15}$

b) $6\sqrt{6} - 6 + 3\sqrt{12} - \sqrt{18}$
$= 6\sqrt{6} - 6 + 6\sqrt{3} - 3\sqrt{2}$

c) $(4\sqrt{2})^2 - (2\sqrt{3})^2$
$= (a + b)(a - b) = a^2 - b^2$
$= 32 - 12$
$= 20$

d) $6\sqrt{36} - 9\sqrt{18} - 4\sqrt{42} + 6\sqrt{21}$
$= 36 - 27\sqrt{2} - 4\sqrt{42} + 6\sqrt{21}$

e) $6x\sqrt{xy} - 2x\sqrt{y^2}$
$= 6x\sqrt{xy} - 2xy$

f) $(a + b)(a - b) = a^2 - b^2$
$= (x\sqrt{x})^2 - (y\sqrt{y})^2$
$= x^3 - y^3$

Q4. a) $\dfrac{(\sqrt{3}-\sqrt{5})(3\sqrt{5}-2\sqrt{3})}{(3\sqrt{5}+2\sqrt{3})(3\sqrt{5}-2\sqrt{3})}$

$= \dfrac{3\sqrt{15}-21+2\sqrt{15}}{33}$

$= \dfrac{5\sqrt{15}-21}{33}$

b) $\dfrac{(\sqrt{6}-\sqrt{2})(2\sqrt{6}+4\sqrt{2})}{(2\sqrt{6}-4\sqrt{2})(2\sqrt{6}+4\sqrt{2})}$

$= \dfrac{12+4\sqrt{12}-2\sqrt{12}-8}{24-32}$

$= \dfrac{-1-\sqrt{3}}{2}$

c) $\dfrac{(2\sqrt{5}-3\sqrt{3})(2\sqrt{5}-3\sqrt{3})}{(2\sqrt{5}+3\sqrt{3})(2\sqrt{5}-3\sqrt{3})}$

$= \dfrac{20-6\sqrt{15}-6\sqrt{15}+27}{-7}$

$= \dfrac{12\sqrt{15}-47}{7}$

d) $\dfrac{3(2\sqrt{x}+3\sqrt{y})}{(2\sqrt{x}-3\sqrt{y})(2\sqrt{x}+3\sqrt{y})}$

$= \dfrac{6\sqrt{x}+9\sqrt{y}}{4x-9y}$

e) $\dfrac{(2-\sqrt{5})(3\sqrt{x}-4\sqrt{y})}{(3\sqrt{x}+4\sqrt{y})(3\sqrt{x}-4\sqrt{y})}$

$= \dfrac{6\sqrt{x}-8\sqrt{y}-3\sqrt{5x}+4\sqrt{5y}}{9x-16y}$

f) $\dfrac{(\sqrt{x}+\sqrt{y})(4\sqrt{x}+2\sqrt{y})}{(4\sqrt{x}-2\sqrt{y})(4\sqrt{x}+2\sqrt{y})}$

$= \dfrac{(4x+2\sqrt{xy}+4\sqrt{xy}+2y)}{16x-4y}$

$= \dfrac{2x+3\sqrt{xy}+y}{8x-2y}$

Q5. a) $\dfrac{5\sqrt{2}+2\sqrt{5}}{10\sqrt{10}}$

$= \dfrac{(5\sqrt{2}+2\sqrt{5})}{10\sqrt{10}} \times \dfrac{\sqrt{10}}{\sqrt{10}}$

$= \dfrac{10\sqrt{5}+10\sqrt{2}}{100}$

$= \dfrac{\sqrt{5}+\sqrt{2}}{10}$

b) $\dfrac{(\sqrt{3})^2-(\sqrt{2})^2}{\sqrt{6}}$

$= \dfrac{1}{\sqrt{6}} \times \dfrac{\sqrt{6}}{\sqrt{6}}$

$= \dfrac{\sqrt{6}}{6}$

c) $\left(\dfrac{\sqrt{35}+2\sqrt{30}}{2\sqrt{42}}\right) \times \dfrac{\sqrt{42}}{\sqrt{42}}$

$= \dfrac{\sqrt{1\,470}+2\sqrt{1\,260}}{84}$

$= \dfrac{\sqrt{1\,470}+4\sqrt{315}}{84}$

d) $\dfrac{(\sqrt{3}-2)-(\sqrt{3}+2)}{(\sqrt{3}+2)+(\sqrt{3}-2)}$

$= \dfrac{-4}{-1} = 4$

e) $\dfrac{3(\sqrt{6}+3)+3(\sqrt{6}-2)}{(\sqrt{6}-2)(\sqrt{6}+3)}$

$= \left(\dfrac{3+6\sqrt{6}}{\sqrt{6}}\right) \times \dfrac{\sqrt{6}}{\sqrt{6}}$

$= \dfrac{3\sqrt{6}+36}{6} = \dfrac{12+\sqrt{6}}{2}$

f) $\dfrac{2(5\sqrt{2}+3)-1(2\sqrt{2}+3)}{(2\sqrt{2}+3)(5\sqrt{2}+3)}$

$= \dfrac{10\sqrt{2}+6-2\sqrt{2}-3}{20+6\sqrt{2}+15\sqrt{2}+9}$

$= \dfrac{(3+8\sqrt{2})}{(29+21\sqrt{2})} \times \dfrac{(29-21\sqrt{2})}{(29-21\sqrt{2})}$

$= \dfrac{249-169\sqrt{2}}{41}$

Q6. $\dfrac{1}{(\sqrt{2}+\sqrt{5})^2} + (\sqrt{2}+\sqrt{5})^2$

$= \dfrac{1}{7+2\sqrt{10}} + (7+2\sqrt{10})$

$= \dfrac{(7-2\sqrt{10})}{(7+2\sqrt{10})(7-2\sqrt{10})} + (7+2\sqrt{10})$

$= \dfrac{(7-2\sqrt{10})+9(7+2\sqrt{10})}{9}$

$= \dfrac{70+6\sqrt{10}}{9}$

Q7. $\dfrac{x\sqrt{x^4}}{2\sqrt{2x^2}}$

$= \dfrac{x^3}{2x\sqrt{2}} \times \dfrac{\sqrt{2}}{\sqrt{2}}$

$= \dfrac{\sqrt{2}\,x^3}{4x}$

$= \dfrac{x^2\sqrt{2}}{4}$

Q8. $(18 + 6\sqrt{6} + 3) - (18 - 6\sqrt{6} + 3)$
$= 12\sqrt{6}$

Q9. $\dfrac{(2\sqrt{3} - 2\sqrt{2})}{(3\sqrt{3} - 3\sqrt{2})} \times \dfrac{(3\sqrt{3} + 3\sqrt{2})}{(3\sqrt{3} + 3\sqrt{2})}$
$= \dfrac{18 - 6\sqrt{6} + 6\sqrt{6} - 12}{9}$
$= \dfrac{6}{9} = \dfrac{2}{3}$ which is a rational number.

LEVEL 5 – Surds and Indices

Q1. a) $\dfrac{y^2}{x^2}$ b) $\dfrac{x^2}{\sqrt{y}}$ c) $\dfrac{\sqrt{y^3}}{x^3}$ d) x^2y^4 e) $\dfrac{y^2 - x^2}{x^2y^2}$ f) $\dfrac{2x}{(x+y)(x-y)}$

Q2. a) 9 b) 512 c) 5 d) $\dfrac{1}{25}$ e) $\dfrac{3}{32}$ f) $\dfrac{4}{81}$

Q3. a) $\dfrac{2}{\sqrt{x}}$ b) $\dfrac{9}{x^{\frac{5}{6}}}$ c) $54x^4$ d) $6\sqrt[3]{x}$ e) $\dfrac{\sqrt{x}}{16}$ f) $\dfrac{5}{\sqrt{x}}$

Q4. a) (i) $(6.022 \times 10^{23}) \div 2 = 3.011 \times 10^{23}$
(ii) $(6.022 \times 10^{23}) \div 4 = 1.506 \times 10^{23}$
(iii) $(6.022 \times 10^{23}) \times \dfrac{20}{12} = 1.004 \times 10^{24}$

b) (i) 1.2×10^{-2} g
(ii) 1.2×10^{-12} g
(iii) 1.2×10^{-17} g
(iv) 1.2×10^{-20} g

Q5. a) (i) $3 \times 10^8 \times 60 = 1.8 \times 10^{10}$ m
(ii) $3 \times 10^8 \times 60 \times 60 = 1.08 \times 10^{12}$ m
(iii) $3 \times 10^8 \times 60 \times 60 \times 24 = 2.592 \times 10^{13}$ m

b) Distance = Speed × Time
$\therefore$ Time $= \dfrac{6\,370\,000}{3 \times 10^8} = 2.1 \times 10^{-2}$ s $= 0.021$ s

c) 1 'light year'= 365 days = $365 \times 24 \times 60 \times 60$ seconds
$= 3.1536 \times 10^7$ seconds
Distance traveled in 1 light year $= 3.1536 \times 10^7 \times 3 \times 10^8$
$= 9.46 \times 10^{15}$ m
$= 9.46 \times 10^{12}$ km
Therefore 4.85 light years $= 4.85 \times 9.46 \times 10^{12}$
$= 4.5881 \times 10^{13}$ km

Q6. a) $\dfrac{12}{4} \div 100 = 0.03$ per quarter
b) For A = 1 000 and $n = 4$, $V = 1\,000\,(1.03)^4 = \$1\,125.51$
c) For the investment to have doubled, its value must be at least \$2 000.
As $1\,000\,(1.03)^{24} = \$2\,032.79$ then it will take 24 quarters to double the investment.
i.e. It will take $24 \times 3 = 72$ months = 6 years.

LEVEL 6 – Surds and Indices

Q1. a) Edge length = $\sqrt{252} = 6\sqrt{7}$ cm b) Diagonal length = $\sqrt{252+252} = \sqrt{504} = 6\sqrt{14}$ cm

c) $(6\sqrt{7})^3 = 1\,512\sqrt{7}$ cm^3 d) $12 \times 6\sqrt{7} = 72\sqrt{7}$ cm

Q2. a) Half base length = 3cm. $h = \sqrt{6^2 - 3^2} = \sqrt{27} = 3\sqrt{3}$ cm, Area = $\frac{1}{2} \times 6 \times 3\sqrt{3} = 9\sqrt{3}$ cm^2

b) (i) Perimeter = $2(\sqrt{18} + \sqrt{48}) = 2(3\sqrt{2} + 4\sqrt{3}) = (6\sqrt{2} + 8\sqrt{3})$ cm

Area = $\sqrt{18} \times \sqrt{48} = 3\sqrt{2} \times 4\sqrt{3} = 12\sqrt{6}$ cm^2

(ii) Perimeter = $2(2\sqrt{72} + 10\sqrt{128}) = 2(12\sqrt{2} + 80\sqrt{2}) = 2 \times 92\sqrt{2} = 184\sqrt{2}$ cm

Area = $2\sqrt{72} \times 10\sqrt{128} = 12\sqrt{2} \times 80\sqrt{2} = 960 \times 2 = 1\,920$ cm^2

Q3. a) (i) $x = (4 + \sqrt{2}) - 3 = 1 + \sqrt{2}$ (ii) $y = (6 + 2\sqrt{3}) - (5 + \sqrt{2}) = 1 + 2\sqrt{3} - \sqrt{2}$

b) (i) Perimeter = $2(4 + \sqrt{2} + 6 + 2\sqrt{3}) = 2(10 + \sqrt{2} + 2\sqrt{3}) = 20 + 2\sqrt{2} + 4\sqrt{3}$

(ii) Area = $(4 + \sqrt{2})(6 + 2\sqrt{3}) - (1 + \sqrt{2})(1 + 2\sqrt{3} - \sqrt{2}) = (2\sqrt{6} + 8\sqrt{3} + 6\sqrt{2} + 24) - (2\sqrt{6} + 2\sqrt{3} - 1) = 6\sqrt{3} + 6\sqrt{2} + 25$

Q4. a) Area of wrapping paper = $2(10\sqrt{6} \times 8\sqrt{3} + 10\sqrt{6}(4\sqrt{2} + 5) + 8\sqrt{3}(4\sqrt{2} + 5))$

$= 2(240\sqrt{2} + 50\sqrt{6} + 80\sqrt{3} + 32\sqrt{6} + 40\sqrt{3}) = (480\sqrt{2} + 164\sqrt{6} + 240\sqrt{3})$cm^2

b) Volume = $10\sqrt{6} \times 8\sqrt{3} \times (4\sqrt{2} + 5) = (1\,200\sqrt{2} + 1\,920)$ cm^3

Q5. a) Fourth term: $2ab^{-2} \times (b\sqrt{a})^3 = 2a^2b\sqrt{a} = 2a^{\frac{5}{2}}b$

Fifth term: $2ab^{-2} \times (b\sqrt{a})^4 = 2a^3b^2$

b) $2ab^{-2} \times 2ab^{-2}(b\sqrt{a}) \times 2ab^{-2}b\sqrt{a})^2 \times 2ab^{-2} \times (b\sqrt{a})^3 \times 2ab^{-2} \times (b\sqrt{a})^4 = 32a^{10}$

LEVEL 1 – Equations and Formulae

Q1. a) $x = -4$ b) $p = 9$ c) $x = \frac{2}{5}$ d) $a = 1\frac{1}{4}$ e) $b = \frac{2}{3}$ f) $a = -4$

g) $x = -5$ h) $m = 8$ i) $y = \frac{5}{9}$

Q2. a) $x = 24$ b) $x = 3\frac{1}{3}$ c) $x = -3$ d) $x = 4$ e) $x = 0$ f) $x = -\frac{1}{4}$

Q3. a) $x < 18$

16 17 18 19

b) $x > 6$

4 5 6 7 8

c) $x \leq \frac{3}{4}$

−2 −1 0 $\frac{3}{4}$ 1 2

d) $x < 5$

3 4 5 6

e) $x < -\frac{1}{2}$

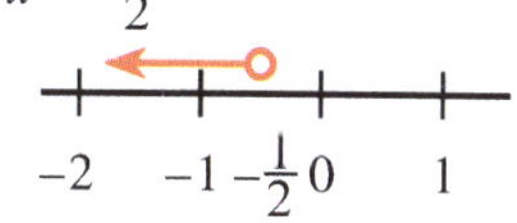

f) $x \geq -3$

−4 −3 −2 −1 0

Q4. a) Let the daughters = x years old ∴ Mother = $2x$ years old

∴ $(x - 5) + (2x - 5) = 62$ ⇨ $x = 24$ ⇨ Present ages are 24 and 48.

b) Let the full bottle = x mL ⇨ $\frac{3}{4}x - \frac{1}{2}x = 275$ ⇨ $x = 1\,100$ mL

c) Let the son = x years old ∴ Father = $5x$ years old

∴ 3 years ago: $5x - 3 = 8(x - 3)$ ⇨ $x = 7$ fathers present age = 35 years old

d) Let H = hire price and k = number of km and d = number of days hired

H = $53d + 0.84k$ ⇨ $349.68 = 53 \times 3 + 0.84k$ ⇨ $k = 227$ km

e) Let x = number of 20¢ coins and y = number 50¢ coins.

∴ $x + y = 112$ – (i) Solve these 2 equations simultaneously

$20x + 50y = 3\,710$ – (ii) ∴ $x = 63$ and $y = 49$

Q5. a) $X = 9$ b) $x = \pm\frac{2}{9}$ c) $y = \pm\frac{2}{3}$

Q6. i) $2as = v^2 - u^2$

$\therefore\ a = \frac{v^2 - u^2}{2s}$

ii) $2as = v^2 - u^2$

$\therefore\ s = \frac{v^2 - u^2}{2a}$

iii) $v = \sqrt{u^2 + 2as}$

iv) $u^2 = v^2 - 2as$

$\therefore\ u = \sqrt{v^2 - 2as}$

LEVEL 2 – Equations and Formulae

Q1. a) $x = -6, y = 16$ b) $x = 3, y = -1$ c) $x = -1, y = 5$ d) $x = -1, y = -5$

e) $x = 1, y = 2\frac{1}{2}$ f) $x = -2, y = -3$

Q2. a) $x = 1, y = -2$ b) $x = 9, y = 12$ c) $x = 2, y = 3$ d) $x = -4, y = 14$

e) $x = 3, y = 5$ f) $x = -3, y = \frac{1}{2}$

Q3. a) $x - y = 11$ ––––– (1)

$2y - 4 = x$ ––––– (2)

$x = 26$ and $y = 15$

b) $6x + 4y = 30$ ––––– (1)

$7x + 2y = 23$ ––––– (2)

$x = 2$ and $y = 4\frac{1}{2}$

$\therefore$ Weight of large marble = 4.5 g

c) $A = x(x + 5)$ ––––– (1)

$A + 50 = (x + 2)(x + 7)$ ––––– (2)

$\therefore\ A = x^2 + 9x - 36$

$\therefore\ x^2 + 9x - 36 = x^2 + 5x$

$\therefore\ 4x = 36$

$x = 9$

breadth = 9 cm

length = (9 + 5) cm

= 14 cm

d) $y = 5x$ ––––– (1)

$y - 4 = 9(x - 4)$ ––––– (2)

$\therefore\ y = 9x - 32$

$\therefore\ x = 8$ and $y = 40$

Son = 8 years old

Father = 40 years old

e) $2a + 2b = 20$ ––––– (1)

$8a + 4b = 48$ ––––– (2)

$a = 2$ and $b = 8$

MNOP = 8 cm × 2 cm

ABCD = 16 cm × 8 cm

f) $(x + 3) + (x + 3) + (x - 2) + (x - 2) = 30$

$\therefore\ 4x + 2 = 30$

$\therefore\ x = 7$

Length of square's side = 7 cm

g) $y = 3x + 2$ ––––– (1)

$y = 7x - 6$ ––––– (2)

$\therefore\ 7x - 6 = 3x + 2$

$\therefore\ 4x = 8$

$\therefore\ x = 2$ and $y = 8$

Point of intersection = (2, 8)

h) $3x + 2y - 6 = 0$ ––––– (1)

$2x - 3y - 17 = 0$ ––––– (2)

$\therefore\ 9x + 6y = 18$ ––––– (1)

$\therefore\ 4x - 6y = 34$ ––––– (2)

$\therefore\ 13x = 52$

$x = 4$ and $y = -3$

Point of intersection = (4, –3)

Q4. $4x - 3y + 23 = 0$ ––––– (1)

$3x + 2y - 4 = 0$ ––––– (2)

$5x + 4y - 10 = 0$ ––––– (3)

Substitute (–2, 5) into (1).

It holds true ➡ the 3 lines are concurrent because they all pass through the same point (–2, 5).

Solve (2) and (3) simultaneously to find points of intersection (–2, 5)

Q5. $2 = 3m + b$ ––––– (1)

$-1 = 2m + b$ ––––– (2)

Solve simultaneously ➡ $m = 3$ and $b = -7$

$\therefore\ y = 3x - 7$

Note: Substitute (3, 2) and (2, –1) into $y = mx + b$

LEVEL 3 – Equations and Formulae

Q1. a) $x = -2, -3$ b) $x = -4, 6$ c) $x = 4$ d) $x = -4, 3$
e) $x = -8, 7$ f) $x = -3, 8$

Q2. a) $2(x - 2)(3x + 2) = 0$
$\therefore\ x = 2, -\frac{2}{3}$

b) $(4x + 1)(2x - 3) = 0$
$\therefore\ x = -\frac{1}{4}, 1\frac{1}{2}$

c) $(4x + 3)(5x + 1) = 0$
$\therefore\ x = -\frac{3}{4}, -\frac{1}{5}$

d) $2(3x + 4)(x - 3) = 0$
$\therefore\ x = -1\frac{1}{3}, 3$

e) $(3x + 7)(4x + 1) = 0$
$\therefore\ x = -2\frac{1}{3}, -\frac{1}{4}$

f) $(9x + 2)(x - 8) = 0$
$\therefore\ x = -\frac{2}{9}, 8$

Q3. a) $(x + 5)(x - 5) = 0$
$\therefore\ x = -5, 5$

b) $x^2 + 6x - 27 = 0$
$(x + 9)(x - 3) = 0$
$\therefore\ x = -9, 3$

c) $10x^2 - x - 2 = 0$
$(5x + 2)(2x - 1) = 0$
$\therefore\ x = -\frac{2}{5}, \frac{1}{2}$

d) $x(x - 1) = 0$
$\therefore\ x = 0, 1$

e) $x(5 - x) = 0$
$\therefore\ x = 0, 5$

f) $x^2 - 9x = 0$
$x(x - 9) = 0$
$\therefore\ x = 0, 9$

g) $x^2 - 1 = 0$
$(x + 1)(x - 1) = 0$
$\therefore\ x = -1, 1$

h) $(3x + 2)(3x - 2) = 0$
$\therefore\ x = -\frac{2}{3}, \frac{2}{3}$

Q4. a) $x = -2 \pm \sqrt{7}$ b) $x = 3 \pm \sqrt{8}$ c) $x = -1 \pm \sqrt{6}$ d) $x = \frac{3 \pm \sqrt{17}}{2}$
e) $x = \frac{7 \pm \sqrt{41}}{2}$ f) $x = \frac{-9 \pm \sqrt{113}}{2}$

Q5. a) $x = \frac{1}{2}, 2$ b) $x = \frac{-5 \pm \sqrt{13}}{4}$ c) $x = \frac{5 \pm \sqrt{73}}{12}$ d) $x = \frac{3 \pm \sqrt{89}}{10}$
e) $x = \frac{3 \pm \sqrt{21}}{6}$ f) $x = \frac{4 \pm \sqrt{7}}{3}$

Q6. a) $x = \frac{9 \pm \sqrt{57}}{6}$ b) $x = \frac{-1 \pm \sqrt{3}}{2}$ c) $x = -1, \frac{-3}{4}$ d) $x = \frac{-11 \pm \sqrt{161}}{4}$
e) $x = \frac{-9 \pm \sqrt{21}}{6}$ f) $x = \frac{3 \pm \sqrt{89}}{10}$

Q7. a) $3x^2 + 5x + 2 = 0$
$x = -\frac{2}{3}, -1$

b) $5x^2 + 8x - 3 = 0$
$x = \frac{-4 \pm \sqrt{31}}{5}$

c) $3x^2 + 12x + 4 = 0$
$x = \frac{-6 \pm 2\sqrt{6}}{3}$

d) $4x^2 + x - 2 = 0$
$x = \frac{1 \pm \sqrt{33}}{8}$

e) $4x^2 - 4x + 1 = 0$
$(2x - 1)(2x - 1) = 0$
$x = \frac{1}{2}$

f) $8x^2 - 14x + 3 = 0$
$(4x - 1)(2x - 3) = 0$
$x = \frac{1}{4}, 1\frac{1}{2}$

Note: In Q7 you must firstly clear all fractions by multiplying through by the LCD. Secondly, rearrange in order to get zero on the right hand side.

Q8. a) $x(x + 2) = 288$
$\therefore\ x^2 + 2x - 288 = 0$
$\therefore\ x = 16, 18$

b) $x + x^2 = 156$
$\therefore\ x^2 + x - 156 = 0$
$x = -13$

c) $\frac{1}{2}x \times (h + 6) \times h = 140$
$\therefore\ h^2 + 6h - 280 = 0$
base = (14 + 6) cm = 20 cm

d) $[x + (x + 1) + (x + 2)]^2 = 441$
$\therefore\ (3x + 3)^2 = 441$
$\therefore$ numbers are 6, 7, 8 or −8, −7, −6

LEVEL 4 – Equations and Formulae

Q1. a) $y = 3x - 1$ b) $x + y = 5$ or $y = 5 - x$ c) $y = \frac{1}{2}x - 1$

Q2. a) $y = \frac{1}{2}x + 1$ b) $x + y = 4$ or $y = 4 - x$ c) $y = -2x + 2$

Q3. a) $6x - 9 = 36x - 189$
$\therefore\ 180 = 30x$
$\therefore\ x = 6$

b) $9x + 12 = 6x - 15$
$\therefore\ 3x = -27$
$\therefore\ x = -9$

c) $10y - 15 = 6y + 2$
$\therefore\ 4y = 17$
$\therefore\ y = 4\frac{1}{4}$

d) $2(x - 3) = 4(3x - 1)$
$\therefore\ 2x - 6 = 12x - 4$
$\therefore\ -10x = 2$
$\therefore\ x = -\frac{1}{5}$

e) $20m + 10 = 18m + 24$
$\therefore\ 2m = 14$
$\therefore\ m = 7$

f) $35 - 21b = 16b - 40$
$\therefore\ 75 = 37b$
$\therefore\ b = 2\frac{1}{37}$

g) $27x - 6 = 15x + 35$
$\therefore\ 12x = 41$
$\therefore\ x = 3\frac{5}{12}$

h) $96 - 60a = 40a + 60$
$\therefore\ 36 = 100a$
$\therefore\ a = \frac{36}{100} = \frac{9}{25}$

Q4. i) $h = \frac{V}{\pi r^2}$

ii) $r^3 = \frac{3V}{4\pi}$
$\therefore\ r = \sqrt[3]{\frac{3V}{4\pi}}$

iii) $r^2 = 3b - 4ac$
$\therefore\ 4ac = 3b - r^2$
$\therefore\ a = \frac{3b - r^2}{4c}$

iv) $x^2 = b^2 - 4ac$
$\therefore\ b^2 = x^2 + 4ac$
$\therefore\ b = \sqrt{x^2 + 4ac}$

v) $am = 4a - 4b$
$\therefore\ a(m - 4) = -4b$
$\therefore\ a = \frac{-4b}{m - 4}$ or $\frac{4b}{4 - m}$

vi) $3w = x^2 - 2$
$\therefore\ x^2 = 3w + 2$
$\therefore\ x = \sqrt{3w + 2}$

Q5. a) $3k^2 = 48$
$\therefore\ k^2 = 16$
$\therefore\ k = \pm 4$

b) $p^2 - 7 = 42$
$\therefore\ p^2 = 49$
$\therefore\ p = \pm 7$

c) $2m^2 = 128$
$\therefore\ m^2 = 64$
$\therefore\ m^2 = \pm 8$

d) $m = \sqrt[3]{-27}$
$\therefore\ m = -3$

e) $m^3 - 9 = 55$
$\therefore\ m^3 = 64$
$\therefore\ m = 4$

f) $4a^3 = -500$
$\therefore\ a^3 = -125$
$\therefore\ a = -5$

Note: When taking the square root of a number, there are always 2 possible answers – a positive number and its negative.

Q6. a) (i) Total surface area = $2 \times 3xh + 2 \times xh + 2 \times 3x^2 = 8xh + 6x^2 = 42$

$8xh = 42 - 6x^2,\ h = \frac{6(7 - x^2)}{8x} = \frac{3(7 - x^2)}{4x}$

(ii) $14x^2 = 42,\ x^2 = 3,\ x = \sqrt{3}$ cm

b) (i) $V = 3x^2h = 3x^2 \times \frac{3(7 - x^2)}{4x} = \frac{9x(7 - x^2)}{4}$

(ii) For $x = 1$, $h = \frac{3 \times 6}{4} = 4\frac{1}{2}$. Width = 1 and length = 3 × 1 = 3. Volume = $4\frac{1}{2} \times 1 \times 3 = 13\frac{1}{2}$ cm^3.

For $x = 2$, $h = \frac{3 \times 3}{8} = 1\frac{1}{8}$. Width = 2 and length = 3 × 2 = 6. Volume = $1\frac{1}{8} \times 2 \times 6 = 13\frac{1}{2}$ cm^3.

Q7. a) $6 \times \frac{1}{2}(8x - 1) = 6 \times \frac{1}{3}(6x + 9)$
$\therefore\ 24x - 3 = 12x + 18$
$\therefore\ 12x = 21$
$\therefore\ x = \frac{21}{12} = 1\frac{3}{4}$

b) $5x - 3 = 6(x + 2)$
$\therefore\ 5x - 3 = 6x + 12$
$\therefore\ x = -15$

c) $(x - 1) - 3(x + 6) = 5(x + 6)$
$\therefore\ -2x - 19 = 5x + 30$
$\therefore\ 7x = -49$
$\therefore\ x = -7$

d) $3(x + 2) + 6(4 - x) = 2(4 - x)$
$\therefore\ 3x + 6 + 24 - 6x = 8 - 2x$
$\therefore\ x = 22$

e) $8(3 - 2x) - 2(2x + 3) = 1(3 - 2x)$
$\therefore\ 24 - 16x - 4x - 6 = 3 - 2x$
$\therefore\ 18x = 15$
$\therefore\ x = \frac{15}{18} = \frac{5}{6}$

f) $2(x - 2) + 3(x + 3) = 6x$
$\therefore\ 2x - 4 + 3x + 9 = 6x$
$\therefore\ x = 5$

g) $9(4x - 3) - 8(3x + 1) = 3 \times 5x$
$\therefore\ 36x - 27 - 24x - 8 = 15x$
$\therefore\ 3x = -35$
$\therefore\ x = -11\frac{2}{3}$

h) $4(3x - 1) - 5(2x + 3) = 20(x - 3)$
$\therefore\ 12x - 4 - 10x - 15 = 20x - 60$
$\therefore\ 18x = 41$
$\therefore\ x = \frac{41}{18} = 2\frac{5}{18}$

i) $5 + x^2 = 9$
$\therefore\ x^2 = 4$
$\therefore\ x = \sqrt{4} = \pm 2$

LEVEL 5 – Equations and Formulae

Q1. a) $0 = (x-4)(x+2)$
Roots are $x = -2, 4$

b) $0 = x(x-7)$
Roots are $x = 0, 7$

c) $0 = (x-5)(x-3)$
Roots are $x = 5, 3$

d) $0 = (4+x)(4-x)$
Roots are $x = -4, 4$

e) $0 = (x-3)(x-3)$
Root is $x = 3$

f) $0 = (5-x)(3+x)$
Roots are $x = -3, 5$

Note: (i) Roots are another name for the x intercepts.
(ii) In part e) the parabola touches the x-axis in only 1 place.
i.e. the x-axis is a tangent to the parabola.

Q2. $x^2 + 5x - 3 = 3x + 5$
$\therefore\ x^2 + 2x - 8 = 0$
$\therefore\ (x-2)(x+4) = 0$
$\therefore\ x = 2$ and $x = -4$
$\therefore\ y = 11$ and $x = -7$

By simple substitution.

Points of intersection are:
(2, 11) and (–4, –7)

Q3. a) Substitute $t = 3$ into $h = 12t - t^2$
$\therefore$ Height $= 12 \times 3 - 3^2 = 27$ m

b) $h = 12t - t$
$\therefore\ 32 = 12t - t^2$
$\therefore\ t^2 - 12t + 32 = 0$
$\therefore\ (t-4)(t-8) = 0$

After 4 seconds and after 8 seconds.

c) At ground level, $h = 0$
$\therefore\ 0 = 12t - t^2$
$\therefore\ 0 = t(12-t)$
$\therefore\ t = 12$
The ball lands on the ground after 12 seconds.

Q4. a) $x = -1 \pm \sqrt{k+1}$

b) $x = -1 \pm \sqrt{1-4k}$

c) $x = -2 \pm 2\sqrt{k+1}$

d) $x = k \pm \sqrt{k^2+1}$

e) $x = -2k \pm \sqrt{4k^2+1}$

f) $x = 3k \pm \sqrt{9k^2+4}$

g) $x = \dfrac{2 \pm 2\sqrt{1-k}}{k}$

h) $x = \dfrac{1 \pm \sqrt{33}}{4}$

i) $x = \dfrac{1 \pm \sqrt{1-32k}}{4}$

Q5. a) The general quadratic formula used to solve the quadratic equation
$ax^2 + bx + c = 0$ is: $x = \dfrac{-b \pm \sqrt{b^2-4ac}}{2a}$.

When the section under the square root, $\Delta = b^2 - 4ac$ is negative then there are no solutions as the square root of a negative number doesn't exist as a real number.
When Δ is positive then the $\pm$ sign is used to provide two real solutions.
When Δ is zero then the $\pm$ sign is redundant and the quadratic formula becomes $x = \dfrac{-b}{2a}$.

b) (i) $\Delta = (-4)^2 - 4(1)(-3) = 28$ so there will be two real solutions.
(ii) $\Delta = (4)^2 - 4(2)(10) = -64$ so there will be no real solutions.
(iii) $\Delta = (1)^2 - 4(3)(8) = -95$ so there will be no real solutions.
(iv) $\Delta = (-4)^2 - 4(1)(4) = 0$ so there will be one real solution.

c) $\Delta = (m)^2 - 4(1)(4) = m^2 - 16$.
(i) $\Delta = m^2 - 16 = 0$, $m = \pm 4$
(ii) For $m^2 - 16 < 0$, $-4 < m < 4$
(iii) For $m^2 - 16 > 0$, $m < -4$ or $m > 4$

d) (i) $\Delta = (m)^2 - 4(1)(m-1) = m^2 - 4m + 4 = (m-2)^2 = 0$, $m = 2$
(ii) As the square for $(m-2)^2$ is positive or zero for any values of m then the equation must always have at least one real solution.

Q6. $y = 9x^2 - 30x + 40$ and $y = 18x - 24$.

$\therefore$ $9x^2 - 30x + 40 = 18x - 24$

$\therefore$ $9x^2 - 48x + 34 = 0$

$\therefore$ Δ = discriminant = $b^2 - 4ac$

$\therefore$ $\Delta = 48^2 - 4(9)(64) = 0$

$\therefore$ Only 1 point of intersection

$\therefore$ The line is a tangent.

Q7. a) $a = 1, b = 4, c = 5$ b) $a = 3, b = 5, c = 2$ c) $a = 0, b = c = 4$

d) $a = 2, b = 12, c = 20$ e) $a = 3, b = 5, c = 7$

Q8. Imagine the square as part of a number plane.

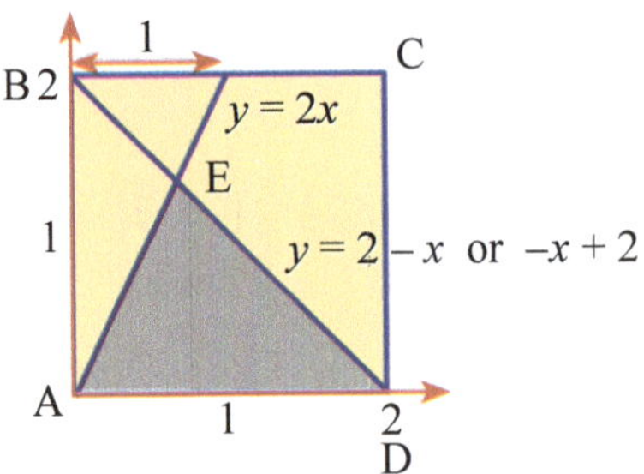

Using $y = mx + b$:

Equation of AE is $y = 2x$ ----- (1)

Equation of BD is $y = -x + 2$ ----- (2)

Solve simultaneously to find the coordinates of E.

$\therefore$ $2x = -x + 2$

$\therefore$ $x = \frac{2}{3}$ and $y = \frac{4}{3}$

$\therefore$ E has coordinates $\left(\frac{2}{3}, \frac{4}{3}\right)$

Area of ΔAED $= \frac{1}{2} \times b \times h = \frac{1}{2} \times 2 \times \frac{4}{3} = 1\frac{1}{3}$ units^2

Area of square ABCD $= 2 \times 2 = 4$ units^2

$\therefore$ Fraction shaded $= \dfrac{1\frac{1}{3}}{4} = \frac{1}{3}$.

Q9.

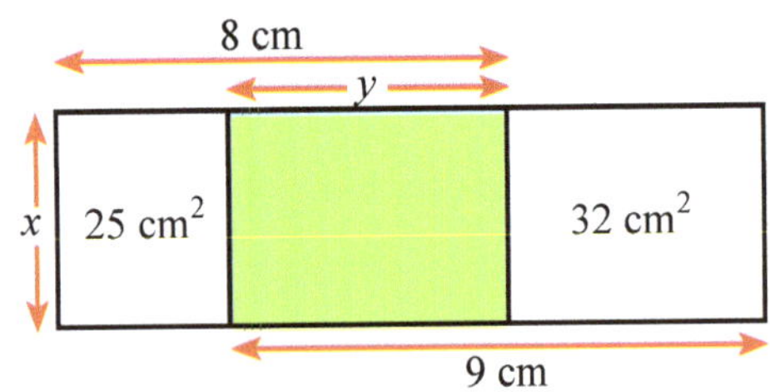

Let the 2 unknown dimensions be x and y as shown in the diagram.

$\therefore$ $9x - xy = 32$ ------ (i)

$\therefore$ $8x - xy = 25$ ------ (ii)

Solve simultaneous equations.

(i) subtract (ii)

$\therefore$ $x = 7$ and $y = \frac{31}{7}$ or $4\frac{3}{7}$

$\therefore$ Green area $= 7 \times \frac{31}{7} = 31$ cm^2

LEVEL 6 – Equations and Formulae

Q1. a) (i) $y = \pm\frac{1}{2}\sqrt{4 - x^2}$ (ii) For $x < -2$ and $x > 2$ then $4 - x^2 < 0$. For $4 - x^2 < 0$, $\sqrt{4 - x^2}$ doesn't exist.

(ii)

x	−2	−1.5	−1	−0.5	0	0.5	1	1.5	2
y	0	±0.661	±0.866	±0.968	±1	±0.968	±0.866	±0.661	0

b) (i) $x = \pm 2\sqrt{1 - y^2}$ (ii) For $y < -1$ and $y > 1$ then $1 - y^2 < 0$. For $1 - y^2 < 0$, $\sqrt{1 - y^2}$ doesn't exist.

(ii)

y	−1	−0.75	−0.5	−0.25	0	0.25	0.5	0.75	1
x	0	±1.323	±1.732	±1.937	±2	±1.937	±1.732	±1.323	0

c) (i) $x = \pm\frac{a}{b}\sqrt{b^2 - y^2}$ (ii) $y = \pm\frac{b}{a}\sqrt{a^2 - x^2}$ (iii) $a = \pm\dfrac{bx}{\sqrt{b^2 - y^2}}$ (iv) $b = \pm\dfrac{ay}{\sqrt{a^2 - x^2}}$

Q2. a) (i) For $n = 0$, $d = \frac{1}{2} \times 0(0-3) = 0$. For $n = 1$, $d = \frac{1}{2} \times 1(0-1) = -\frac{1}{2}$.

For $n = 2$, $d = \frac{1}{2} \times 2(2-3) = -1$. Having a shape with $-\frac{1}{2}$, -1 diagonals are not possible.

(ii) For $n = 0$ with no sides there is no shape. For $n = 1$ this is a line and not a closed shape. For $n = 2$ with two sides this is not a closed shape.

b) (i) $d = \frac{1}{2}n(n-3)$, $n(n-3) = 2d$, $n^2 - 3n = 2d$, $n^2 - 3n - 2d = 0$

$$n = \frac{3 \pm \sqrt{9 - 4 \times 1 \times (-2d)}}{2} = \frac{3 \pm \sqrt{9 + 8d}}{2}$$

(ii) For $d = 2 : n = \frac{3 \pm \sqrt{9 + 8 \times 2}}{2} = \frac{3 \pm \sqrt{25}}{2} = \frac{3 \pm 5}{2} = -1, 4$. Reject $n = -1$, Accept $n = 4$

For $d = 5 : n = \frac{3 \pm \sqrt{9 + 8 \times 5}}{2} = \frac{3 \pm \sqrt{49}}{2} = \frac{3 \pm 7}{2} = -2, 5$. Reject $n = -2$, Accept $n = 5$

For $d = 9 : n = \frac{3 \pm \sqrt{9 + 8 \times 9}}{2} = \frac{3 \pm \sqrt{81}}{2} = \frac{3 \pm 9}{2} = -3, 6$. Reject $n = -3$, Accept $n = 6$

For $d = 14 : n = \frac{3 \pm \sqrt{9 + 8 \times 14}}{2} = \frac{3 \pm \sqrt{121}}{2} = \frac{3 \pm 11}{2} = -4, 7$. Reject $n = -4$, Accept $n = 7$

c) (i) For $d = \{2, 5, 9, 14\}$ starting with the first value, 2, 3 is added to 2 to get 5, then 4 is added to get 9, then 5 is added to get 14. So one more than before is added to the next number to give the next term. The next four terms in the pattern are {20, 27, 35, 44}.

(ii) $d = \frac{1}{3}t(t+3)$

(iii) For $t = 1$, $d = \frac{1}{2} \times 1(1+3) = 2$, For $t = 2$, $d = \frac{1}{2} \times 2(2+3) = 5$,

For $t = 3$, $d = \frac{1}{2} \times 3(3+3) = 9$, For $t = 4$, $d = \frac{1}{2} \times 4(4+3) = 14$,

For $t = 5$, $d = \frac{1}{2} \times 5(5+3) = 20$, For $t = 6$, $d = \frac{1}{2} \times 6(6+3) = 27$,

For $t = 7$, $d = \frac{1}{2} \times 7(7+3) = 35$, For $t = 8$, $\mathrm{d} = \frac{1}{2} \times 8(8+3) = 44$

Q3. a) $I = 2\,000 \times 1.05^4 - 2\,000 = \$431.01$

b) (i) $P = \dfrac{I}{\left(1 + \frac{R}{100}\right)^n - 1}$

(ii) $P = \dfrac{1\,000}{(1.08)^4 - 1} \approx \$2\,774$

c) (i) $P\left(1 + \frac{R}{100}\right)^n - \mathrm{P} = P$, $P\left(1 + \frac{R}{100}\right)^n = 2P$, $\left(1 + \frac{R}{100}\right)^n = 2$

(ii) For $R = 10$, $1.1^{\mathrm{n}} = 2$. Using trial and error, $n \approx 7.3$ years or 88 months.

Q4.

a) (i) $a + 1$ (ii) $2 + 1 = 3$ (iii) $a + 1 + 1 = a + 2$

b) (i) $2a$ (ii) $2 \times 3 = 6$ (iii) $2(a - 1) = 2a - 2$

c) (i) $a^2 + 1$ (ii) $4^2 + 1 = 17$ (iii) $(a + 1)^2 + 1 = a^2 + 2a + 1 + 1 = a^2 + 2a + 2$

d) (i) $2a^2$ (ii) $2 \times 5^2 = 50$ (iii) $2(a - 1)^2 = 2(a^2 - 2a + 1) = 2a^2 - 4a + 2$

e) (i) $a^2 + a$ (ii) $6^2 + 6 = 42$ (iii) $(a + 1)^2 + a + 1 = a^2 + 2a + 1 + a + 1 = \mathrm{a}^2 + 3a + 2$

LEVEL 1 – Coordinate Geometry

Q1. a) A(6, 3) B(–3, 5) C(–7, 0) D(–3, –6) E(2, –2) F(0, 0) G(7, –6) H(5, 0) I(–5, –2) J(0, –3)

Q2. a) (4, 6) b) (3, 3) c) $\left(3\frac{1}{2}, 2\right)$ d) (2, 1) e) $\left(\frac{1}{2}, 1\right)$

f) $\left(-1\frac{1}{2}, -5\frac{1}{2}\right)$

Q3. a) $2\sqrt{10}$ units b) $\sqrt{97}$ units c) $\sqrt{153}$ units d) $\sqrt{10}$ units e) $2\sqrt{61}$ units

f) $\sqrt{73}$ units

Q4. a) $\frac{1}{2}$ b) $-1\frac{2}{3}$ c) $\frac{1}{9}$ d) $-2\frac{1}{2}$ e) $-\frac{9}{7}$ or $-1\frac{2}{7}$

f) $-\frac{4}{13}$

Q5. a)

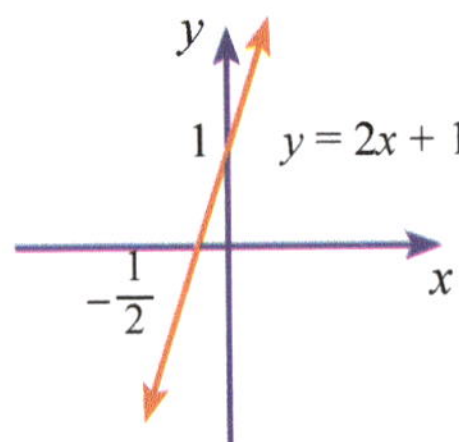

b)

y
x = 5
5
x

c)

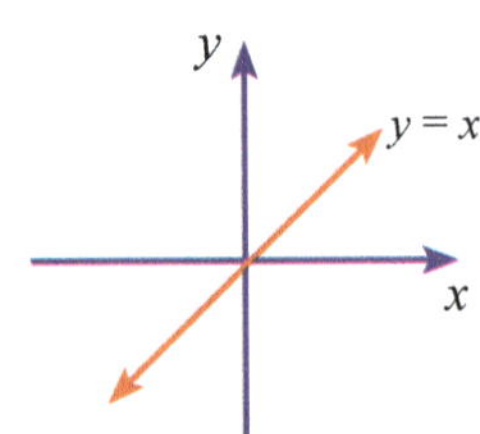

d)

y
3
y = 3 − x
3
x

e)

y
y = 4x − 5
5/4
x
−5

f)

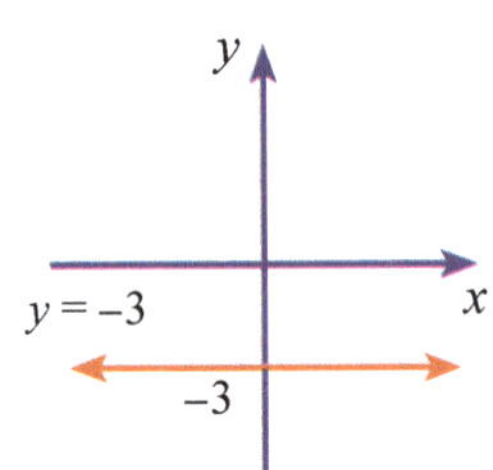

Q6. a) $m = 5, b = -2$ b) $m = -2, b = +3$ c) $m = \frac{3}{2}, b = -4$

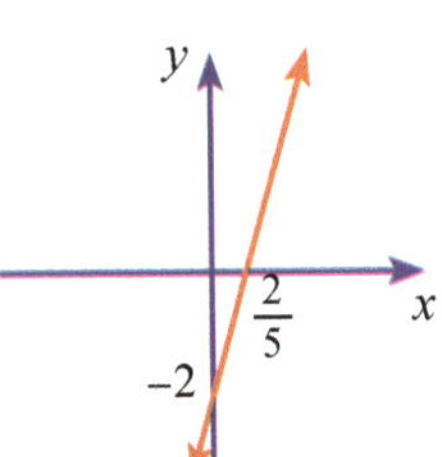

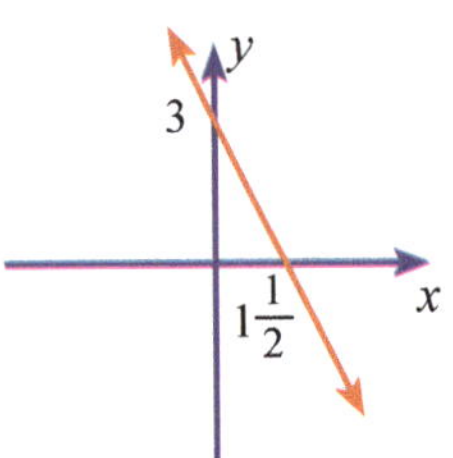

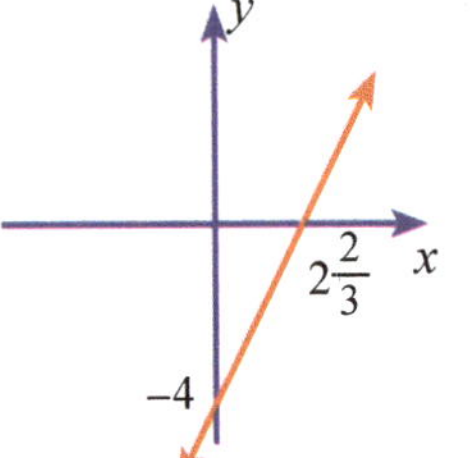

d) $m = -\frac{1}{2}, b = 1$ e) $m = -1, b = 0$ f) $m = -1, b = \frac{1}{2}$

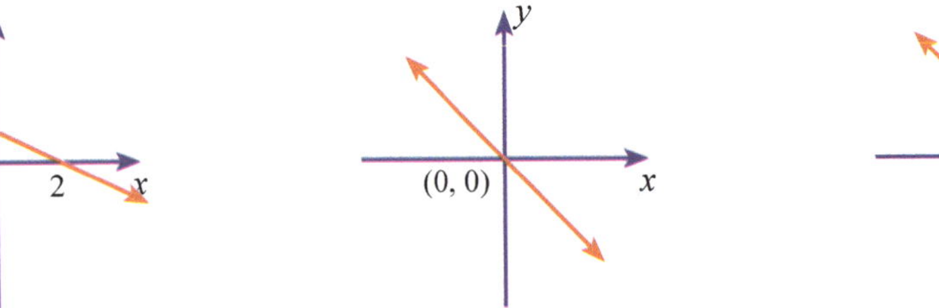

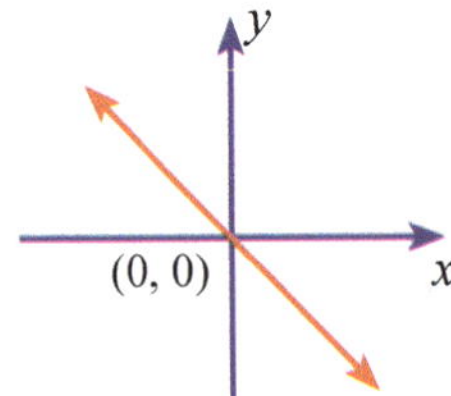

LEVEL 2 – Coordinate Geometry

Q1. a) $x + y = 7$ or $y = 7 - x$ b) $y = 3x - 1$ c) $y = \frac{1}{2}x + 5$

d) $y = x^2 - 4$ e) $y = \frac{1}{2}x - 1\frac{1}{2}$ or $x - 2y - 3 = 0$ f) $y = 2x - 17$

Q2. a) C b) D c) F d) A

e) E f) G g) H h) B

Q3.

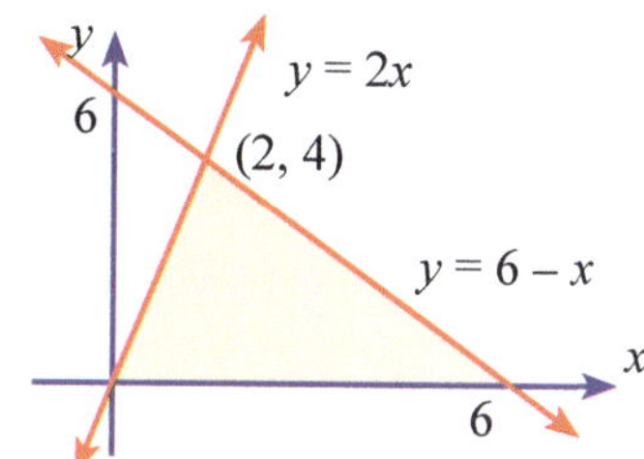

The 2 lines intersect at the point (2, 4).

Area shaded is a triangle.

Area $= \frac{1}{2}bh = \frac{1}{2} \times 6 \times 4 = 12$ units^2

Q4. a) $y = x^2 - 2x - 8$

b) c)

x	–3	–2	–1	0	1	2	3	4	5
y	7	0	–5	–8	–9	–8	–5	0	7

d) The 2 points of intersection are:
(–1, –5) and (5, 7).

Note: The lowest point (or minimum) is at (1, –9).

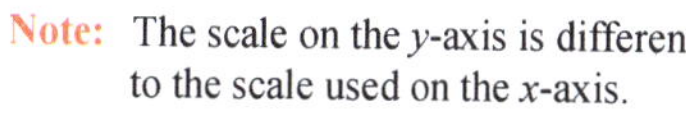
Note: The scale on the y-axis is different to the scale used on the x-axis.

The minimum is at (1, –9).

Q5. a) $y = x^2 - 4$

x	–3	–2	–1	0	1	2	3
y	5	0	–3	–4	–3	0	5

The 2 points of intersection are:
(–1, –3) and (3, 5).

Note: The lowest point (or minimum) is at (0, –4).

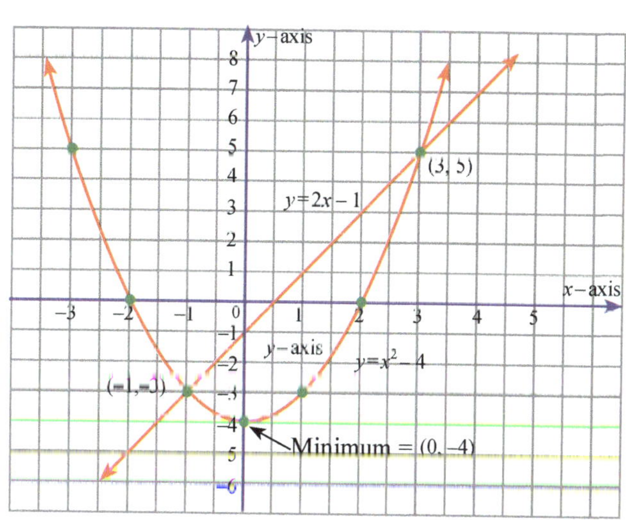

b) $y = x^2 - x - 6$

x	–3	–2	–1	0	1	2	3	4
y	6	0	–4	–6	–6	–4	0	6

The 2 points of intersection are:
(–3, 6) and (3, 0).

Note: The lowest point (or minimum) is at $(\frac{1}{2}, -6\frac{1}{4})$.

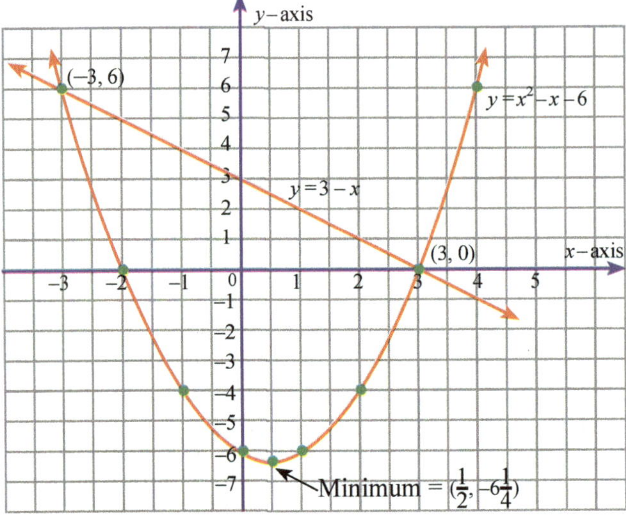

c) $y = 4x - x^2$

x	–1	0	1	2	3	4	5
y	–5	0	3	4	3	0	–5

The 2 points of intersection are:
(–1, –5) and (3, 3).

Note: The highest point (or maximum) is at (2, 4).

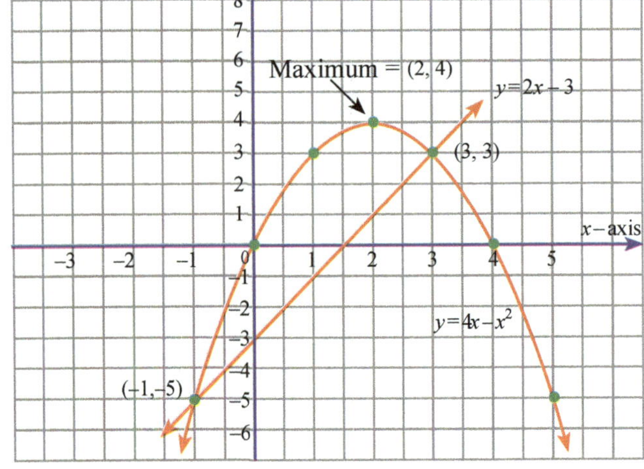

All parabolas are symmetrical. They all have one line of symmetry.

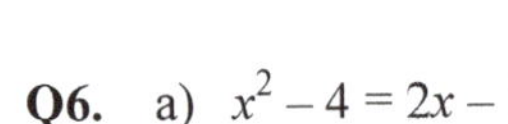

Q6. a) $x^2 - 4 = 2x - 1$
$\therefore\ x^2 - 2x - 3 = 0$
$\therefore\ (x + 1)(x - 3) = 0$
$\therefore\ x = -1$ or $x = 3$
$\therefore$ (–1, –3) and (3, 5)

b) $x^2 - x - 6 = 3 - x$
$\therefore\ x^2 - 9 = 0$
$\therefore\ (x + 3)(x - 3) = 0$
$\therefore\ x = -3$ or $x = 3$
$\therefore$ (–3, 6) and (3, 0)

c) $4x - x^2 = 2x - 3$
$\therefore\ 0 = x^2 - 2x - 3$
$\therefore\ (x + 1)(x - 3) = 0$
$\therefore\ x = -1$ or $x = 3$
$\therefore$ (–1, –5) and (3, 3)

Q7. a) $y = x^2 - 2x - 8$

y, x, –2, 4, (1, –9)

b)

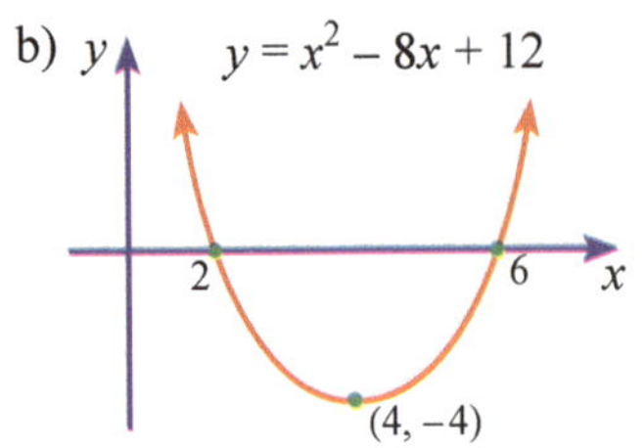

c)

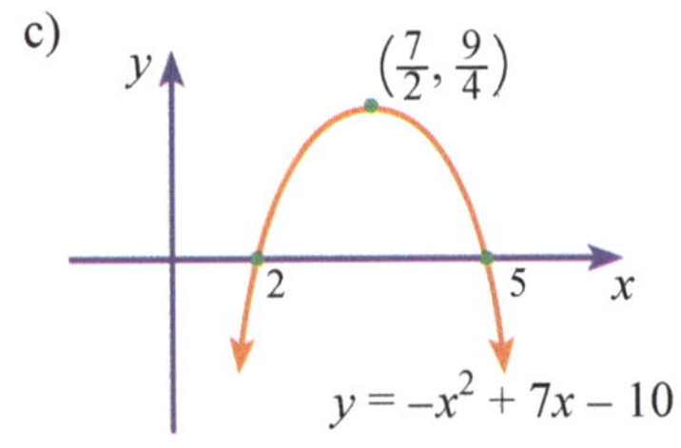

d)

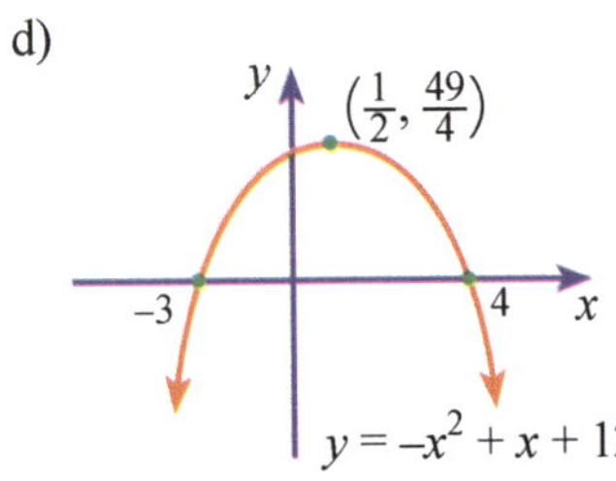

e)

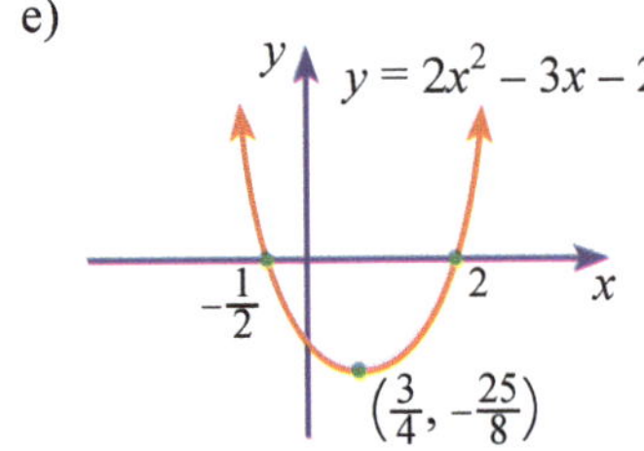

f)

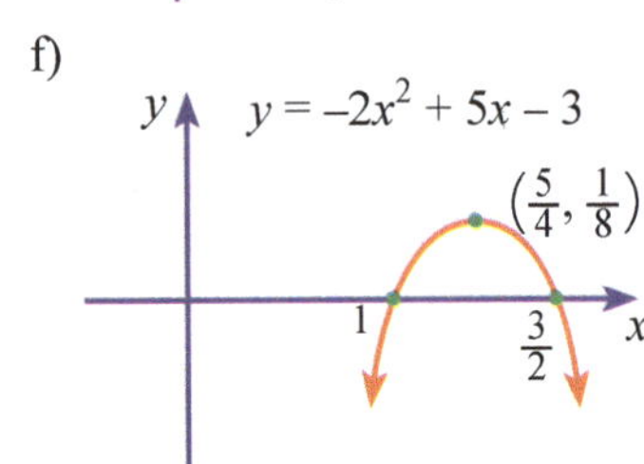

LEVEL 3 – Coordinate Geometry

Q1. a) $m = 2, b = 3$ b) $m = -2, b = 3$ c) $m = -1, b = -1\frac{2}{3}$ d) $m = -\frac{1}{5}, b = 0$

e) $m = \frac{2}{3}, b = \frac{1}{3}$ f) $m = \frac{1}{4}, b = -1$

Q2. a) $y = 3x - 9$ b) $y = -2x + 11$ c) $y = 5x + 20$ d) $y = -\frac{1}{2}x + \frac{1}{2}$

e) $y = \frac{2}{3}x + 5\frac{2}{3}$

(Note: Answers may also be written in general form.)

Q3. a) $y = 2x - 3$ b) $y = -2x + 5$ c) $y = -4x + 8$ d) $y = 3x + 2$

e) $y = -\frac{3}{2}x + 4$ f) $y = \frac{1}{2}x - 2\frac{1}{2}$

(Note: Answers may also be written in general form.)

Q4. $2x - y + 1 = 0$ and $4x - 2y + 3 = 0$

Q5. $y = 4x - 3$ or $4x - y - 3 = 0$

Q6. $3y - x + 3 = 0$ and $6y - 2x + 4 = 0$

Q7. $y = -\frac{3}{2}x + 5$ or $3x + 2y - 10 = 0$

Q8. (3, 15), (8, 35) and (4, 19)

Q9. all the points lie on the line $x - 2y - 6 = 0$

Q10. a) $5x - y - 8 = 0$ b) $2x - 3y + 21 = 0$ c) $12x - 15y - 5 = 0$

Q11. a)

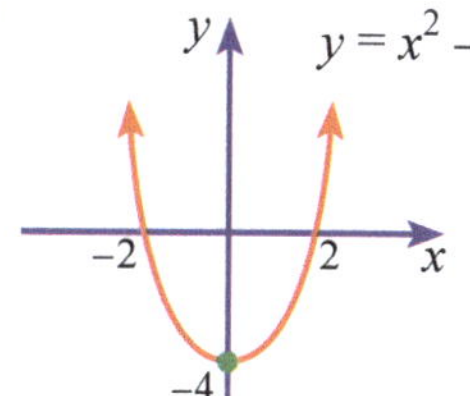

b)

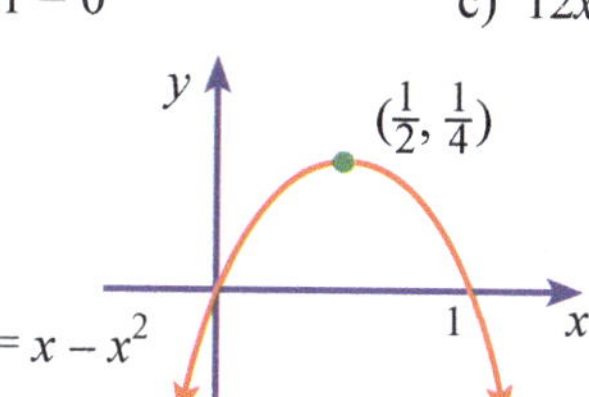

c)

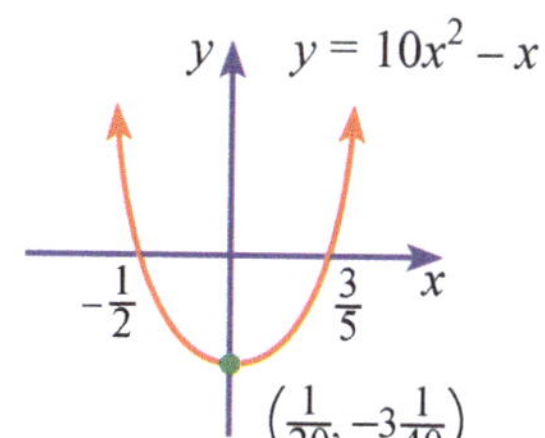

d)

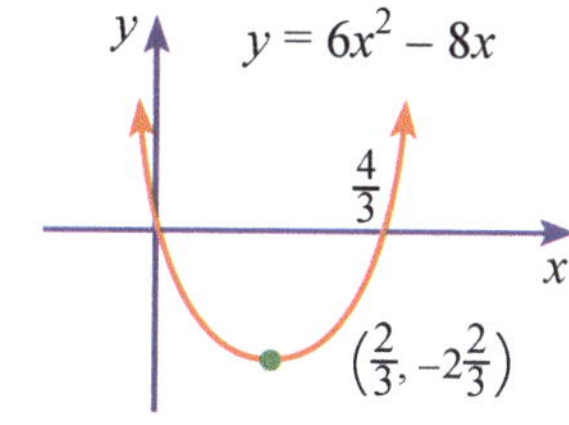

e)

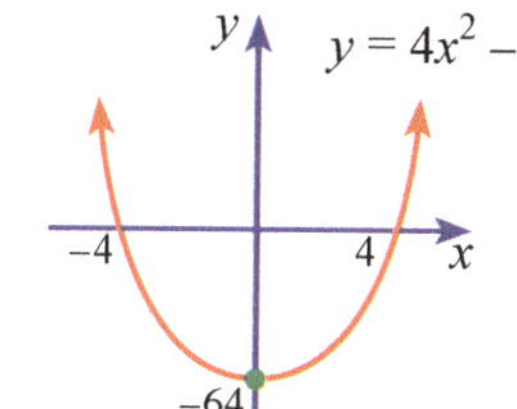

f)

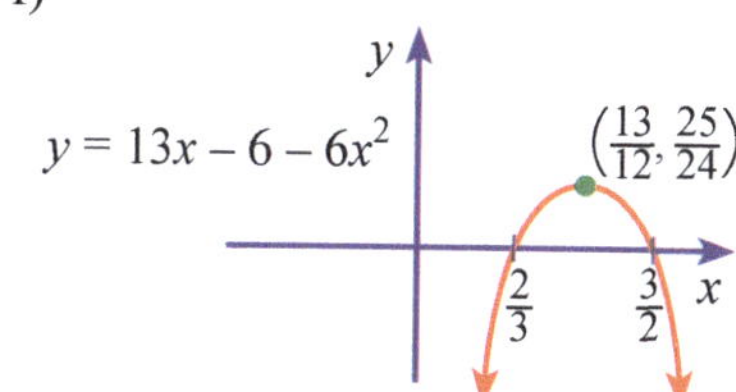

g)

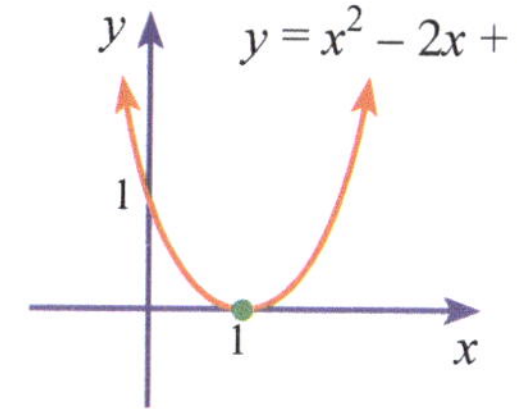

h)

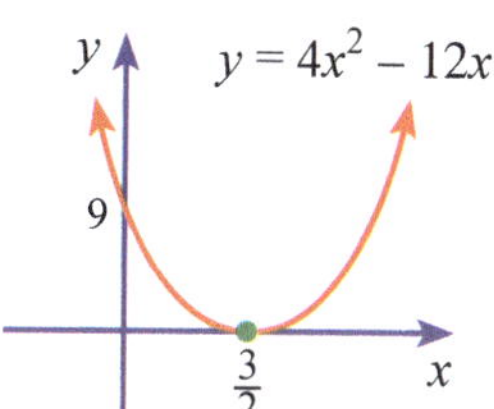

i)

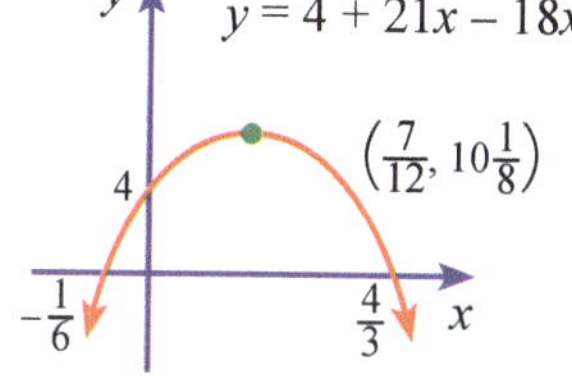

LEVEL 4 – Coordinate Geometry

Q1. a) Using $y = mx + b$ where $m = \frac{2}{3}$, $b = 2$ $\therefore$ $y = \frac{2}{3}x + 2$ or $2x - 3y + 6 = 0$

b) Using $y = mx + b$ where $m = -\frac{1}{2}$, $b = -1$ $\therefore$ $y = -\frac{1}{2}x - 1$ or $x + 2y + 2 = 0$

c) Using $y = mx + b$ where $m = 4$, $b = -2$ $\therefore$ $y = 4x - 2$ or $4x - y - 2 = 0$

Q2. Midpoint $= \left(\frac{6-2}{2}, \frac{2+4}{2}\right) = (2, 3)$ Gradient $m = \frac{y_2 - y_1}{2} = \frac{3-0}{2-0} = \frac{3}{2}$

Using $y_2 - y_1 = m(x_2 - x_1)$ $\therefore$ $y - 0 = \frac{3}{2}(x - 0)$ $\therefore$ $y = \frac{3}{2}x$ or $3x - 2y = 0$

Q3. $\left(\frac{m+8}{2}, \frac{n+5}{2}\right) = (2, 4)$ ⇨ $m = -4$ and $n = 3$

Q4. $M = \left(\frac{3}{2}, 1\right)$ and $B = (6, -2)$ $\therefore$ $BM = \sqrt{\left(6 - \frac{3}{2}\right)^2 + (-2 - 1)^2} = \sqrt{20\frac{1}{4} + 9} = \sqrt{\frac{117}{4}} = \frac{\sqrt{117}}{2}$ units

Q5. $2x - 3y + 3 = 0 \quad \therefore\ 3y = 2x + 3 \quad \therefore\ y = \frac{2}{3}x + 1 \Rightarrow m = \frac{2}{3}$

Using $y - y_1 = m(x - x_1) \Rightarrow y + 4 = \frac{2}{3}(x - 6) \Rightarrow 2x - 3y - 24 = 0$

Q6. Gradient of PQ $= m = \frac{y_2 - y_1}{x_2 - x_1} = \frac{2a - a}{a - 0} = \frac{a}{a} = 1$

Using $y - y_1 = m(x - x_1) \Rightarrow y - a = 1(x - 0) \Rightarrow x - y + a = 0$

Q7. $5x + 2y - 1 = 0 \quad \therefore\ 2y = -5x + 1 \quad \therefore\ y = -\frac{5}{2}x + \frac{1}{2} \Rightarrow m_1 = -\frac{5}{2}$

$m_1 \times m_2 = -1 \Rightarrow -\frac{5}{2} \times m_2 = -1 \Rightarrow m_2 = \frac{2}{5}$

Using $y - y_1 = m(x - x_1) \Rightarrow y - 7 = \frac{2}{5}(x - 2) \Rightarrow 2x - 5y + 39 = 0$

Q8.

A(3, 6)

C(−5, 2)

B(−2, −4)

It would be useful to first draw a rough sketch. We could prove that gradient of AC is negative reciprocal of the gradient of CB. i.e. $m_1 \times m_2 = -1$. Or we can verify that Pythagoras' Theorem holds true.

$AC = 4\sqrt{5} \quad BC = 3\sqrt{5} \quad AB = 5\sqrt{5}$

$AC^2 + BC^2 = 80 + 45 = AB^2 = 125$

$\therefore$ Right angled because Pythagoras' holds true.

Area $\Delta ABC = \frac{1}{2} \times BC \times AC = \frac{1}{2} \times 3\sqrt{5} \times 4\sqrt{5}$

$= 30 \text{ units}^2$

Q9. $AB = CD = \sqrt{5}$ units; $BC = AD = \sqrt{17}$ units; $m_{AB} = m_{DC} = \frac{1}{2} \quad \therefore AB \parallel DC$

$m_{BC} = m_{AD} = -\frac{1}{4} \quad \therefore BC \parallel AD$

Q10. All sides are $5\sqrt{2}$ units; $m_{AB} = m_{CD} = \frac{1}{7} \quad \therefore AB \parallel CD; \quad m_{BC} = m_{AD} = 7 \ \therefore BC \parallel AD.$

Area of rhombus $= \frac{1}{2} \times$ (product of 2 diagonals) $= \frac{1}{2} AC \times BD$

$= \frac{1}{2} \times 6\sqrt{2} \times 8\sqrt{2} = 48$ units

Q11. Solve $4x - y - 1 = 0$ and $2x - y + 5 = 0$ simultaneously to find their point of intersection at (3, 11).

Gradient of line joining (3, 11) and (−2, 1) $= m = \frac{11 - 1}{3 + 2} = 2$

Using Using $y - y_1 = m(x - x_1) \quad \therefore\ y - 11 = 2(x - 3) \Rightarrow y = 2x + 5$

Q12. A(4, 5) B(1, −1) c(x, y)

$\frac{x + 4}{2} = 1 \Rightarrow x = -2 \qquad \frac{y + 5}{2} = -1 \Rightarrow y = -7$

$\therefore$ Coordinates of C = (−2, −7)

Q13. Solve $5x - 3y + 2 = 0$ and $x + 3y + 1 = 0$ simultaneously to find the point of intersection at $(-\frac{1}{2}, -\frac{1}{6})$.

The gradient of $2x + 3y + 3 = 0$ is $m = \frac{-2}{3}$.

Using Using $y - y_1 = m(x - x_1) \quad \therefore\ y + \frac{1}{6} = \frac{-2}{3}(x + \frac{1}{2}) \Rightarrow 4x + 6y + 3 = 0$

LEVEL 5 – Coordinate Geometry

Q1. a)

A B D M

0 1 2 3 4 5 6 7

4 3 2 1 -1 -2 -3

b) (i) $m_{AB} = \frac{3 - 1}{3 - 0} = \frac{2}{3}$

(ii) $m_{AD} = \frac{3 - 2}{3 - 6} = -\frac{1}{3}$

(iii) $m_{BD} = \frac{2 - 1}{6 - 0} = \frac{1}{6}$

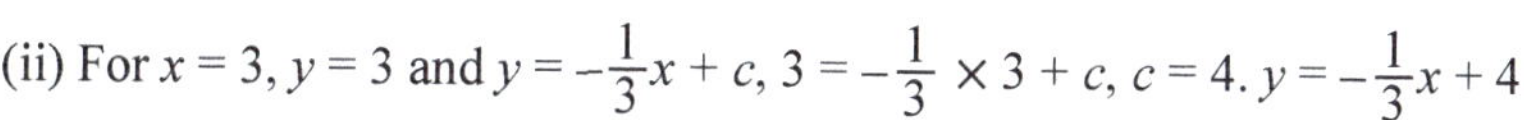

c) (i) For $x = 3$, $y = 3$ and $y = \frac{2}{3}x + c$, $3 = \frac{2}{3} \times 3 + c$, $c = 1$ which can be seen on the graph as the y-intercept of the line AB. $y = \frac{2}{3}x + 1$

(ii) For $x = 3$, $y = 3$ and $y = -\frac{1}{3}x + c$, $3 = -\frac{1}{3} \times 3 + c$, $c = 4$. $y = -\frac{1}{3}x + 4$

(iii) For $x = 0$, $y = 1$ and $y = \frac{1}{6}x + c$, $1 = \frac{1}{6} \times 0 + c$, $c = 1$ which can be seen on the graph as the y-intercept of the line BD. $y = \frac{1}{6}x + 1$

d) (i) $m_{AC} = -\frac{1}{m_{BD}} = -\frac{1}{\frac{1}{6}} = -6$

(ii) For $x = 3$, $y = 3$ and $y = -6x + c$, $3 = -6 \times 3 + c$, $c = 21$. $y = -6x + 21$

e) $y = -6x + 21 = \frac{1}{6}x + 1$, $6\frac{1}{6}x = 20$, $x = 3\frac{9}{37}$ and $y = \frac{1}{6} \times 3\frac{9}{37} + 1 = 1\frac{20}{37}$. Coordinates of M are $(3\frac{9}{37}, 1\frac{20}{37})$.

f) For $x = 4$, $y = -6 \times 4 + 21 = -3$.

g) Length of $AC = \sqrt{(3-4)^2 + (3-(-3))^2} = \sqrt{37}$

Q2. a)

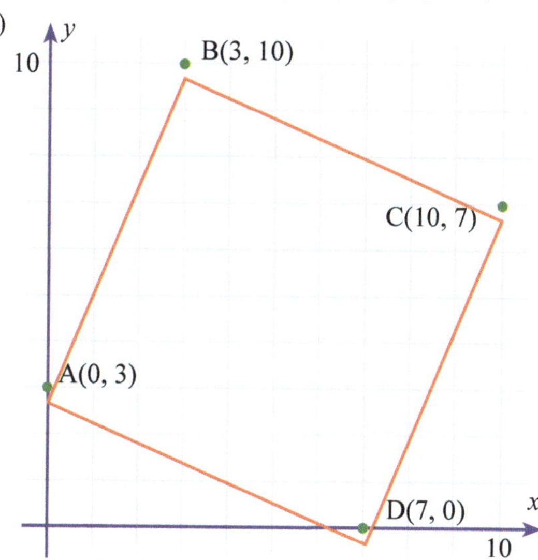

b) (i) $AB = \sqrt{(3-0)^2 + (10-3)^2} = \sqrt{58}$

(ii) $BC = \sqrt{(3-10)^2 + (10-7)^2} = \sqrt{58}$

(iii) $CD = \sqrt{(10-7)^2 + (7-0)^2} = \sqrt{58}$

(iv) $AD = \sqrt{(7-0)^2 + (0-3)^2} = \sqrt{58}$

c) $m_{AB} = \frac{7}{3}$, $m_{AD} = -\frac{3}{7}$. As $m_{AB} = -\frac{1}{m_{AD}}$ then AB is perpendicular to AD. The shape is a square.

d) (i) $m_{AB} = \frac{7}{3}$, $m_{BC} = -\frac{3}{7}$. (ii) They are perpendicular to each other.

e) Area $= \sqrt{58} \times \sqrt{58} = 58$ square units.

f) $A'(0, -3)$, $B'(3, -10)$, $C'(10, -7)$, $D'(7, 0)$.

g) $A''(0, 3)$ $B''(-3, 10)$ $C''(-10, 7)$ $D''(-7, 0)$.

Q3. a)

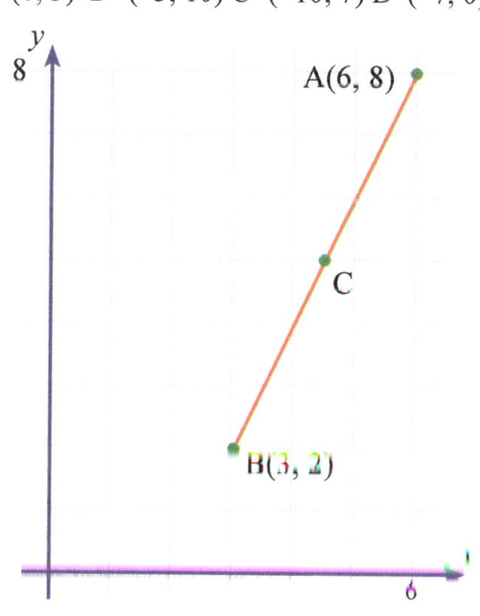

b) Coordinates of C: $\left(\frac{3+6}{2}, \frac{2+8}{2}\right) = (4\frac{1}{2}, 5)$

c) Coordinates of D: $\left(\frac{3\times2+6\times1}{3}, \frac{2\times2+8\times1}{3}\right) = (4, 4)$

d) Length of $CD = \sqrt{(4\frac{1}{2}-4)^2+(5-4)^2} = \frac{\sqrt{5}}{2}$

e) Length of $AB = \sqrt{(6-3)^2+(8-2)^2} = 3\sqrt{5}$,

Ratio of lengths AB to $CD = \frac{3\sqrt{5}}{\frac{\sqrt{5}}{2}} = \frac{6}{1}$

LEVEL 6 – Coordinate Geometry

Q1. a) (i) $\frac{8--12}{-2-6} = -2\frac{1}{2}$ (ii) $y = -2\frac{1}{2}x + 3$ b) Distance $AB = \sqrt{(6--2)^2+(-12-8)^2} = 4\sqrt{29}$

c) $M = \left(\frac{-2+6}{2}, \frac{8-12}{2}\right) = (2, -2)$. Distance $AM = \sqrt{(2--2)^2+(-2-8)^2} = 2\sqrt{29}$

Distance $BM = \sqrt{(2-6)^2+(-2--2)^2} = 2\sqrt{29}$, so the distance of AM is the same as the distance BM.

d) $y = -2\frac{1}{2}x + 10$

Q2. a) Equation for A: $y = \frac{1}{3}x + \frac{1}{3}$ so the gradient for A is $\frac{1}{3}$. b) -3 c) $y = -3x + 14$

d) $(4\frac{1}{10}, 1\frac{7}{10})$

Q3. a) $A(-1, 0)$ $B(2, 4)$ $C(6, 1)$ $D(3, -3)$

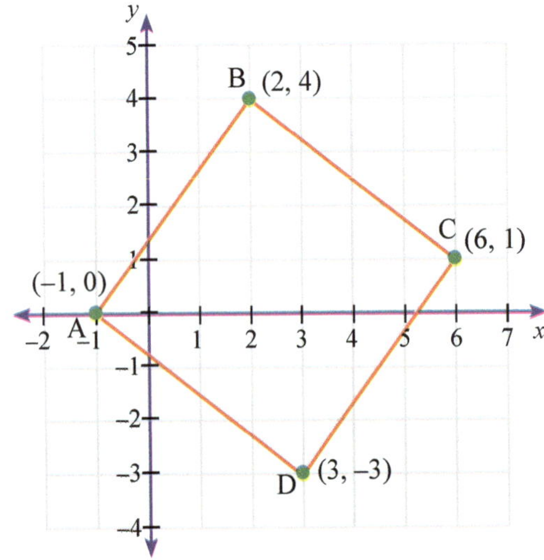

b) (i) $AB = \sqrt{(2--1)^2+(4-0)^2} = 5$ (ii) $BC = \sqrt{(6-2)^2+(1-4)^2} = 5$

(iii) $CD = \sqrt{(6-3)^2+(1--3)^2} = 5$ (iv) $AD = \sqrt{(3--1)^2+(-3-0)^2} = 5$

$ABCD$ could be a rhombus or a square.

c) (i) $m_{AB} = \frac{4-0}{2--1} = \frac{4}{3}$ (ii) $m_{BC} = \frac{4-1}{2-6} = -\frac{3}{4}$ (iii) $m_{CD} = \frac{4--3}{6-3} = \frac{4}{3}$ (iv) $m_{AD} = \frac{0--3}{-1-3} = -\frac{3}{4}$

As $m_{AB} = m_{CD} = -\frac{1}{m_{BC}} = -\frac{1}{m_{AD}}$ then $\angle ABC = \angle BCD = \angle CDA = \angle DAB = 90°$, hence $ABCD$ is a square.

d) Area $= 5^2 = 25$ square units

e) (i) $y = \frac{4}{3}x + \frac{4}{3}$ (ii) $y = -\frac{3}{4}x + \frac{11}{2}$ (iii) $y = \frac{4}{3}x - 7$ (iv) $y = -\frac{3}{4}x - \frac{3}{4}$

f) Midpoint $AB = M = (\frac{1}{2}, 2)$. Midpoint $BC = N = (4, 2\frac{1}{2})$.

Midpoint $CD = P = (4\frac{1}{2}, -1)$. Midpoint $AD = Q = (1, -1\frac{1}{2})$.

Finding the lengths:

$MN = \sqrt{(2\frac{1}{2} - 2)^2 + (4 - \frac{1}{2})^2} = \frac{5\sqrt{2}}{2}$, $NP = \sqrt{(2\frac{1}{2} - -1)^2 + (4 - 4\frac{1}{2})^2} = \frac{5\sqrt{2}}{2}$

$PQ = \sqrt{(-1 - 1\frac{1}{2})^2 + (4\frac{1}{2} - 1)^2} = \frac{5\sqrt{2}}{2}$, $MQ = \sqrt{(2 - -1\frac{1}{2})^2 + (\frac{1}{2} - 1)^2} = \frac{5\sqrt{2}}{2}$

Gradients of lines:

$m_{MN} = \frac{2\frac{1}{2} - 2}{4 - \frac{1}{2}} = \frac{1}{7}$, $m_{NP} = \frac{2\frac{1}{2} - -1}{4 - 4\frac{1}{2}} = -7$, $m_{PQ} = \frac{-1 - -1\frac{1}{2}}{4\frac{1}{2} - 1} = \frac{1}{7}$, $m_{MQ} = \frac{2 - -1\frac{1}{2}}{\frac{1}{2} - 1} = -7$

As $m_{MN} = m_{PQ} = -\frac{1}{m_{NP}} = -\frac{1}{m_{MQ}}$ then $\angle MNP = \angle NPQ = \angle PQM = \angle QMN = 90°$,

hence $MNPQ$ is a square.

Q4. a)

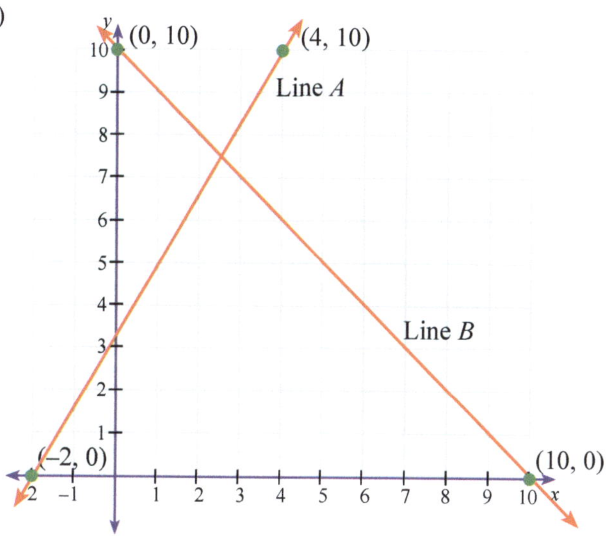

b) (i) Line A: $y = \frac{5}{3}x + \frac{10}{3}$

(ii) Line B: $y = 10 - x$

c) Point of intersection: $(2\frac{1}{2}, 7\frac{1}{2})$

d) Area $= \frac{1}{2} \times 7\frac{1}{2} \times 12 = 45$ square units

e) (i) Line A: $y = \frac{5}{3}x + \frac{16}{3}$, Line B: $y = 12 - x$

(ii) Point of intersection: $(2\frac{1}{2}, 9\frac{1}{2})$

(iii) Area $= \frac{1}{2} \times 15\frac{1}{5} \times 9\frac{1}{2} = 72\frac{1}{5}$ square units

LEVEL 1 – Further Trigonometry

Q1.	a) 0.772	b) 0.138	c) 0.116	d) 2.494	e) 0.929	f) 0.593
Q2.	a) 81° 5′	b) 43° 33′	c) 23° 59′	d) 67° 58′	e) 59° 59′	f) 36° 41′
Q3.	a) 16.06 cm		b) 20.06 m		c) 9.31 cm	
Q4.	a) $\theta = 58°\,24'$		b) $\theta = 36°\,4'$		c) $\theta = 43°\,48'$	
Q5.	a) 6.15 cm		b) 14.96 m		c) 15.59 m	
Q6.	a) 11.96 cm		b) 14.05 m		c) 6.90 cm	
Q7.	a) 41.8 m²		b) 57.9 cm²		c) 143 m²	
Q8.	a) 153°		b) 236°		c) 338°	

LEVEL 2 – Further Trigonometry

Q1.	a) ∠F = 64° 41′	b) a = 10.03 cm	c) ∠X = 29° 03′
Q2.	a) m = 16.7 cm	b) ∠C = 54° 36′	c) g = 4.6 cm
Q3.	a) 156.03 cm²	b) 202.90 cm²	c) 3 873.55 cm²
Q4.	401 m	**Q5.** 71.97 km (correct to 2 d.p.)	

Q6. 1.1 km from A, 1.9 km from B

Q7. 1.6 km

Q8. a) 46.3 cm b) 102 cm^2

Q9. a) $\tan 60^\circ = \sqrt{3}$ b) $\cos 30^\circ = \frac{\sqrt{3}}{2}$ c) $\sin 45^\circ = \frac{1}{\sqrt{2}}$

d) $\cot 30^\circ = \sqrt{3}$ e) $\operatorname{cosec} 60^\circ = \frac{2}{\sqrt{3}}$ f) $\sec 30^\circ = \frac{2}{\sqrt{3}}$

Q10. a) Q.II and Q.III b) Q.III and Q.IV c) Q.II and Q.IV

Q11. a) $\tan 150^\circ = -\tan 30^\circ = \frac{-1}{\sqrt{3}}$ b) $\cos 210^\circ = -\cos 30^\circ = \frac{-\sqrt{3}}{2}$ c) $\sin 135^\circ = \sin 45^\circ = \frac{1}{\sqrt{2}}$

LEVEL 3 – Further Trigonometry

Q1. a) $\sin 210^\circ = -\sin 30^\circ = \frac{-1}{2}$ b) $\cos 300^\circ = \cos 60^\circ = \frac{1}{2}$

c) $\tan 300^\circ = -\tan 60^\circ = -\sqrt{3}$ d) $\tan 150^\circ = -\tan 30^\circ = \frac{-1}{\sqrt{3}}$

Q2. a) $\theta = 60^\circ$ or 120° b) $\theta = 120^\circ$ or 300°

c) $\theta = 150^\circ$ or 210° d) $\theta = 225^\circ$ or 315°

Q3.

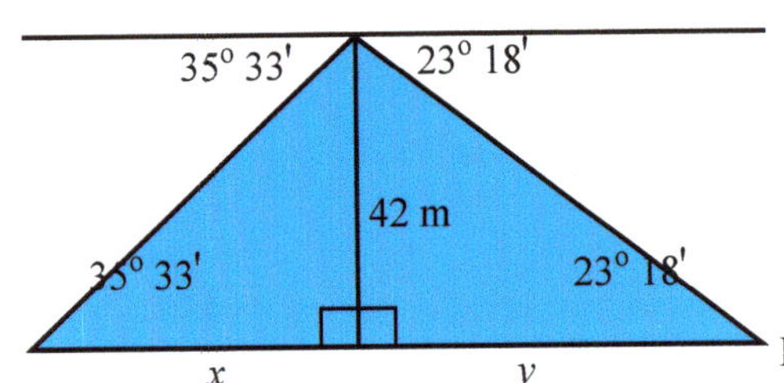

$x = \frac{42}{\tan 35^\circ 33'}$ $y = \frac{42}{\tan 23^\circ 18'}$

$\therefore\ x + y = 156.2 \approx 156$ m

Q4.

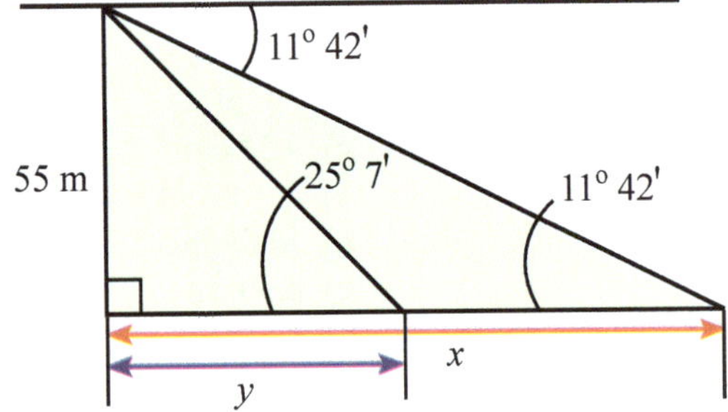

$d = x - y$

$= \frac{55}{\tan 11^\circ 42'} - \frac{55}{\tan 25^\circ 7'} \approx 148$ m

Speed = 30×148 m/h = 4.44 km/h

$= 4.44 \div 1.852 \approx 2$ nautical miles per hour

Q5. a) $AC^2 = BC^2 + AB^2$ (Pythagoras)

$\therefore$ $AC = \sqrt{72} = 8.49$ cm

b) $EC^2 = EF^2 + FC^2$ (Pythagoras)

$\therefore$ $EC = \sqrt{9^2 + 4.245^2} = 9.95$ cm

b) $\sin E\hat{C}F = \frac{9}{9.95}$

$\therefore$ $E\hat{C}F = 64^\circ 46'$

d) $\cos E\hat{C}B = \frac{3}{9.95}$

$\therefore$ $E\hat{C}B = 72^\circ 27'$

Q6.

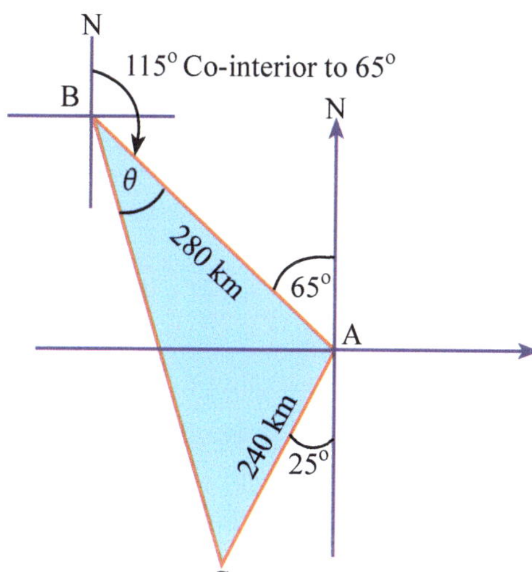

$\tan\theta = \dfrac{240}{280}$

$\therefore\quad \theta = 40°\,36'$

$\therefore\quad \theta =$ Bearing of C from B

$= 115° + 40°\,36'$

$= 155°\,36'$

Note that it is a right angled triangle $\angle BAC = 90°$.

Q7.

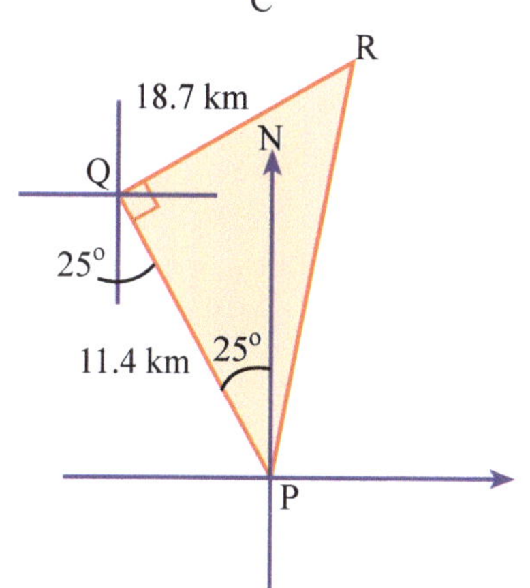

Using geometry theorems, the angle in the triangle at Q = 90° as shown.

a) $RP^2 = 11.4^2 + 18.7^2$

$RP = \sqrt{479.65}$

$= 21.9$ km

b) $\tan Q\hat{P}R = \dfrac{18.7}{11.4}$

$\therefore\quad Q\hat{P}R = 58°\,38'$

$\therefore$ Bearing of R from P

$= 58°\,38' - 25°$

$= 33°\,38'$

Q8.

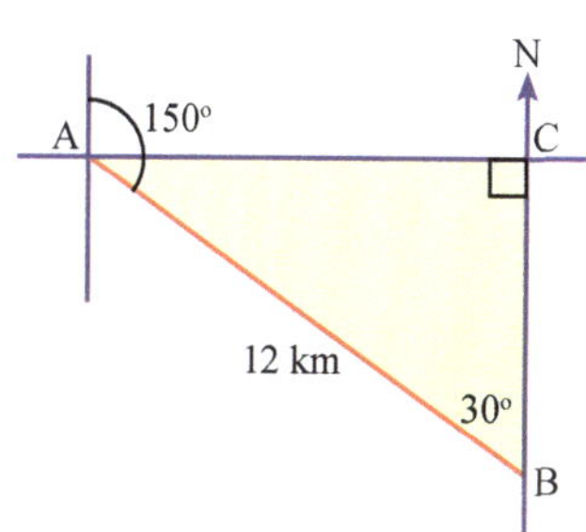

a) $\angle ABC = 30°$

b) (i) $\sin 30° = \dfrac{AC}{12}$ ⇨ $AC = 12 \times \sin 30° = 6$ km

(ii) $\tan 8° = \dfrac{H}{6}$ ⇨ $H = 6 \times \tan 8° = 0.843245$ km $= 843.25$ m

c) (i) $\tan 30° = \dfrac{6}{CB}$ ⇨ $CB = 10.392$ km $= 10\,392$ m

(ii) $\tan\theta = \dfrac{843.25}{10\,395}$ ⇨ $\theta = 4.64°$

LEVEL 4 – Further Trigonometry

Q1. a) $a = 14\tan(46°\,32') = 14.8$ cm, $b = \dfrac{14\sin(32.4°)}{\cos(46°\,32')} = 10.9$ cm, $c = 16\sin(32.4) = 8.6$ cm,

$d = \dfrac{14\cos(32.4°)}{\cos(46°\,32')} - 16 = 1.2$ cm, $e = \dfrac{14}{\cos(46°\,32')} - 16\cos(32.4°) = 6.8$ cm

b) $\alpha = \cos^{-1}\left(\dfrac{40}{50}\right) = 36.9°$, $\beta = \tan^{-1}\left(\dfrac{50}{8}\right) = 80.9°$, $\theta = \tan^{-1}\left(\dfrac{8}{50}\right) = 9.1°$

Note: The right angled Δ has sides of 30 cm, 40 cm and 50 cm from Pythagoras'.

Q2. a) (i) $\sin(\theta) = \dfrac{\sin(30°)}{16} \times 24$, $\theta = \sin^{-1}\left(\dfrac{\sin(30°)}{16} \times 24\right) = 48.6°$ and $\theta = 180° - 48.6° = 131.4°$

(ii) For $\theta = 48.6°$, $\alpha = 180° - (48.6° + 30°) = 101.4°$

For $\theta = 131.4°$, $\alpha = 180° - (131.4° + 30°) = 18.6°$

(iii) For $\alpha = 101.4°$, $AB = \sqrt{24^2 + 16^2 - 2 \times 24 \times 16 \cos(101.4^o)} = 31.37$ m

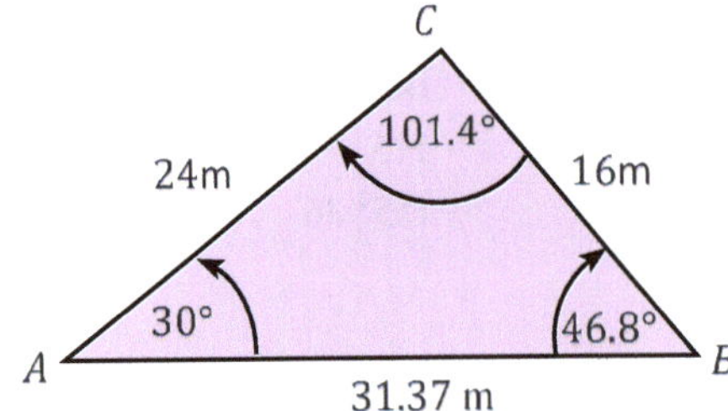

For $\alpha = 18.6°$, $AB = \sqrt{24^2 + 16^2 - 2 \times 24 \times 16 \cos(18.6°)} = 10.20$ m

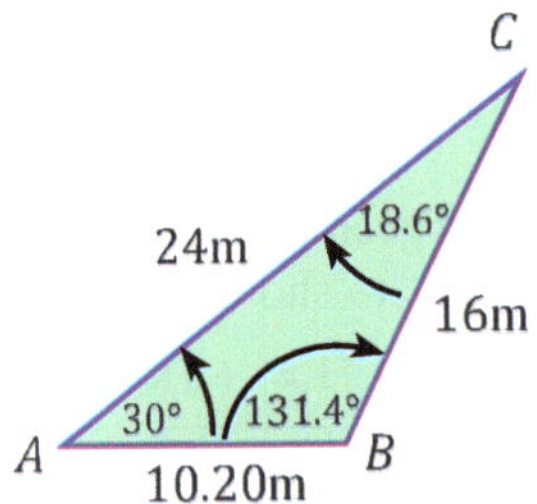

(iv) Area for triangle: AC = 24 m, BC = 16m and $\angle BAC = 101.4° = 24 \times 16 \sin(101.4°) = 376.4$ m^2
Area for triangle: AC = 24 m, BC = 16m and $\angle BAC = 18.6° = 24 \times 16 \sin(18.6°) = 122.5$ m^2

(b) (i) $\sin(\theta) = \frac{\sin(34^o)}{18} \times 20$, $\theta = \sin^{-1}\left(\frac{\sin(34^o)}{18} \times 20\right) = 38.4°$ and $\theta = 180° - 38.4° = 141.6°$

(ii) For $\theta = 38.4°$, $\alpha = 180° - (38.4° + 34°) = 107.6°$
For $\theta = 141.6°$, $\alpha = 180° - (141.6° + 34°) = 4.4°$

(iii) For $\alpha = 107.6°$, $AB = \sqrt{20^2 + 18^2 - 2 \times 20 \times 18 \cos(107.6^o)} = 30.69$ m

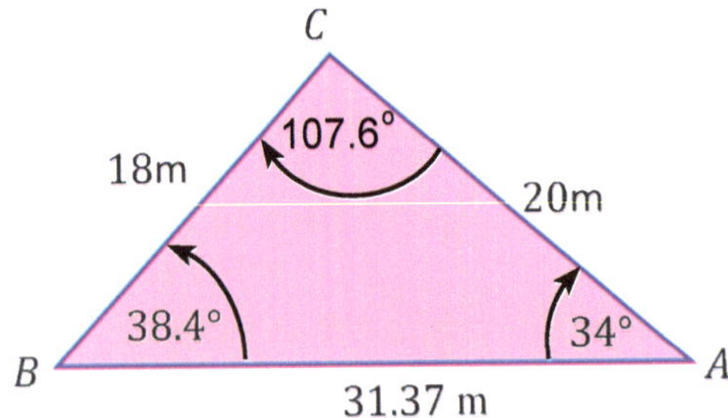

For $\alpha = 4.4°$, $AB = \sqrt{20^2 + 18^2 - 2 \times 20 \times 18 \cos(4.4^o)} = 2.47$ m

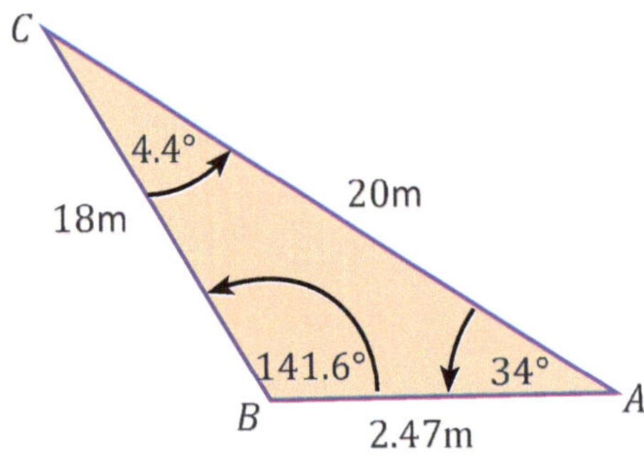

As you can see from this question there are 2 possible answers for θ and therefore there are 2 possible different triangles.

(iv) Area for triangle: AC = 20 m, BC = 18 m and $\angle ACB = 107.6°$
Area = $20 \times 18 \sin(107.6°) = 343.1$m^2
Area for triangle: AC = 20 m, BC = 18 m and $\angle ACB = 4.4°$
Area = $20 \times 18 \sin(4.4°) = 27.6$ m^2

Q3. a) BC = $\sqrt{12^2 + 18^2 - 2 \times 12 \times 18 \cos(100^\circ)}$ = 23.3 cm

b) (i) $\angle DCB = \sin^{-1}\left(\frac{12\sin(100^\circ)}{23.3}\right) = 30.5^\circ$ (ii) $\angle ACB = \sin^{-1}\left(\frac{24\sin(70^\circ)}{23.3}\right) = 75.4^\circ$ (iii) $\angle ACD = 105.9^\circ$

c) $\angle ABD = 360^\circ - 70^\circ - 100^\circ - 105.9^\circ = 84.1^\circ$

d) $\angle ABC = 180^\circ - (70^\circ + 75.4^\circ) = 34.6^\circ$, AC = $\left(\frac{24\sin(34.6^\circ)}{\sin(75.4^\circ)}\right)$ = 14.1 cm

Area of ABDC = $\frac{1}{2} \times 18 \times 12 \times \sin 100^\circ + \frac{1}{2} \times 24 \times 14.1 \times \sin 70^\circ$

= 265.35 cm^2

Q4. a) $\angle ADB = 180^\circ - (24^\circ + 18^\circ) = 138^\circ$, AD = $\left(\frac{36\sin(18^\circ)}{\sin(138^\circ)}\right)$ = 16.63 m

b) BD = $\left(\frac{36\sin(24^\circ)}{\sin(138^\circ)}\right)$ = 21.88 m

c) $\angle ADE = 180^\circ - 138^\circ = 42^\circ$, AD = $\left(\frac{16.63\sin(42^\circ)}{\sin(70^\circ)}\right)$ = 11.84 m

d) $\angle DAE = 180^\circ - (70^\circ + 42^\circ) = 68^\circ$, ED = $\left(\frac{16.63\sin(68^\circ)}{\sin(70^\circ)}\right)$ = 16.41 m

e) $\angle BDC = 180^\circ - 138^\circ = 42^\circ$, $\angle CBD = 180^\circ - (42^\circ + 95^\circ) = 43^\circ$

BC = $\left(\frac{21.88\sin(42^\circ)}{\sin(95^\circ)}\right)$ = 14.70 m

f) CD = $\left(\frac{21.88\sin(43^\circ)}{\sin(95^\circ)}\right)$ = 14.98 m

ABOUT THE AUTHOR

Ian Bull B.Sc., Dip. Ed., BFA, Grad Dip attended Ivanhoe Grammar School in Melbourne, Australia winning a number of academic prizes as well as representing the school in a variety of sports. He won a scholarship to complete his first degree in Applied Science (B.Sc.) majoring in Chemistry at the Royal Melbourne Institute of Technology.

He completed his Diploma of Education (Dip.Ed.) at the University of Melbourne and has continued to teach in an uninterrupted career for more than forty years. He has taught Mathematics to the highest levels in both Government and Private schools as well as being a Senior Examination marker in Victoria for many years. While teaching he completed a Fine Arts Degree part time at the Royal Melbourne Institute of Technology majoring in Printmaking and then completed a Graduate Diploma in Mathematics at the University of Melbourne.

With a wide diversity of interests, he is currently in charge of the Gifted and Talented Education programs at a prestigious private school in Melbourne. He has coached teams and lead them to compete at the International level of 'Tournament of Minds'. He has presented papers both at the Mathematical Association of Victoria annual conference as well as the Australian Association of Mathematics Teachers conference for many years. He has also presented papers at the Asian Pacific Conference for Gifted and Talented Students both in Sydney and Dubai.

Ian was the lead author in the Pearson Mathematics series, Mathematics Dimensions for years 7 to 10 as well as writing the senior text General Maths for the Victorian Certificate of Education units 1&2. He is a widely published and prolific author having written the titles of Revision and Practice books 1 to 4, Preparing for Secondary School Mathematics, Extension Mathematics books 1 and 2, Maths Assessment Tasks for CSF levels 5 and 6, Maths Practice and Tests for years 7 and 8, the Primary Resource Book as well as Maths Problem Solving for Higher Achieving Students.

He has been committed to the sharing of his education experiences and has continued to write a range of appealing and popular resources to cutting-edge educational issues.

'Education is not the learning of facts, but the training of the mind to think'

Albert Einstein
(1879 - 1955)

ABOUT THE AUTHOR

Warwick Marlin B.Sc., Dip.Ed. went to St Bees school in the North of England where he completed 'A' levels in Maths, Physics and Chemistry. He obtained his Bachelor of Science (B.Sc.) at the University of Natal (South Africa) in 1970, majoring in Pure Maths and Physics.

He completed his Diploma in Education (Dip.Ed.) at Christchurch (NZ). He has had over 30 years of teaching experience mainly in Australia, but also in South Africa and New Zealand.

He was the Author, Editor and Publisher of all the books in the original Understanding Maths Series, which were first published in 1988.

In 1986 he was also the Founder and Senior Director of Australian Youth Challenge – a program designed to help teenagers in the key areas of self esteem, motivation, goal setting, communication and study skills.

He was the Coordinator and Principal for the July Holiday Math courses which ran successfully for 15 years at St Andrews Cathedral School in Sydney.

In the latest series of *Understanding Maths*, he is the author of the Year 3, 4, 5, 6, 7, 8, 9 and 10 books. He is also the author of all 8 books in the *Important Facts and Formulas* series.

He is the co-author of the 6 elementary books in the '*Essential Exercises Series*'.

He is the editor of the 6 books in the '*Understanding Comprehension Series*'.

In 2010, he moved and relocated to the Philippines full time, and he transferred the publishing and distribution of his books in Australia to *Five Senses Education Pty. Ltd.* He is still very active in writing and updating books on behalf of *Five Senses Education*.

During the last two years he has developed and written new ADVANCED versions of the 'Understanding Maths Series' (for books Year 3, 4, 5 and 6).

During 2020, he completely updated the books for Years 9 and 10. Previously, he had written a combined Year 9 and 10 book for the Intermediate course and a combined Year 9 and 10 book for the Advanced course. There are now 4 books: Year 9 and Year 9 Advanced Edition, Year 10 and Year 10 Advanced Edition. These new updated and improved books contain additional theory and exercises, more challenging exercises and also more 'problem solving' exercises.

These 4 books are based on the NSW and Australian curriculum.

'The best teacher is not the one who knows the most, but the one who is most capable of reducing knowledge to that simple compound of the obvious and wonderful...'

H.L. Menken